Lecture Notes in Artificial Intelligence 13048

Subseries of Lecture Notes in Computer Science

Series Editors

Randy Goebel
University of Alberta, Edmonton, Canada

Yuzuru Tanaka
Hokkaido University, Sapporo, Japan

Wolfgang Wahlster
DFKI and Saarland University, Saarbrücken, Germany

Founding Editor

Jörg Siekmann
DFKI and Saarland University, Saarbrücken, Germany

More information about this subseries at http://www.springer.com/series/1244

Víctor Rodríguez-Doncel · Monica Palmirani ·
Michał Araszkiewicz · Pompeu Casanovas ·
Ugo Pagallo · Giovanni Sartor (Eds.)

AI Approaches to the Complexity of Legal Systems XI-XII

AICOL International Workshops 2018 and 2020:
AICOL-XI@JURIX 2018, AICOL-XII@JURIX 2020,
XAILA@JURIX 2020
Revised Selected Papers

 Springer

Editors
Víctor Rodríguez-Doncel ⓘ
Technical University of Madrid
Madrid, Spain

Michał Araszkiewicz ⓘ
Jagiellonian University
Krakow, Poland

Ugo Pagallo ⓘ
University of Turin
Turin, Italy

Monica Palmirani ⓘ
University of Bologna
Bologna, Italy

Pompeu Casanovas ⓘ
La Trobe University
Melbourne, VIC, Australia

Autonomous University of Barcelona
Barcelona, Spain

Giovanni Sartor ⓘ
University of Bologna
Bologna, Italy

ISSN 0302-9743   ISSN 1611-3349 (electronic)
Lecture Notes in Artificial Intelligence
ISBN 978-3-030-89810-6   ISBN 978-3-030-89811-3 (eBook)
https://doi.org/10.1007/978-3-030-89811-3

LNCS Sublibrary: SL7 – Artificial Intelligence

This Springer imprint is published by the registered company Springer Nature Switzerland AG
The registered company address is: Gewerbestrasse 11, 6330 Cham, Switzerland

Preface

We are delighted to introduce a set of contributions to today's state of the art in Law and Technology in this volume, the fifth in a series on the AI Approaches to the Complexity of Legal Systems (AICOL). Since the first AICOL workshop held in Beijing in 2009, many other have followed, at the crossroads of artificial intelligence, the web of (linked) data, political science, legal theory, and legal studies. Some AICOL workshops, with the same cluster of topics, have taken place in conjunction with EU-funded FP7 and H2020 projects. But most of them have been launched jointly with conferences devoted to AI and law, such as the International Conference of AI and Law (ICAIL) and the International Conference on Legal Knowledge and Information Systems (JURIX).

AICOL I-II, published in 2010, included papers from the first AICOL workshop (co-located with the 24th IVR World Congress, September 15–20, Beijing, China) and the follow up held later that year (co-located with JURIX 2009, November 16–18, Rotterdam, The Netherlands). AICOL III, published in 2012, resulted from the third AICOL workshop (co-located with the 25th IVR World Congress, August 15–20, 2011, Frankfurt, Germany) along with AICOL-IV@IVR (July 21–27, 2013, Belo Horizonte, Brazil), and AICOL-V@SINTELNET-JURIX (December 11, 2013, Bologna, Italy). The fourth volume included papers from AICOL-VI@JURIX 2015, AICOL-VII@EKAW 2016, AICOL-VIII@JURIX 2016, AICOL-IX@ICAIL 2017, and AICOL-X@JURIX 2017. The present volume includes AICOL-XI@JURIX2018, held in Groningen, the Netherlands, on December 12, 2018, and AICOL-XII@JURIX2020, held online via Brno, Czech Republic, on December 9, 2020.

The AICOL proceedings are usually published every two years in the form of a selection of the best papers. The present edition is no exception, although COVID-19 has made things a bit more difficult. We faced this situation by joining forces with Technologies for Regulatory Compliance (TERECOM, from the H2020 Project LYNX H2020-780602) and EXplainable & Responsible AI in Law (XAILA), also held jointly with JURIX in 2020, as detailed in the Introduction. It is worth mentioning that the LAST-JD-RIoE Marie Skłodowska-Curie Innovative Training Network provided some papers with promising research as well.

The present volume contains 21 contributions (17 long papers and 4 short papers). All contributions were submitted to a double peer-review process — first to be accepted at the scientific workshops, and then to be resubmitted, reviewed again, and published. The contributions have been organized into the following blocks: (i) Knowledge Representation; (ii) Logic, Rules, and Reasoning; (iii) Explainable AI in Law and Ethics; (iv) Law as a Web of Linked Data and the Rule of Law; (v) Data Protection, Privacy Modeling, and Reasoning. This time, the volume reflects the ethical turn that has occurred in the field of AI during the past three years, also fostering a new reflection on legal rights, legal governance, and the rule of law.

As always, we thank the reviewers for their excellent work, and we encourage the reader to explore the outcomes of research having in mind what they can observe by themselves, in the ever changing brave new world of digital regulation.

August 2021

Víctor Rodríguez-Doncel
Monica Palmirani
Michał Araszkiewicz
Pompeu Casanovas
Ugo Pagallo
Giovanni Sartor

AICOL Organization

Program Committee Chairs

Víctor Rodríguez-Doncel   Universidad Politécnica de Madrid, Spain
Pompeu Casanovas   La Trobe University, Australia/IDT-UAB, Spain
Monica Palmirani   University of Bologna, Italy
Ugo Pagallo   University of Turin, Italy
Giovanni Sartor   University of Bologna/EUI, Italy
Danièle Bourcier   CNRS, Paris 2 Panthéon-Assas University, France

Program Committee

Laura Alonso Alemany   National University of Córdoba, Argentina
Michal Araszkiewicz   Jagiellonian University, Poland
Guido Boella   University of Turin, Italy
Marcello Ceci   University College Cork, Ireland
Luigi di Caro   University of Turin, Italy
Tom van Engers   University of Amsterdam/TNO, The Netherlands
Enrico Francesconi   ITTIG, Italy/EurLex, Luxembourg
Michael Genesereth   Stanford University, USA
Jorge González-Conejero   Autonomous University of Barcelona, Spain
Guido Governatori   Data61, CSIRO, Australia
Davide Grossi   University of Liverpool, UK
John Hall   Model Systems, UK
Renato Iannella   Airservices Australia, Brisbane, Australia
Beishui Liao   Zhejiang University, China
Arno Lodder   Vrije University, The Netherlands
Marco Manna   University of Calabria, Italy
Martin Moguillansk   Universidad Nacional del Sur, Argentina
Pablo Noriega   IIIA-CSIC, Spain
Paulo Novais   University of Minho, Portugal
Adrian Paschke   Freie Universität Berlin, Germany
Silvio Peroni   Universtity of Bologna, Italy
Ginevra Peruginelli   ITTIG, Florence, Italy
Enric Plaza   IIIA-CSIC, Spain
Marta Poblet   RMIT, Australia
Martín Rezk   Free University of Bozen-Bolzano, Italy
Antoni Roig   Autonomous University of Barcelona, Spain
Livio Robaldo   University of Luxembourg, Luxembourg
Piercarlo Rossi   University of Piemonte Orientale, Italy
Antonino Rotolo   University of Bologna, Italy

Carles Sierra	IIIA-CSIC, Spain
Barry Smith	University of Buffalo, USA
Clara Smith	UNLP/UCALP, Argentina
Said Tabet	RuleML Initiative, USA
Daniela Tiscornia	ITTIG, Italy
Leon van der Torre	University of Luxembourg, Luxembourg
Raimo Tuomela	University of Helsinki, Finland
Anton Vedder	Tilburg University, The Netherlands
Serena Villata	Inria, France
Fabio Vitali	University of Bologna, Italy
Adam Wyner	University of Aberdeen, UK
Radboud Winkels	University of Amsterdam, The Netherlands
John Zeleznikow	La Trobe University, Australia

XAILA Organization

Program Committee Chairs

Michał Araszkiewicz	Jagiellonian University, Poland
Martin Atzmueller	Osnabrück University, Germany
Grzegorz J. Nalepa	Jagiellonian University, Poland
Bart Verheij	University of Groningen, The Netherlands

Program Committee

Kevin Ashley	University of Pittsburgh, USA
Floris Bex	Utrecht University, The Netherlands
Szymon Bobek	Jagiellonian University, Poland
Jörg Cassens	University of Hildesheim, Germany
David Camacho	Universidad Autonoma de Madrid, Spain
Pompeu Casanovas	La Trobe University, Australia/IDT-UAB, Spain
Enrico Francesconi	ITTIG, Italy/EurLex, Luxembourg
Paulo Novais	University of Minho, Portugal
Tiago Oliveira	National Institute of Informatics, Japan
Martijn von Otterlo	Tilburg University, The Netherlands
Adrian Paschke	Freie Universität Berlin, Germany
Juan Pavón	University Complutense de Madrid, Spain
Monica Palmirani	University of Bologna, Italy
Radim Polčák	Masaryk University, Czech Republic
Víctor Rodríguez-Doncel	Universidad Politécnica de Madrid, Spain
Ken Satoh	National Institute of Informatics, Japan
Jaromír Šavelka	Carnegie Mellon University, USA
Erich Schweighofer	University of Vienna, Austria
Michal Valco	Constantine the Philosopher University of Nitra, Slovakia
Bart Verheij	University of Groningen, The Netherlands
Tomasz Żurek	Maria Curie-Skłodowska University of Lublin, Poland

Additional Reviewers for this AICOL-XAILA Edition

Paulo H. C. Alves
M. Luisa Alvite-Díaz
Ilia Chalkidis
Rajaa El Hamdani
Nicoletta Fornara
Alessio Fiorentino
Carles Górriz-López
Mustafa Hashmi
Llio Humphreys

Mercedes Martínez-González
Harshvardhan J. Pandit
Juan Pavón
Radim Polčák
Shashishekar Ramakrishna
Ariana Rossi
Jaromir Savelka
Piotr Skrzypczynski
Gijs van Dijck

Contents

Introduction: A Hybrid Regulatory Framework and Technical Architecture for a Human-Centered and Explainable AI

Víctor Rodríguez-Doncel[1]([⊠]) [iD], Monica Palmirani[2] [iD], Michał Araszkiewicz[3] [iD], Pompeu Casanovas[4,5] [iD], Ugo Pagallo[6] [iD], and Giovanni Sartor[2] [iD]

[1] Universidad Politécnica de Madrid, 28040 Madrid, Spain
vrodriguez@fi.upm.es
[2] University of Bologna, 1088 Bologna, Italy
{monica.palmirani,giovanni.sartor}@unibo.it
[3] Jagiellonian University, 31-007, Kraków, Poland
michal.araszkiewicz@uj.edu.pl
[4] La Trobe University, Bundoora, VIC 3086, Australia
p.casanovasromeu@latrobe.edu.au
[5] Autonomous University of Barcelona, 08193 Barcelona, Spain
pompeu.casanovas@uab.cat
[6] University of Turin, 10124 Torino, Italy
ugo.pagallo@unito.it

Abstract. This introduction presents the fifth volume of a series started twelve years ago: the AI Approaches to the Complexity of Legal Systems (AICOL). The introduction revises the recurrently addressed topics of technology, Artificial Intelligence and law and presents new challenges and areas of research, such as the AI ethical and legal turn, hybrid and conflictive intelligences, regulatory compliance and AI explainability. Other domains not yet fully explored include the regulatory models of the Web of Data and the Internet of Things that integrate legal reasoning and legal knowledge modelling.

Keywords: AICOL workshops · Artificial intelligence and law · Semantic web · LegalXML · Web of linked data · Internet of Things · Ethics · Human rights · Privacy · Rule of law

1 Introduction

This is the Introduction to the fifth volume of *AI Approaches to the Complexity of Legal Systems* (AICOL). During the past twelve years, AICOL editions have been consistently showing the evolution of the research carried out in the field of technology, Artificial Intelligence, and law. In this edition, we point out the main trends and developments of the last three years, in a time marked by the pandemic, but also by a growing and sustained heed in the regulatory models of the Web of linked Data and the Internet of Things integrated with legal reasoning and legal knowledge modelling. It would be a case

V. Rodríguez-Doncel et al. (Eds.): AICOL-XI 2018/AICOL-XII 2020/XAILA 2020, LNAI 13048, pp. 1–11, 2021.
https://doi.org/10.1007/978-3-030-89811-3_1

of increasing returns, as many outcomes of research are just a step ahead, not having been implemented yet. In this scenario we are assisting to the cross-fertilization of different research fields: subsymbolic AI should be combined with symbolic AI and Semantic Web in order to provide better instruments to implement the concept of *explicability* and the ethics principles. Human-centered AI needs integrated approach where technical, social, legal and ethics approaches are used together for supporting the Human-in-the-loop principle. This is a matter of time, as more young researchers are joining and are advancing and figuring out new approaches to cope with the problems raised in digital scenarios that are becoming the natural ecosystems of our time. But it is also a matter of will, as many solutions require the agreement and collaborative effort of citizens, companies, governments, and social and economic institutions. As we will state in this volume, we are facing *hybrid* and conflictive environments, in which, the simultaneous emergence of new technologies is changing the notion of what it means to be human in the digital age [1]. The reminder of the Introduction will briefly address these renewed challenges, focusing on its ethical and legal components, before briefly introducing the contents of the volume.

2   A Latest AI Ethical and Legal Turn

Scholars and institutions have extensively discussed the ethical (and legal) principles of AI over the past years. So far, we have got more than a hundred of such declarations, recommendations, or charters[1]. Since the early 2020s, this kind of exercise has somehow turned out to be more precise from a legal viewpoint with a series of initiatives brought about by the European Commission. The Commission has proposed a number of acts that directly affect the design, production or use of AI systems. This is the case of the Digital Services Act package—which includes the Digital Services Act and the Digital Markets Act—much as the proposal for a new AI regulation from 21 April 2021. The final version of these texts remains of course an open issue. For example, the final form of the GDPR which was approved in 2016 was very different from the initial proposal of the European Commission in 2014. We need no prophetic powers, however, to suspect that most of this new set of legal provisions are here to stay and will affect some of the ways in which AI can be used and even designed.

Some of the new set of legal rules proposed by the European Commission are already the focus of some chapters in this book. After all, the proposal for a new AI regulation in Europe adopts several recommendations that your AICOL editors have advanced over the past years, be it Art. 53 of the AI Regulation on sandboxes [2], or Art. 56 on the European AI Board [3]. On the other hand, according to a recurrent technique of EU regulations, the legal bar is often set in connection with the state-of-the-art. In the phrasing of Recital no. 49 of the AI Act, "High-risk AI systems should perform consistently throughout their lifecycle and meet an appropriate level of accuracy, robustness, and cybersecurity in accordance with the generally acknowledged state of the art."

Likewise, the regulation hinges on the state of the art in Art. 9(3) on risk management systems, and Art. 10(5) on data and data governance. In this light it is fundamental also

---

[1] To be more precise, as of August 2021, the online inventory of AI Ethics guidelines gathered 173 documents. https://inventory.algorithmwatch.org/.

to investigate the Data Governance Act proposal of the European Commission where a new form of consent is supposed to open new intensive debates among scholars: "data altruism" permits to contribute personal data to recognized organizations for "general interest, such as scientific research purposes or improving public services". The Data Strategy is oriented to encourage big data sharing as "common good" and it is frequently clashing with the human rights framework devoted to asking for transparency, accountability, explicability and knowability of dataset, AI modelling, parameters, and algorithm processing. The next challenge of the ongoing digital transformation is to govern the groundbreaking technologies through a harmonized regulation, capable of correctly balancing all of the following: economic growth of digital economy, unit-trust policy for a fair competition, data sovereignty for protecting society and fundamental rights (e.g., face recognition EU approach).

3   A Hybrid and Conflictive Environment

Hybrid intelligence (HI) has been recently defined as follows: "We define HI as the combination of human and machine intelligence, augmenting human intellect and capabilities instead of replacing them, to make meaningful decisions, perform appropriate actions, and achieve goals that were unreachable by either humans or machines alone" [4]. The authors have also identified the main HI research challenge for the times to come: "how to build adaptive intelligent systems that augment rather than replace human intelligence, leverage our strengths, and compensate for our weaknesses while taking into account ethical, legal, and societal considerations."

This might certainly be an undisputed goal for general AI and for regulatory designs as well, putting AI at the center of these ethical and legal challenges. From AICOL, we could suggest an additional turn. Stemming from the same perspective, the regulatory lens should be adjusted to promote this approach, adapting it to an evolving and changing legal environment. We deem the authors right. Without a deep social knowledge of the economic and political conditions in which the technology is created and implemented, we barely will reach this 'noble dream' (borrowing from Hart this qualification). The technology we will live by in the immediate future is not neutral and can be populated, designed, and used for different and opposite purposes in an increasingly conflicting world [5].

But just think through legal lenses. Even if the systems built are *cooperative* (working in synergy with humans), *adaptive* (they learn from and are adapted to humans), *responsible* (behaving ethically), and *explainable* (sharing with humans their rationale), they still will be disruptive and will generate controversies, disputes and eventually lawsuits and conflicts that are susceptible to escalate. *Plus ça change...* In sum, law, legal instruments, what we have been understanding as counting as law so far, should also be rebuilt to give way to a common hybrid relationship both with machines and also *with humans dealing with machines*. This second order relationship is even more relevant than a simple M-H-M contextualisation of possible scenarios, because our humanity is being transformed through this experience. This is the reason why rethinking legal change and new legal toolkits is so important.

As shown by some of the papers that we are publishing here, there are new ways of regulating that are more cooperative and open to dialogue. We have already mentioned the use of sandboxes to test innovation. We could add many other instruments in which agents cooperate within multi-stakeholder governance systems (standards, protocols, agreements, soft law, among many others). However, even though, the balance between what is (or should be) mandatory or forbidden and what is (or should be) dialogical or directly allowed and secured is an open question. Fostering trust and ensuring secure and safe transactions on the web might not cover all possible venues. There are many other conceptual tools to be put in place and implemented *in-between* enforcement and agreement technologies: business (regulatory) and legal compliance; strict compliance and relative (or partial) compliance; centralized or decentralized governance systems; integrated or distributed management systems, are open issues for digital currencies and platform-driven economies, urging drafters, regulators, and legislators to revise the architecture of the formal rule of law, the place of national jurisdictions, and the substantive enactment of rights.

At the same time, red flag issues about populist extremism, hate speech, organised crime, and violent social behaviour, to mention only a few problems, raise new concerns and lead us to undertake an in-depth discussion on the shift that is currently occurring in the digital age and the technological way to respond to these challenges.

For instance, the use of AI tools for filtering, unlawful or objectionable online content and activities has become possible due to recent technological achievements (e.g., in the domain of NLP), and is indeed needed for preventing online abuses and ensuring effective moderation. This opportunity, however, has raised well-grounded worries concerning the extent to which to online freedoms may consequently be curtailed, through censorship and further measures against the concerned individuals. Only by enhancing technologies and better complementing and integrating them with human skills can this challenge be met.

Further issues pertain the extent to which AI expands opportunities for surveillance, control, and differential assessment of individuals. Technologies such as face recognition enable people being tracked in physical environments, complementing the technologies for online tracking. At the same time, the collected data can be processed through AI technologies for the purposes of assessing individuals, assigning them classifications that may link to detrimental consequences. Defining criteria for fair and accurate assessments of individuals, by combining automated analysis of relevant data (collected and processed in compliance with data protection law) with human oversight and complementary analysis, is a key legal and political challenge for the future.

4 A Hybrid Technological Framework

The non-symbolic AI is rapidly evolving, and it is everyday more evident that a hybrid technical framework could produce better results in the legal domain, combining machine learning and deep learning based on stochastic technologies, with semantic knowledge modelling, legal reasoning, and a symbolic rule-based approach [6, 7]. The main problem in the current applications of non-symbolic AI (ML/DL) in the legal domain is the lack of contextual information that affects the capacity to create useful relationships between the different annotations, classifications, clustering, correlation, and regression elaborated by ML/DL technologies. Interconnecting all the extracted legal knowledge permits a better interpretation [8] by human beings and the implementation of the Human-in-the loop, Human-on-the-loop, Human-in-command principles[2] [9].

There are four main problems in the current state of the art of the ML/DL applications of the legal documents: 1) the ML/DL works without logic and semantics, and many contexts information included in the legal document are neglected with an evident lower capacity of interpretation; 2) the legal citation is a consolidated best practice in legal disciplines to delegate to external textual resources some important meta-role (e.g., definitions, derogations, modifications, integration of prescriptiveness, penalty, conditions, etc.). This means that ML/DL should also consider the cited text because especially some algorithms (e.g., similitude, grouping) can find similar the text "art. 3" and "art. 13" when the content is completely different. For this reason, the network of norms through citations should be included in the baseline of the experiments; 3) temporal parameters are fundamental for creating a robust ML/DL dataset. Case-law based on repealed legislation should have lower weight respect case-law based on new legislation even if these datasets are less frequent. So, frequency, probabilistic, temporal series should be mitigated with relevance and legal validity criteria; 4) logic and semantic web annotation should be integrated with ML/DL in order to understand the type and meaning of relationship that connect different sentences in the text (e.g., obligation and penalty, obligation and derogation).

For this reason, a hybrid architecture is necessary to integrate the ML/DL legal knowledge extraction with Semantic Web annotation and legal deontic logic modelling, and to achieve a better, sound result for the legal discipline [8]. Moreover, this hybrid approach is functional also with regard to the explicability principle that is included in the proposal of the European Commission about Artificial Intelligent Act (AIA[3]) in arts. 13 and 14. The hybrid approach is also fundamental for implementing co-regulation through standardization bodies (based on the New Legislative Framework - NLF[4]) according to Recital (61)[5] AIA [10]. It is essential for supporting interoperability within the large landscape of the Artificial Intelligence domain, to minimize fragmentation, overcome technical and organizational barriers, and set different benchmarking criteria and legacy systems [11].

---

[2] "Human oversight helps ensuring that an AI system does not undermine human autonomy or causes other adverse effects. Oversight may be achieved through governance mechanisms such as a human-in-the-loop (HITL), human-on-the-loop (HOTL), or human-in-command (HIC) approach. HITL refers to the capability for human intervention in every decision cycle of the system, which in many cases is neither possible nor desirable. HOTL refers to the capability for human intervention during the design cycle of the system and monitoring the system's operation. HIC refers to the capability to oversee the overall activity of the AI system (including its broader economic, societal, legal and ethical impact) and the ability to decide when and how to use the system in any particular situation.", on 8 April 2019, the High-Level Expert Group on AI presented Ethics Guidelines for Trustworthy Artificial Intelligence, pag. 16.

[3] European Commission, Proposal for a Regulation of the European Parliament and of the Council laying down harmonised rules on artificial intelligence (Artificial Intelligence Act) and amending certain Union legislative acts (COM(2021) 206 final) (hereafter AIA).

[4] https://ec.europa.eu/growth/single-market/goods/new-legislative-framework_en.

[5] "[s]tandardization should play a key role to provide technical solutions to providers to ensure compliance with this Regulation".

5   AICOL, XAI, and XAILA and TERECOM Workshops

Against this framework, we are delighted to introduce a set of vibrant contributions to today's state of the art in this volume, the fifth in a series of books on the *AI Approaches to the Complexity of Legal Systems (AICOL)*. The book contains 21 contributions. Some of them were presented in the AICOL workshops of 2018 (co-located within the JURIX conference in Groningen, The Netherlands) and 2020 (co-located with the online JURIX conference in Brno, Czech Republic). We also selected some papers that were submitted first to workshops such as TERECOM (on *technologies for regulatory compliance*, in 2018, Groningen), and XAILA (*eXplainable AI and law*, in the 2018, 2019 and 2020 editions, as we will precise below).

Contributions have been logically organised into the following blocks: (i) Knowledge representation; (ii) Logic, rules, and reasoning; (iii) Explainable AI in Law and Ethics; (iv) Law as Web of linked Data and the Rule of Law; (v) Data Protection and Privacy Modelling and Reasoning.

The chapters included into these sections are diverse in nature, and they cover both theoretical reflections (as the thoughts on rule of law and compliance) and practical problems (as the representation of Italian law in a machine-readable form), spanning from the description of complete legal information systems to fine-grain algorithms. Let's elaborate on top of that.

During recent years, the ubiquitous character of AI-based tools and solutions, many of which provide the basis of automated or semi-automated decision making relevant from the point of view of human rights, prompted the social interest in understandability of the algorithms' operation. This topic has been discussed under the heading of Explainable AI (XAI) and the related concepts such as transparency or simulatability. The European approach towards the development of AI emphasizes the ethical and the legal aspects of intelligent systems, directed towards building trustworthy AI. This direction has already been reflected in documents issued by the different European bodies, including the High-Level Expert Group on Artificial Intelligence (AI-HLEG), the European Parliament, and the European Commission, as discussed above. Therefore, it is not surprising that the topics of XAI have gained importance in the AI and Law community and prompted the idea of explainable computational models of legal reasoning and of the understandability of the operation of Machine Learning models supporting the performance of legal tasks.

As advanced above, among the recent events dedicated to this topic, the three consecutive XAILA (eXplainable AI and Law) workshops have taken place as accompanying events of JURIX 2018 in Groningen, JURIX 2019 in Madrid and the online JURIX 2020 organized by the Masaryk University in Brno, Czech Republic. The list of workshops' co-organizers encompasses Grzegorz J. Nalepa, Martin Atzmueller, Michał Araszkiewicz, Paulo Novais and Bart Verheij.

In parallel, along with the development of the European Project LYNX, two Workshops on regulatory and legal compliance were held at JURIX 2017 in Luxembourg, and JURIX 2018 in Groningen. Víctor Rodríguez-Doncel, Elena Montiel, Jorge González-Conejero, and Pompeu Casanovas served as organizers and co-chairs.

6 Contents of the Present Volume

This volume includes revised and extended versions of selected papers. The contents of these papers provide an instructive overview of the topics and approaches recognized as relevant in the AICOL and XAILA communities. Clustering is always a difficult task that we have carried out in a flexible way, as many papers might fall into two or three sections. They have not been distributed according to their provenance but rather following a reasonable order, considering their topic, homogeneity, methodology and theoretical approach. This will contribute to facilitate their readability, allowing the reader to make easier connections to papers with similar contents. It is worth mentioning here hat long papers have been placed first in their correspondent section. Short and position papers come later.

In the first section, on knowledge representation, we have placed three long papers, followed by a short one. In "Identification of Legislative Errors through Knowledge Representation and Interpretive Argumentation", Michał Araszkiewicz and Tomasz Zurek, in line with some positions previously referred to in this Introduction, assert that high-quality legislative drafting requires not only linguistic fluency, but also in-depth legal expertise. They address the latter issue, and develop a frame-based, semi-formal model, useful in the identification and discussion of potential legislative errors. Monica Palmirani in "Lexdatafication: Italian Legal Knowledge Modelling in Akoma Ntoso" focuses on legal information retrieval. She offers the results of Lexdatafication, a project for modelling Italian legal knowledge information and Constitutional Court decisions through Akoma Ntoso and the existing XML and metadata provided by the Open Data portal. In "A critical reflection on ODRL" Milen Girma Kebede, Giovanni Sileno and Tom van Engers address the issue of representation of rights into policy languages. They reflect on the pros and cons of the Open Digital Rights Language (ODRL), a well-known W3C Recommendation by now. It is worth noting that the authors' observations stem from their own practical experience of trying ODRL as a potential solution for representing obligations, prohibitions, permissions, rights and duties in data-sharing regulation. Finding a suitable policy specification language to represent legal fundamental concepts present in normative sources in a computational form is the subject matter of the last (short) paper of this section, "Automating normative control for healthcare research", by Milen Girma Kebede. Her contribution addresses the challenges of developing an access control model capable of specifying and operationalizing policies from legal and other institutionally relevant sources, artifacts or events, and reasoning over them.

The second section is devoted to logic, rules and reasoning. As already stated, some papers under this category could also fit nicely into the previous one. This is the case of "Principles and Semantics: Modelling Violations for Normative Reasoning", by Silvano Colombo Tosatto, Guido Governatori and Antonino Rotolo; and "Towards a Formal Framework for Partial Compliance of Business Processes", by Ho-Pun Lam, Mustafa Hashmi and Akhil Kumar. The first paper proposes a structural operational semantics for explicitly representing in force obligations and violations as events in a temporal framework. The second one builds an evaluation framework to quantify the degree of compliance of business processes across different levels of abstraction and across multiple dimensions for each task. Partial compliance might replace the binary yes/no compliance checking, being able to better reflect the complexity of regulatory processes. Both papers are written at a high level of formality but having practical legal cases in mind. Likewise, in their contribution, Heng Zheng, Davide Grossi and Bart Verheij apply classical logic to enable case-based reasoning in a setting cognate to but in many respects diverging from the classical dimension of factor-based approach discussed in the earlier literature. Their comparison approach identifies and defines analogies, distinctions, and relevances, and is applied to the temporal dynamics of case-based reasoning using also a model of real-world cases. The last paper in this section, "Reasoning over Knowledge Graphs in an Intuitionistic Description Logic", by Bernardo Alkmim, Edward Haeusler and Daniel Schwabe, present a way to model and reason over Knowledge Graphs (KG) via an Intuitionistic Description Logic called iALC. The authors introduce a natural deduction system to reason on KG, and they apply this modelling to a case study, supporting the definition of trust, privacy and transparency. The two latter papers originally are an illustration of the relevance of symbolic representation for the purposes of developing of explainable computational legal reasoning models.

Consequently, the papers included into the third section are devoted to exploring explainable AI in Law and Ethics. To begin with a general perspective, the contribution by Martijn van Otterlo and Martin Atzmueller, "A Conceptual View on the Design and Properties of Explainable AI Systems for Legal Settings", discusses the two existing approaches concerning explainable AI for legal settings. The first approach aims at designing legal requirements, while the second one deals with a strategy of ensuring ethical requirements with explainability as a core element. The next paper, "Towards Grad-CAM Based Explainability in a Legal Text Processing Pipeline", by Łukasz Górski, Shashishekar Ramakrishna and Jędrzej M. Nowosielski introduces a technique for producing explanations of legal AI systems with the help of adapted Grad-CAM metrics, and it shows the interplay between the choice of embeddings, its consideration of contextual information, and their effect on downstream processing. A different approach is proposed in the third paper under this section, titled "The Difference between Making Things Explainable and Explaining: Requirements and Challenges under the GDPR" co-authored by Francesco Sovrano, Fabio Vitali and Monica Palmirani, which elaborates on the user-centric approach to the explainability of Automated Decision Systems. The authors introduce the approach of explanatory AI (YAI), which includes the idea of holding pragmatic explanatory discourses aiming at providing explanations tailored to the users' needs. The contribution is discussed against the background of GDPR requirements. The third paper, "Explaining Arguments at the Dutch National Police", by Anne Marie Borg and Floris Bex, shows how the standard approach towards computational argumentation modeling—that is, argumentation frameworks—can be used to generate explanations on top of the argumentation systems held by the Dutch National Police. The authors take as example the system in use to assist in the processing of complaints on online trade fraud. Finally, Giovanni Sileno, Alexander Boer, Geoff Gordon and Bernhard Rieder discuss in their contribution, "Like Circles in the Water: Responsibility as a System-Level Function Explainable AI systems", the foundational topic of what eventually determines the semantics of algorithmic decision-making. They contend that it is not the program artefact, nor—if applicable—the data used to create it, but the preparatory (enabling) and consequent (enabled) practices holding in the environment (computational and human) in which such algorithmic procedure is embedded. They relate this idea to the structure of the notion of responsibility.

Section four delves into the rule of law and the web of linked data. There are two ways of facing it, depending on the place we stand by. In a first interpretation, the protections of the substantive rule of law (the content of norms and rights) are expressed on actionable and interoperable web languages. In a second one, the rule of law refers to a set of working (hybrid) practices to protect individuals and vulnerable social groups. In both cases its scope and intention are deemed transnational, i.e. broader than its link to national jurisdictions. In "The Rule of Law and Compliance: Legal Quadrant and Conceptual Clustering", Pompeu Casanovas, Mustafa Hashmi and Louis de Koker present part of the toolkit they have built to carry out an extended survey on legal compliance in the frame of the Australian Data to Decisions CRC project. They describe the design of a legal quadrant to map and navigate the rule of law, the clustering of legal concepts, and the methodology and metrics followed to select the papers and handle their content. The second paper of the section, "Legal Ecosystems to Address Violent Extremism Fuelled by Hate Speech in Social Media", by Andre Oboler and Pompeu Casanovas, explores the social and political conditions in which the protections and constraints of the rule of law should be sustainably deployed and implemented on the web to address such a problem. It contends that a combination of regulatory instruments, incentives, training, proactive self-awareness and education can be effective to create legal ecosystems to improve the present situation. The third paper of the section, "SPIRIT: Semantic and Systemic Interoperability for Identity Resolution in Intelligence Analysis" addresses the dimension of law enforcement to fight organised crime, a binding dimension of the rule of law. Costas Davarakis, Eva Blomqvist, Marco Tiemann and Pompeu Casanovas introduces an identity resolution service that has been designed to learn about identity patterns, to build up a social graph related to them, and thereby facilitate Law Enforcement Agent's investigation work. Finally, from a technical side, Victor Rodriguez-Doncel and Maria Navas-Loro present in this section a short paper titled "TimeLex: a Suite of Tools for Processing Temporal Information in Legal Texts". TimeLex includes different systems able to process temporal information from legal texts—lawORdate, Añotador, and WhenTheFact. These are tools to handle different aspects of this temporal dimension on legal documents.

The last section of the volume refers to data protection and privacy modelling and reasoning. Its first paper, "Inferring the meaning of non-personal, anonymised, and anonymous data", by Emanuela Podda and Monica Palmirani, explores the problems linked to having two mutually exclusive definition of personal and non-personal data in the legal framework in force, provided by the European General Data Protection Regulation (GDPR), and the Free Flow Data Regulation (FFDR). It discusses the two main data processing tools provided by GDPR, anonymization and pseudonymization. "Challenges in the implementation of Privacy Enhancing Semantic Technologies (PESTs) supporting GDPR", by Rana Saniei, proposes the use of Semantic Technologies in PETs in the form of an Intelligent Compliance Agent (ICA) to support data controllers in carrying out a Data Protection Impact Assessment (DPIA). It contends that models and ontologies representing entities involved in the DPIA process can help data controllers determine the risk of their processing activities. "Publication of court records: circumventing the privacy-transparency trade-off", by Tristan Allard, Louis Béziaud, and Sébastien Gambs, addresses the problem of the sensitive nature of legal decisions being published and accessed as linked data. They carefully analyse the different venues and propose a strawman multimodal architecture for a privacy preserving legal data publishing system. In "Challenges in the Digital Representation of Privacy Terms", Beatriz Esteves discusses the challenges of the implementation of a service based on decentralised Web technologies and Semantic Web standards and specifications to facilitate the communication between data subjects and data controllers. Finally, "The use of Decentralized and Semantic Web Technologies for Personal Data Protection and Interoperability", by Mirko Zichichi, Victor Rodriguez-Doncel and Stefano Ferretti, brings to the discussion on privacy and data protection the role of Distributed Ledger Technologies (DLTs) and smart contracts in handling personal data in trustless scenarios.

To end up our Introduction, it is worth mentioning that at the time this volume was being edited, the first draft for an AI regulation in Europe had been published and a number of AI-related challenges were being addressed in the most recent legislation. If this series of books is about *AI approaches to Legal System*, we can fairly say that this time *Legal Systems* are approaching *AI*. If this volume, published 3 years after GDPR is applicable has collected many papers related to GDPR and data protection, it would be no surprise if the next AICOL volume devotes a number of contributions to address the challenges posed by the new AI regulation.

References

1. Plummer, D., et al.: Gartner's top strategic predictions for 2020 and beyond: technology changes the human condition. In: Gartner ID G00450595, 29 October (2019)
2. Pagallo, U.: Algo-rhythms and the beat of the legal drum. Philos. Technol. **31**(4), 507–524 (2017). https://doi.org/10.1007/s13347-017-0277-z
3. Pagallo, U., Casanovas, P., Madelin, R.: The middle-out approach: assessing models of legal governance in data protection. Artif. Intell. Web Data Theory Pract. Legis. **7**(1), 1–25 (2019)
4. Akata, Z., et al.: A research agenda for hybrid intelligence: augmenting human intellect with collaborative, adaptive, responsible, and explainable artificial intelligence. Computer **53**(8), 18–28 (2020)
5. Casanovas, P., Wishart, D., Chen, J.: The digital world we will live by. Law Context Socio-Legal J. **37**(1), 1–5 (2020)
6. Palmirani, M., Bincoletto, G., Leone, V., Sapienza, S., Sovrano, F.: Hybrid refining approach of pronto ontology. In: Kő, A., Francesconi, E., Kotsis, G., Tjoa, A.M., Khalil, I. (eds.) EGOVIS 2020. LNCS, vol. 12394, pp. 3–17. Springer, Cham (2020). https://doi.org/10.1007/978-3-030-58957-8_1
7. Ashley, K., et al.: Proceedings of the Fourth Workshop on Automated Semantic Analysis of Information in Legal Text held online in conjunction with the 33rd International Conference on Legal Knowledge and Information Systems, ASAIL@JURIX 2020, 9 December 2020. CEUR Workshop Proceedings 2764 (2020)
8. Deakin, S., Markou, C.: Is Law Computable? Critical Perspectives on Law and Artificial Intelligence. Hart Publishing (2020)
9. Monarch, R.M.: Human-in-the-Loop Machine Learning: Active Learning and Annotation for Human-Centered AI. Simon and Schuster (2021)

10. Ebers, M.: Standardizing AI - the case of the european commission's proposal for an artificial intelligence act. In: The Cambridge Handbook of Artificial Intelligence: Global Perspectives on Law and Ethics, 6 August 2021
11. Veale, M., Zuiderveen Borgesius, F.: Demystifying the draft EU artificial intelligence act. Comput. Law Rev. Int. **22**(4), 97–112 (2021)

Knowledge Representation

Identification of Legislative Errors Through Knowledge Representation and Interpretive Argumentation

Michał Araszkiewicz[1]([⊠])[iD] and Tomasz Zurek[2][iD]

[1] Department of Legal Theory, Faculty of Law and Administration,
Jagiellonian University, Ul. Gołębia 24, 31-007 Kraków, Poland
michal.araszkiewicz@uj.edu.pl
[2] Institute of Computer Science, Maria Curie Sklodowska University,
Ul. Akademicka 9, 20-033 Lublin, Poland
tomasz.zurek@poczta.umcs.lublin.pl

Abstract. High-quality legislative drafting requires not only linguistic fluency, but also in-depth legal expertise. In this paper we focus on the second problem and we develop a frame-based, semi-formal model useful in identification and discussion of potential legislative errors. We present a partial systematization of (potential) legislative errors in terms of the types of expert knowledge necessary to ascertain them. The systematization is based on the anticipated interpretation of the statutory text.

Keywords: Frames · Legal interpretation · Legislative errors · Knowledge representation

1 Introduction

This paper[1] presents how (1) frame-based knowledge that represents statutory norms and (2) argumentation schemes that represent justification of interpretive hypotheses—two methods widely employed in symbolic AI & Law—may be fruitfully used to identify errors and potential errors in statutory text. The order of investigations is as follows. In Sect. 2, we present an informal, theoretical discussion of the notion of legislative errors. In Sect. 3, we provide an informal systematization of legislative errors and potential legislative errors in terms of the types of knowledge that are required to identify them. Section 4 presents a semi-formal model that may be used for the sake of the identification of (potential) legislative errors, particularly in the process of statutory drafting. We demonstrate the usefulness of the model with an example. Section 5 discusses results and suggests the directions of future research.

---

[1] It presents the results of the project UMO-2018/29/B/HS5/01433 entitled Legislative Errors and Comprehensibility of Legal Text.

© Springer Nature Switzerland AG 2021
V. Rodríguez-Doncel et al. (Eds.): AICOL-XI 2018/AICOL-XII 2020/XAILA 2020, LNAI 13048, pp. 15–30, 2021.
https://doi.org/10.1007/978-3-030-89811-3_2

2   The Notion of Legislative Error

Statutory texts are drafted in natural language; therefore, a certain degree of indeterminacy is their essential feature. The sources of this indeterminacy are widely debated in the legal-theoretical literature [9]. Let us enumerate some of them.

- Ambiguity. If a word or expression is ambiguous, then it is possible to assign more than one different meanings to it. We distinguish semantic and syntactic ambiguity. If an expression is semantically ambiguous, then it has at least two different, potentially mutually exclusive, meanings. A classical example of such word is "key" which has at least three different meanings: a device used to open the door, an element of a keyboard or a musical key (there are 12 major and 12 minor keys in the diatonic system). Syntactic ambiguity is caused by the composition of a complex phrase where, for instance, it may be not clear to which part of a sentence another part of a sentence refers to. Other type of syntactic ambiguity may follow from the use of connectives. Ambiguity has been extensively investigated in the logical research on law for decades [1,2].
- Open texture. Although the term "open texture" is often used to refer to any type of indeterminacy in language, it must be stressed that according to the original account presented by Waismann, who introduced the term [18], its meaning should be clarified in the following manner. For any predicate P of a language L (excluding the language of mathematics), it is possible that a situation S will occur such that either (i) it will be necessary to modify the existing meaning of the predicate P to include S in its extension, or (ii) it will be necessary to coin another predicate, Q, to include S in its extension. Therefore, basically no predicate in natural language (except from the mathematical predicates), even though its meaning may appear to be settled, is safe from potential doubts if an untypical situation occurs.
- Vagueness. Of the many theories of vagueness, the one referring to the pragmatic concept of the native speaker is particularly useful. In this account, a predicate P in language L is vague if and only if its fringe is non-empty. The fringe of the predicate is a set of such objects that a native speaker of the language L doubts regarding whether they belong to the extension of P or non-P. The "native speaker" in this context is a theoretical construct, an idealized person who has complete knowledge of the language and does not make linguistic mistakes [10]. Therefore, even full linguistic competence does preclude doubts with regard to the use of some terms. Common terms of this type are "large" or "significant". The use of obviously vague terms is sometimes desirable in legal provisions to attain the flexibility of regulation.
- Contextual dependence. Certain terms are of such a nature that their proper use is relative only to a particular context. For instance, a person may be considered "tall" in the context of the general population but not in the context of a basketball team. In law, a classical term possessing this feature is "fault"; for instance, a psychophysical behavior may be considered to be

at fault in the context of the ascription of civil liability but not at fault in the context of evaluation from the point of view of criminal liability (or vice versa). This is because different contrast classes are taken into account in the assessment of the behavior. The contrast class is a set of objects or states of affairs that creates a context for the assessment of whether a given object or state of affairs may be included into the extension of a predicate.

- Value judgment sensitivity. Although it may be claimed that legal classification always requires value judgments, certain linguistic expressions call for it explicitly, such as "fair trial", "reasonable deadline", or "good faith". Value judgment sensitivity may be perceived as a subtype of contextual dependence; however, the application of such terms requires specific evaluation regarding the promotion or demotion of legally relevant values.

The abovementioned phenomena are considered "natural" elements of ethnic, and thus legal, language. However, the presence of ambiguity is generally assessed as an error in legislation; therefore, efforts should be taken to eliminate it in the drafting process. In particular, syntactic ambiguity should be eliminated to the highest degree possible, and semantically ambiguous expressions should rather be avoided. However, if the latter expressions are actually used in the legal text, the ambiguity may still be eliminated through interpretation: if one and the same expression is used in different provisions of the text, it should be assigned with the same meaning, unless there are important reasons for other decision. Other phenomena discussed above cannot be eliminated in their entirety, and thus, the occurrences of such expressions cannot be labeled as legislative errors, except for the case where a vague term in used in regulation requiring a high degree of precision. It should be emphasized that the use of vague or highly contextual-dependent expressions may be perfectly justified in order to achieve the possibility of adaptation of the regulation to different circumstances. A question arises regarding the features of legal texts that should be taken into account, excepting obvious grammatical errors or errors in punctuation, as potential errors in legislative drafting.

In the process of legislative drafting, the author of the text should take into account how the statute will be interpreted and applied. In consequence, it is recommended to formulate the text in a way that minimizes the risk of interpretive doubt. As we have noted above, a certain degree of the occurrence of problems leading to such doubt is not eliminable. Moreover, the elimination of such properties of a legal text is sometimes not even desirable. A significant measure of the scope of such doubts may be removed in legal practice by the application of rules of interpretation. However, if the anticipated scope of such problems exceeds a tolerable level, the statute drafter should amend the proposed wording. Of course, the criterion of "tolerability" is in itself vague; therefore, the distinction between legislative error and mere imperfect drafting is blurred and may be subject to debate in concrete situations. Our model may serve as a clarification tool to facilitate the discussion in such doubtful situations. However, we also indicate a category of situations that qualify as definitive legislative errors.

In such cases, the drafting of a statute determines a violation of at least one rule of interpretation in any possible understanding of the statutory text.

3   A Systematization of (Potential) Legislative Errors

Here, we present a systematization of legislative errors and potential legislative errors in terms of the type of knowledge that is necessary to ascertain them. We are particularly interested in the category of potential errors, i.e., situations in which it is unclear whether a part of a legal text is actually an error or whether it is just complex or encumbered with some natural linguistic features, discussed in the previous section. For instance, the presence of (a high degree of) vagueness does not necessary have to be assessed as an error of the text, because the use of a vague expression may be justified in the light of the purposes of the regulation.

Let us begin with the category of situations that may be straightforwardly referred to as errors in legal texts. The first category encompasses a set of syntactic errors (i.e., situations in which the legal text violates the rules of syntax of language under which it is formulated). Some examples of such errors would include

(1) Incomplete provisions: *The seller ought to transfer the property to the.* (the recipient description is missing)
(2) Other grammatically incorrect provisions: *The buyer ought to transfers the property to the sellers.*

The erroneous character of the above examples seems obvious on the one hand, but it may also be discussed in terms of legal interpretation theory on the other hand. Each act of legal interpretation should be carried out in conformity with the rules of the syntax of the language used to write the legislation. Therefore, we may conclude that the knowledge needed to ascertain the erroneous character of the above examples is the knowledge of English syntax.

The second category encompasses semantically incorrect provisions (i.e., syntactically properly constructed provisions, which are, however, classified as semantic nonsense):

(3) *Colorless green ideas ought to sleep furiously*[2].

Here, the knowledge necessary to identify the error is basic knowledge of English semantics. Let us, however, note that the category of semantic nonsense should be treated with caution in the domain of law, taking into account the possibility of existence of stipulative definitions, which are abstracts of the actual, common usage of the terms. In certain definitional settings, the provision (3) could be assessed as meaningful and correct.

However, the abovementioned categories of errors are not very common, as they are relatively easily identifiable in the process of legislative drafting (if they occur in the first place).

The third category of errors in legal text encompasses logical incompatibilities between provisions, as in

---

[2] Example adapted from [8].

(4) *Smoking in the university buildings is forbidden.*
(5) *Smoking in the university buildings is allowed.*

We assume here that the conflict between the provisions cannot be eliminated by, for example, a chronological collision rule (lex posterior derogat legi priori). This would mean that there are inconsistent (via elementary deontic logic) provisions in the legal system. Here, the domain of knowledge that enables us to evaluate the situation as a legislative error is logic (in this example, elementary deontic logic) and the rationality assumption concerning the lawmaker, according to which

A. The rational lawmaker does not enact contradictory provisions.

However, in actual legislative drafting, the errors typically have a much more subtle character, which has connections with the remaining assumptions concerning the rationality of the lawmaker:

B. The rational lawmaker does not enact regulation with legal gaps.
C. The rational lawmaker does not enact redundant provisions.
D. The rational lawmaker creates a coherent regulation (especially, axiologically coherent).

The legal expert knowledge related to the abovementioned assumptions of rationality plays a crucial role in connection with the evaluation of a legislative draft. As noted, the author of the legislative text should anticipate how the act will be interpreted and applied. The rationality assumptions are extensively used in the process of legal interpretation and they form the basis of interpretive arguments widely used in contemporary legal cultures [12]. However, as the concepts used here (legal gaps, redundancy, and coherence) are themselves ambiguous, vague, and subject to debate, it is obvious why the boundary between the category of legislative error and merely imperfect drafting or ineliminable indeterminacy is not clear. Let us show some sources of the possible disagreement in connection with the assessment of the following examples, each of which is related to one of the rationality postulates B and C. As the problems of axiological coherence of legislative text tend to be particularly complex, we do not discuss them in this paper due to space limitations.

(6) *The mayor is authorized to collect from the owner the calculated payment resulting from the increase of the value of real property.*

At first look, the provision (6) seems not to generate any interpretive doubts. However, closer scrutiny reveals that before the payment is collected, it must first be calculated. The mayor is explicitly authorized only to collect the payment; we assume here that no other provisions identify an authority competent to calculate it. The question arises regarding whether one might infer that the mayor is also authorized to calculate the payment in order to make the provision (6) operative. A similar provision has actually been assessed by Polish administrative courts and has been evaluated as a genuine legal gap, requiring the amendment of the statute because no authority has been explicitly authorized to calculate the payment.

(7) *During the partnership, the partner may not require the debtor to pay his share in the company's receivables or submit the company's receivables to his creditor for set-off.*

The above example is an actual provision 36.1 of the Polish Code of Commercial Companies, commonly criticized for its redundant character. The repeal of this provision would not alter the position of the partner of the partnership because the latter is a separate legal entity; therefore, it is obvious that the partners do not have the abovementioned claims against the debtor or the creditor. On the other hand, it is argued that this provision has a clarifying character, and though it is normatively redundant, it is useful, because it removes any potential doubt concerning the scope of the partners' rights. The concept of coherence is one of the most debated legal notions [3,5]. Axiological coherence of a legal regulation may be accounted for as the following set of property of legal provisions: (1) the states of affairs which involve similar values should render similar legal consequences and (2) of a state of affairs A promotes (or demotes) a legal value V to the degree D and a state of affairs B promotes (or demotes) a legal value V to the degree E, where E > D, then legal consequences attached to states of affairs A and B should appropriately reflect this difference. These conditions encompass the notions of legally relevant similarity (1) and proportionality (2). An example of an arguably incoherent regulation is discussed in [11]. According to the Polish civil code (art. 480§1):

(8) *In the event of a default of the debtor in performing the obligation to perform, the creditor may, while retaining the claim for compensation for damage, demand authorization by the court to perform the activities at the debtor's expense.*

The above provision enables the creditor to recover expenses from the debtor, however only in the event of default of the debtor. On the other hand, if the creditor claims the performance of an obligation in kind (the behavior to the performance of which the debtor is primarily obliged to) and the debtor fails to do so, then in the course of the enforcement proceedings the creditor may claim the costs of substitute performance. This effect holds even of the debtor is not responsible for the non-performance of an obligation. The relevant rule reconstructed from the provisions of the Civil Code and the Code of Civil Proceedings may be represented as follows:

(9) *If the creditor sues the debtor for performance in kind, even if the debtor is not liable for the default, the creditor will be able to claim the costs of substitute performance in the enforcement proceedings.*

The said regulation should be assessed as incoherent one, because, on the one hand, the art. 480§1 of the Polish civil code enables the creditor to recover the expenses of substitute performance only if the debtor is in default, but in the other hand the rules of civil procedure grant the possibility of recovery if the creditor sues for a performance in kind, even if the debtor is not responsible for the situation. Therefore, although the two situations are similar with regard to the claims of the creditor, in one of them it is necessary for the default condition to be fulfilled, while in another one it is possible to obtain the incurred costs even if the debtor was not at any fault.

4   The Model

The key point of our analysis is, therefore, in the observation that many important legislative errors do not have a purely logical character but instead appear in the specific context of commonsense and legal knowledge. The detection of such errors requires a deep understanding of the context of a norm, case, and whole legal and extra-legal environment. However, we believe that there are specific patterns which can be expressed by means of logical formalism on the basis of which potential errors can be detected, especially some patterns violating principles B, C, and D, presented in the previous section. Due to space limitations, we do not introduce a fully formalized model of knowledge representation and interpretive argumentation, but we remain on the semi-formal, conceptual level. We focus on the violation of the postulates B and C which concern completeness of regulation and the lack of redundancy.

4.1   Basics

Legal provisions usually do not take into consideration particular instances of agents or objects but they operate on the sets of entities instead. Accordingly, we assume a set of sets of entities regulated by legal provisions. For the sake of simplicity, we assume a minimal necessary set of such sets. In order to model a wider class of legal provisions in an adequate way, the more sophisticated models of legal norms and legal relations can be used (for example [7,14,16,17]).

Definition 1 (Agents). *Let AGENTS be a set of types of agents whose behavior is regulated by legal norms.*

Definition 2 (Objects). *Let OBJECTS be a set of types of objects regulated by legal norms.*

Definition 3 (Relations). *Let RELATIONS be a set of the types of relations (between agents or between agent and objects) established by legal norms.*

Definition 4 (Domains). *Let DOMAINS be a set of sets including RELATIONS, OBJECTS, AGENTS. Other sets may also be included in the set DOMAINS possibly other domains regulated by legal norms.*

We use frames that will serve as knowledge representation tools, enabling us to represent relations reconstructed from legal provisions.

Definition 5 (Frames). *Let FRAMES be a set of frames, that is, sets of slots. Each slot may include an element from a subset of DOMAINS*

Frame-based knowledge modeling was introduced in [13]. Here, intuitively, frames represent the content of legal provisions. For instance, a frame may indicate what type of agent may do with regard to a certain object. Here "What an agent may do" refers to a relation between an agent and an object. We assume that slot *relation* can contain the element of the set *RELATIONS*, slots *agent* can contain subsets of *AGENTS*, and slot *object* can contain the subset of set *OBJECTS* etc.

In the following exposition we will typically use frames where at least two agents exist; hence the next slot *counterpart_agent* will also be used. Other slots may represent additional parameters of the represented content (spatial, temporal, procedural, conditions of norm application, etc.). It is important to notice that some slots can be empty and there exist two senses of emptiness of a slot. (1) Empty slot (slot without any content) means that there are no restrictions on the content of this slot. For instance, for some provisions, the temporal aspect of a given action may be irrelevant, which means that this action may be performed at any time; therefore, the slot representing the temporal aspect of the action may be omitted in the frame. This type of emptiness is not problematic from the point of view of the evaluation of a regulation. (2) If the content of a slot is $\emptyset$, it means that it is not possible for the slot to be fit by any element from any set included in *DOMAINS*. This type of situation may be indicative of a legislative error.

By $relation(rel_x) = Y$ we denote that Y is the type of relation represented by frame rel_x, respectively by $agent(rel_x) = A$ we denote that slot *agent* of relation rel_x contains elements of set A (set A is a subset of *AGENTS*), etc. By $slot(rel_x) = \{relation(rel_x) \cup agent(rel_x), ...\}$ we denote a set of sets representing the content of all slots in the frame (rel_x).

For example, by $relation(rel_x) = may\_sell, agent(rel_x) = only\_licensed\_pubs$, $object(rel_x) = beer$, and $counterpart\_agent(rel_x) = adult$, we denote that only licensed pubs can sell beer to adults.

By $sameType(X, Y)$ we denote that two slots represent the same type of slot, for example if X and Y are *agent* slots, then $sameType(X, Y)$).

Definition 6 (Provisions). *Let* $P = \{P_1, P_2, ...\}$ *be a set of legal provisions.*

Legal decisions are not made on the sole basis of legal provisions. The reconstruction of a normative basis of a law-applying decision requires different types of knowledge, including the commonsense knowledge and the various other types of knowledge which provide the content of premises used in legal interpretation. Therefore, we assume.

Definition 7 (Knowledge). *Let* K *be the set of propositions representing relevant knowledge.*

Definition 8 (Interpretive canons). *Let* CAN *be a set of interpretive canons. By interpretive canons, we understand the patterns of reasoning concerning the ascription of meaning to, or the determination of the scope of, legal terms.*

We do not introduce any particular structure of interpretive canons, although we note that they are naturally reconstructed as argumentation schemes (see [4,19], or [6]). We only assume that the result of use of an interpretive canon is a certain determination of the content of the frame's slots.

Definition 9 (Interpretation) *Let a three-tuple* $(can, (P_y \cup K), rel_x)$*, where* $P_y \subseteq P$ *and* $rel_x \in FRAMES$ *and* $can \in CAN$ *be an interpretation, about which we understand that on the basis of the elements of sets* P_y*,* K*, and interpretative canon* can*, it is possible to derive a frame* rel_x*.*

Thus, we intend to represent a result-oriented account of legal interpretation: a result of application of a particular interpretive canon to a legal provision. Typically it is necessary to use also other types of premises in interpretive reasoning and these premises are contained in the set K. Let INT be a set of all interpretations.

Let us note that the expression P_y denotes a set of provisions, and not necessarily a single provision. Therefore, an interpretation may be built on the basis of more than one provision. In such cases the sums of sets expressed in particular slots will be represented in an interpretation. Moreover, in some cases we may consider interpretation of a provision in the light of an interpretation of another provision. For instance, let us consider P_i and P_j. Let us assume that we have an interpretation $int_1 = (can, (P_j \cup K), rel_b)$. In such situation, an interpretation of P_i in the light of P_j taking into account int_1 may be equivalently represented as follows: $(can, (\{P_i, P_j\} \cup K), rel_a) = (can, (P_i \cup rel_b \cup K), rel_a)$

In the set of interpretive canons (CAN) we identify the specific subset thereof, to which we refer as constraining interpretive rules (CIR). The set CIR encompasses the interpretive canons that aim to strictly delimit the scope of acceptable assignment of meaning or scope to the expressions of the legal text. The "constraining" character of these canons of interpretation consists in the fact that they set the boundaries of acceptability of interpretive statements. The CIR are often based on premises referring to the rationality assumptions concerning the legislator (as discussed above). For instance, we have a constraining interpretive rule which prohibits interpretation leading to ascertaining of a legal gap, or a rule which prohibits an interpretation which renders a portion of a legal text unnecessary (the "per non est" rule).

Definition 10 (Constraining interpretive rules). *Let* CIR *be set of formulae using elements from* K*,* N*,* $FRAME$*,* CAN *and variables (which can be substituted by elements or subsets of* K*,* N*,* $FRAME$*, and* CAN*).*

4.2   The Process of Analysis of Legislation

The model presented here has a semi-formal and static character. We introduce neither reasoning nor argumentation mechanism, but we focus on the comparison of the results of interpretive reasoning. A question may arise why the following model aims to represent the results of interpretation, and not the content of the legislation as such. The reason for this is that any formalization of the legislative text, which is primarily expressed in natural language, is already a result of interpretation. In cases of determining the "literal" interpretation this problem is not easily visible, however, we think that it is worthwhile to emphasize this important observation. We believe that in the context of legal knowledge engineering it is not proper to discuss the representation of legal provisions "as such".

The process of detecting potential errors in legal drafts will be divided into 2 main steps.

First Step of Analysis

In this part we illustrate the analysis on the basis of two types of potential errors related to the problem of redundant provisions.

- The detection of generally inapplicable provisions
  A given provision will be potentially generally inapplicable when it cannot be applied in any real life situation because the legal text containing this provision is erroneously drafted. Such a potential error may occur if in the at least one interpretation at least one slot will contain only $\emptyset$ (there does not exist any possibility to substitute it by any element from any set included in $DOMAINS$):
  - $(can, (P_y \cup K), rel_a) \in INT$ and:
  - $\exists_{X \in slot(rel_a)}(X = \emptyset)^3$ (at least one slot will contain only $\emptyset$. E.g.: there is no agent to whom the provision is addressed, etc.)

- The detection of overlapping provisions
  The non-trivial legislative errors often occur when the scopes of at least two provisions overlap. Therefore, we indicate the following conditions that increase the risk of the occurrence of a legislative error. The potential legislative errors will appear if:
  - $(can, (P_y \cup K), rel_a) \in INT$, $(can, (P_x \cup K), rel_b) \in INT$, $P_x \neq P_y$, where:
  - $relation(rel_a) = relation(rel_b)$ (both relations have the same type) and
  - $\exists_{X \in slot(rel_a), Y \in slot(rel_b)}(X \cap Y \neq \emptyset \wedge sameType(X, Y))$ (the scope of of at least one parameter (slot) in both relations are, at least partially, overlapping. E.g.: overlapping agents, objects etc.)

The above conditions represent situations in which two interpretations (of different provisions) concern the same type of relation and the scopes of at least one pair of slots overlap at least partially which means that the interpretations "cover", at least partially, the same agents, objects, counterpart agents etc.

The above can be illustrated by the example:
We represent the interpretations of two legal provisions P_y, P_x:
$(can, (P_y \cup K), rel_a) \in INT$, $(can, (P_x \cup K), rel_b) \in INT$, where:
$relation(rel_a) = may\_sell$, $agent(rel_b) = licensed\_pub$, $object(rel_1) = beer$,
$relation(rel_a) = may\_sell$, $agent(rel_b) = licensed\_pub$,

---

3 Note that we quantify over the slots.

$object(rel_b) = alcoholic\_beverages$

The first interpreted provision allows licensed pubs to sell beer, the second one allows licensed pubs to sell alcoholic beverages. This raises the issue of redundancy of legal provisions: since beer is an alcoholic beverage, why there are two different provisions? Such a phenomenon may be the symptom of potentially erroneous drafting of a legal text encompassing both P_y and P_x. In such case, a legislator may consider either amending the draft of a bill, or anticipate such an interpretation of any of these provisions that could restore the rationality assumption C: the lack of redundancy. However, in the discussed example it would be difficult, because beer, in any standard interpretation of the term, is an alcoholic beverage. However, the discussed set of provisions could be assessed as a rational one if the draft contained a provision that for the sake of this regulation, "beer" means alcohol-free beer only.

Second Step of Analysis

The second step contains the analysis of potentially erroneous provisions in light of all applicable interpretive rules. Let us add that we can speak of legislative error if there is a violation of a CIR, rather than the situation of an interpretive doubt, which is a natural feature of law. This account encompasses an intuition that the drafting of the legal provision in question (represented as the result of literal interpretation) is fallacious, because it leads to violation of the constraints imposed on the process of interpretation.

4.3   Example 1. Informal Approach

P_1 Licensed pubs may sell alcoholic beverages to adult customers.
P_2 Licensed pubs may sell alcoholic beverages to customers between 18 and 24 years of age.

We assume here that an adult is legally defined as a person who has reached 18 years of age. Therefore, the second provision is apparently redundant because it does not add anything to the normative content of the first provision. This situation violates the rule of interpretation "per non est", according to which no part of a normative text should be considered unnecessary. The above example seems to be a clear legislative error because the second provision could be eliminated from the draft without any modification to the content of the regulation. Note, however, that our assessment would differ were the term "adult" defined as a person who has reached 24 years of age or not legally defined at all. In the latter situation, we would argue that for the sake of the regulation, an adult means a person who has reached 24 years of age, because this interpretation makes the second provision meaningful.

4.4   Example 1. Semi-formal Approach

Let $P_{1,2}$ be a set the above provisions $\{P_1, P_2\}$, K will be the knowledge of the reasoner.

First Step of Analysis

The literal (ling) interpretations of the analyzed provisions may be represented as follows:

$(ling, (P_1 \cup K), rel_1) \in INT$ and $(ling, (P_2 \cup K), rel_2) \in INT$, where:
$relation(rel_1) = may\_sell$, $agent(rel_1) = licensed\_pubs$,
$object(rel_1) = alcoholic\_beverages$,
$counterpart\_agent(rel_1) = adult\_customers$
$relation(rel_2) = may\_sell$, $agent(rel_2) = licensed\_pubs$,
$object(rel_1) = alcoholic\_beverages$,
$counterpart\_agent(rel_2) = (customers\_between\_18 - 20)$
Since all slots are overlapping: $agent(rel_1) = agent(rel_2)$, $object(rel_1) = object(rel2)$ and $counterpart\_agent$ of rel_2 is subset of $counterpart\_agent$ of rel_1:
$counterpart\_agent(rel_1) = adult\_customers$
$counterpart\_agent(rel_2) = customers\_between\_18 - 20$
$customers\_between\_18 - 20 \subset adult\_customers$
then two provisions $\{P_1, P_2\}$ are potentially erroneous.

Second Step of Analysis

The identified overlap between the sets $counterpart\_agent(rel_1)$ and $counterpart\_agent(rel_2)$ generates doubts concerning the possibility of interpreting P_1 and P_2 in such a way that both provisions are considered necessary in the legislative draft (the rule "per non est" in CIR). However, it is clear per our assumptions that as any instantiation of $counterpart\_agent(rel_2)$ is also an instantiation of $counterpart\_agent(rel_1)$, it is impossible to interpret the said set of provisions to avoid the conclusion of redundant character of P_2. Therefore, we conclude that the set of provisions amounts to a definitive legislative error and it should be amended (perhaps by deletion of P_2).

4.5 Example 2. Informal Approach

P_1 Licensed pubs may sell alcoholic beverages to adult customers.
P_2 Licensed restaurants may sell alcoholic beverages to adult customers.
P_3 The mayor may grant a license to sell alcoholic beverages to restaurants.

We assume here that the mayor is the only authority competent to grant a license to sell alcoholic beverages. The example is immediately qualified as a potential legislative error because it leads to serious interpretive problems. Should "pubs" be interpreted as a subtype of restaurants for the sake of interpretation of the third provision? Such interpretation would, however, lead to violation of an interpretive argument based on plain ordinary meaning. In order to sustain a plain ordinary language distinction between restaurants and pubs, the law-applying entity might consider an analogous application of provision 3 to conclude that the mayor is authorized to grant a license to sell alcoholic beverages to pubs, too. However, such inference is precluded by the prohibition to reconstruct the scope of public authority competence through analogy. The scope of competence of

public authorities should follow clearly from the text of legal provisions. Therefore, it should be concluded that provision 1 cannot be applied to any situation,
because, according to the regulation, no pub may be granted with a license to
sell alcoholic beverages. The situation may be referred to as an example of a
specific type of legislative gap because the competence of the mayor to grant
a license to pubs is apparently lacking, and at the same time, the rationality
assumptions require making provision 1 operative.

4.6   Example 2. Semi-formal Approach

This example is more complex than the previous one. In order to properly detect
errors, first we have to analyze provision 3, which can have two different interpretations:

First Step of Analysis
$(ling, (P_3 \cup K), rel_1) \in INT, (analogy, (P_3 \cup K), rel_2) \in INT$
where:
$relation(rel_1) = grant, agent(rel_1) = mayor, object(rel_1) = license,$
$counterpart\_agent(rel_1) = restaurant$
$relation(rel_2) = grant, agent(rel_2) = mayor, object(rel_2) = license,$
$counterpart\_agent(rel_2) = restaurant \cup pubs$
In the next stage, we analyze provisions P_1 and P_2 in light of the above interpretations of provision P_3. We assume that there can be 2 different interpretations
of P_1 in the light of P_3 and one interpretation of P_2 in the light of P_3:
$(ling, (P_{1,3} \cup K), rel_3) \in INT,$
$(analogy, (P_{1,3} \cup K), rel_4) \in INT,$
$(ling, (P_{2,3} \cup K), rel_5) \in INT,$
where:
$relation(rel_3) = may\_sell, agent(rel_3) = \emptyset, object(rel_3) = alcoholic\_beverages,$
$counterpart\_agent(rel_3) = adult\_customers$
(Since the mayor has no right to grant a license for selling alcohol to pubs, pubs
cannot obtain such a license.)
$relation(rel_4) = may\_sell, agent(rel_4) = licensed\_pubs,$
$object(rel_4) = alcoholic\_beverages, counterpart\_agent(rel_4) = adult\_customers$
$relation(rel_5) = may\_sell, agent(rel_5) = licensed\_restaurants, object(rel_5) =$
$alcoholic\_beverages, counterpart\_agent(rel_5) = adult\_customers$
Because rel_3 shows that provision P_1 is generally inapplicable: $(agent(rel_3) = \emptyset)$,
we recognize the analyzed norms as potentially erroneous.

Second Step of Analysis
Above, we have indicated the alternative sets of frames representing knowledge
reconstructed from the provisions $P_1 - P_3$. The question arises whether any of
these interpretations of P_3 makes it possible to grant licenses to pubs. Even if this
would be the case, any such interpretation would violate the rule (from set CIR)
according to which the scope of competence of authorities should be based on
explicit legal text. Another possibility consists in concluding that the extension

of the term "licensed pub" is empty in the context of the regulation, but this leads to the conclusion of the occurrence of a genuine legal gap. Therefore, the analyzed set of provisions should be considered erroneous and it should be amended.

5   Conclusions

The topic of formal or semi-formal modeling of legislative errors lacks an in-depth comprehensive analysis. The present paper introduces such a discussion, using frame-based knowledge representation, commonly used in the earlier literature of the subject but used also in the contemporary research [15]. Frame-based approach is particularly useful where high-level legal expertise is required to solve a particular task. The detection of potential and actual legislative errors, other than simple linguistic issues, is not possible without a deep understanding of the legal and commonsense knowledge, so we believe that it is worthwhile to discuss the specific features of legal provisions that indicate the possibility of an error. In this paper, we have introduced a semi-formal model and analysis of specific types of legislative errors that appear when one norm or a group of norms have an overlapping scope of agents. Importantly, we explicated an assumption that the knowledge-based representation of a legal text is always a result of interpretive activity. Therefore, we did not focus on the representation of the content of legislation "as such", because we are strongly convinced that such an approach is theoretically flawed. We always represent an interpretation of a legal text, even if it is referred to as "literal" or "obvious" interpretation. The introduction of an interpretive layer to the representation of legislation is in our opinion necessary to develop more realistic and adequate models. In the process of legislative drafting it is necessary to take into account the possible interpretations of the drafted provisions. Some of the interpretive rules constrain the scope of admissible interpretations. We have referred to these rules as Constraining Interpretive Rules. This leads to the formulation of a tentative definition of a legislative error: a provision is erroneously drafted if any possible interpretation thereof violates at least one Constraining Interpretive Rule. We have discussed this phenomenon on the basis of examples. Of course, as legal interpretation is a domain of defeasible reasoning, the ascertaining of an error in the text may always be subject to debate. However, because legal interpretation is naturally modeled as argumentation, the semantics developed in the field of computational argumentation may be fruitfully used to determine which conclusion, concerning the existence of error or lack thereof, is stronger. We believe that our model can be an introduction to the discussion of the computer-supported analysis of legal drafts and detection of (potential) legislative errors.

In the future work, we will introduce a comprehensive typology of legislative errors with a formal analysis of their specific features, especially with the structure of the Constraining Interpretive Rules, alongside presenting a fully fledged formal model of argumentation concerning evaluation of legislative drafts. The formal analysis will require incorporating into our model some existing models of legal norms, interpretation, and argumentation.

References

1. Allen, L.E., Lysaght, L.J.: Modern logic as a tool for remedying ambiguities in legal documents and analyzing the structure of legal documents' contained definitions. In: Logic in the Theory and Practice of Lawmaking, pp. 383–407 (2015)
2. Allen, L.E., Saxon, C.S.: More IA needed in AI: interpretation assistance for coping with the problem of multiple structural interpretations. In: Proceedings of the Third International Conference on Artificial Intelligence and Law, ICAIL 1991, Oxford, England, 25–28 June 1991, pp. 53–61 (1991)
3. Amaya, A.: The Tapestry of Reason: An Inquiry into the Nature of Coherence and its Role in Legal Argument. Bloomsbury (2015)
4. Araszkiewicz, M.: Towards systematic research on statutory interpretation in AI and law. In: Legal Knowledge and Information Systems - JURIX 2013: The Twenty-Sixth Annual Conference, 11–13 December 2013, University of Bologna, Italy, pp. 15–24 (2013)
5. Araszkiewicz, M., Savelka, J.: Coherence: Insights From Philosophy, Jurisprudence and Artificial Intelligence. Springer, Dordrecht (2013). https://doi.org/10.1007/978-94-007-6110-0
6. Araszkiewicz, M., Zurek, T.: Comprehensive framework embracing the complexity of statutory interpretation. In: Legal Knowledge and Information Systems - JURIX 2015: The Twenty-Eighth Annual Conference, Braga, Portual, 10–11 December 2015, pp. 145–148 (2015)
7. Benjamins, V.R., Casanovas, P., Breuker, J., Gangemi, A. (eds.): Law and the Semantic Web. LNCS (LNAI), vol. 3369. Springer, Heidelberg (2005). https://doi.org/10.1007/b106624
8. Chomsky, N., Lightfoot, D.W.: Syntactic Structures. De Gruyter (2009)
9. Endicott, T.: Vagueness in Law. Oxford University Press, Oxford (2000)
10. Gizbert-Studnicki, T.: Types of vagueness. In: Krawietz, W., Summers, R., Weinberger, O., von Wright, G.H. (eds.) The Reasonable as Rational? On Legal Argumentation and Justification. Festschrift for Aulis Aarnio, pp. 135–144. Duncker & Humblot (2000)
11. Karasek-Wojciechowicz, I.: Roszczenie o wykonanie zobowiązania z umowy zgodnie z jego treścią (The claim to perform a contractual obligation in accordance with its content). Wolters Kluwer (2014)
12. MacCormick, N., Summers, R.: Interpreting Statutes. A Comparative Study. Ashgate, Dartmouth (1991)
13. Minsky, M.: A framework for representing knowledge. Technical report, USA (1974)
14. Sovrano, F., Palmirani, M., Vitali, F.: Legal knowledge extraction for knowledge graph based question-answering. In: Legal Knowledge and Information Systems - JURIX 2020: The Thirty-third Annual Conference, Brno, Czech Republic, 9–11 December 2020, pp. 143–153 (2020)
15. van Doesburg, R., van Engers, T.: The false, the former, and the parish priest. In: Proceedings of the Seventeenth International Conference on Artificial Intelligence and Law, ICAIL 2019, pp. 194–198, New York, NY, USA. Association for Computing Machinery (2019)
16. van Kralingen, R.W.: A Conceptual Frame-Based Ontology for the Law, pp. 15–22. University of Melbourne (1997)
17. Visser, P.R.S., van Kralingen, R.W., Bench-Capon, T.J.M.: A method for the development of legal knowledge systems. In: Proceedings of the 6th International Conference on Artificial Intelligence and Law, ICAIL 1997, pp. 151–160, New York, NY, USA. Association for Computing Machinery (1997)

18. Waismann, F.: Verifiability. Supplementary Volume XI (3), 119–150 (1945)
19. Walton, D., Sartor, G., Macagno, F.: Statutory interpretation as argumentation. In: Bongiovanni, G., Postema, G., Rotolo, A., Sartor, G., Valentini, C., Walton, D. (eds.) Handbook of Legal Reasoning and Argumentation, pp. 519–560. Springer, Dordrecht (2018). https://doi.org/10.1007/978-90-481-9452-0_18

Lexdatafication: Italian Legal Knowledge Modelling in Akoma Ntoso

Monica Palmirani[✉] [iD]

CIRSFID-ALMA AI, University of Bologna, via Galliera 3, 40121 Bologna, Italy
monica.palmirani@unibo.it

Abstract. This paper presents the result of the research project Lexdatafication that aims to model the legal knowledge information of Italy in Akoma Ntoso. The University of Bologna, in cooperation with IPZS, the Official Gazette entity, developed a framework capable to exploit the existing legacy databases of Normattiva in Akoma Ntoso. Additionally, Constitutional Court decisions were converted in Akoma Ntoso using the existing XML and metadata dataset provided by the open data portal. This output was linked to the legislative information. The collection of the documents in AKN constitutes a great annotated corpus in machine-consumable format capable to produce relevant legal data analytics applications and visualizations to support both practitioners and citizens in legal information retrieval.

Keywords: Akoma Ntoso · eLegislation · eJustice · Legal data analytics · Visualization

1 Introduction: Lexdatafication Project

This paper presents the findings of the project Lexdatafication, developed by University of Bologna in collaboration with different actors (e.g., IPZS and Constitutional Courts) and co-funded by the Presidency of Council of the Ministers of Italy through the Digital Transformation Team[1]. The objectives of this project were to convert all the Official Gazette database used for the official open portal Normattiva in Akoma Ntoso in compliance with the new ordinance of the Government that suggests promoting legal open data using a set of standards, also including Akoma Ntoso and ELI[2] for Italy[3].

Additionally, the project intended to apply the principles of interoperability between different legal knowledge information, also converting the Constitutional Court decisions and to connect the judgments to the consolidated legislative corpus. This permits to provide a complete legal information including interpretations, modifications, annulments deliberated by the Constitutional Court activity. This produces material useful to visualize a graph of the correlations between legislative acts and constitutional decisions. The final goal was to have a serialization in a unique standard, Akoma Ntoso

---

[1] https://teamdigitale.governo.it/en/ in charge for two years 2018 and 2019.

[2] European Legislation Identifier, https://eur-lex.europa.eu/eli-register/about.html.

[3] https://eur-lex.europa.eu/eli-register/italy.html.

© Springer Nature Switzerland AG 2021
V. Rodríguez-Doncel et al. (Eds.): AICOL-XI 2018/AICOL-XII 2020/XAILA 2020, LNAI 13048, pp. 31–47, 2021.
https://doi.org/10.1007/978-3-030-89811-3_3

(Vitali 2011, 2019; Palmirani 2011, 2012), gifted to apply Data Analytics Framework for deducting important information concerning the quality of the law, the certainty of the norms, the effectiveness of the regulation, the impact that the Constitutional Court has on the legislative law-making process.

The University of Bologna has developed a framework of software pipeline capable to convert data in Akoma Ntoso, to monitor the validation against Akoma Ntoso XML schema, to add metadata coming from the original sources, to provide API[4] to use via RESTful techniques and so reusable by everyone. The project underlines also the differences between the original XML standard used (NormeInRete – NIR) and Akoma Ntoso, improving the expressiveness of the original NIR-XML provided by Normattiva that is limited to article level, with some errors (e.g., in the preamble and in the conclusions), with missing information concerning the consolidated versions and in the normative references. Finally, the project produces a prototype of visualization through a web portal that shows how to use the information converted in an integrated manner.

The paper starts with the presentation of the research questions and the background of the project, also analysing the obstacles encountered; the third paragraph presents the methodology adopted for proceeding in a scientific manner; the fourth paragraph presents the approaches analysed; the fifth paragraph presents the technical solutions; the sixth paragraph presents the validation analysis; the seventh paragraph presents integration of the legislative information with the Constitutional Court[5] decisions and a visualization of the results; the last paragraphs presents the conclusions, the limits of the current outcomes and the future works.

2 Background: Normattiva and LegalXML Management in Italy

In 2000, the Italian Government, through AIPA[6] ordinances[7], defined a set of XML standards for modelling legislative documents (NormInRete) and its related naming convention (URN:NIR) with the aim to use Semantic Web technologies for legal information and to publish in an official Web portal all the Italian Law in easy, accessible, reusable manners in accordance with the European Directive Public Sector Information[8]. The portal, Normattiva[9], was settled in 2010 with the cooperation of many institutions (e.g., Senate of Italy, Chamber of Deputies, Court of Cassation, Presidency of the Council of Ministers) and thanks to the efforts of the implementer IPZS, the same official body in

---

[4] API - Application programming interface.

[5] http://bach.cirsfid.unibo.it/ldms-cortecostituzionale/.

[6] Autorità per l'informatica nella pubblica amministrazione (AIPA).

[7] Ordinance on 6 November 2001, n. AIPA/CR/35, *urn:nir:autorita.informatica.pubblica. amministrazione:circolare:2001–11-06;35*, https://www.agid.gov.it/sites/default/files/reposi tory_files/circolari/circolare-aipa-6-11-01_0.pdf; Circular on 22 April 2002 n. AIPA/CR/40, *urn:nir:autorita.informatica.pubblica. amministrazione:circolare:2002–04-22;40*, https:// www.agid.gov.it/sites/default/files/repository_files/circolari/circolare-aipa-22-04-02_0.pdf.

[8] Directive 2003/98/EC, modified with Directive 2013/37 https://eur-lex.europa.eu/legal-content/ en/ALL/?uri=CELEX:32013L0037 and recasted with Directive 2019/1024 https://eur-lex.eur opa.eu/legal-content/EN/TXT/?uri=CELEX%3A32019L1024.

[9] https://www.normattiva.it/.

charge of publishing the Official Gazette of Italy. In 2019 a new ordinance[10] emitted by AGID[11] abrogated the two precedents ordinances and introduced Akoma Ntoso XML standard as the official new format for the legal information in Italy, enlarging the scope with respect to the precedent standard NIR to all the legal information, including parliamentary acts, judiciary documentations and drafting legislative material. This adoption has been conditioned to the emission of official guidelines for implementing Akoma Ntoso for Italy that nowadays, despite the relevant outcomes of the Lexdatafication project, are not still emitted[12], with delay respect the deadline emitted in the ordinance 2/2019 (24 months from the date of enter into force).

In 2019 the IPZS and the DAGL[13] of the Presidency of the Council of Ministries signed a new contract for the reengineering of the Normattiva portal that also includes the conversion of all the formats used in Akoma Ntoso[14], but no reference can be found to the open data initiative[15,16] brought forward to create a portal easily reusable through bulk by companies, banks, insurances, citizenships, associations. Other countries already made this conversion for improving transparency, good governance,

---

[10] Circolare n. 2/2019 titled "Adozione di standard per la rappresentazione elettronica e l'identificazione univoca del patrimonio informativo di natura giuridica e istituzione del Forum Nazionale per l'informazione giuridica"; https://trasparenza.agid.gov.it/moduli/downloadFile. php?file=oggetto_allegati/19203145612OO__OCircolare+2-2019+standard+patrimonio+inf ormativo+di+natura+giuridica.pdf.

[11] Agenzia per l'Italia digitale (AGID).

[12] From the Ordinance 2/2019 AGID: "7. Disposizioni transitorie e finali La presente Circolare sostituisce la Circolare del 22 aprile 2002 n. AIPA/CR/40 e la Circolare del 6 novembre 2001 n. AIPA/CR/35 ed entra in vigore il giorno della sua pubblicazione sul sito istituzionale dell'Agenzia per l'Italia digitale. Dalla data di pubblicazione della presente Circolare la realizzazione di ogni nuova banca dati giuridica è effettuata secondo gli standard adottati. Sul medesimo sito istituzionale l'Agenzia per l'Italia digitale:1. pubblica la documentazione necessaria ed utile alla conoscenza e all'applicazione degli standard adottati in forza della presente Circolare;2. rende disponibili gli strumenti open source per supportare le attività di utilizzo degli standard e di conversione delle banche dati esistenti.Le attività di conversione delle banche dati giuridiche già esistenti agli standard adottati con la presente Circolare, sono completate entro 24 mesi dalla data di entrata in vigore della Circolare medesima."".

[13] DAGL, Department of the Legal and Legislative Affairs of the Government, http://presidenza. governo.it/dagl/.

[14] "d) l'adeguamento ad Akoma Ntoso dei formati di marcatura per la pubblicazione di tutti gli atti in XML;" and "f.1.3. la marcatura automatica degli atti nel formato AKOMA NTOSO;" from the contract.http://presidenza.governo.it/AmministrazioneTrasparente/Provvedimenti/Provve dimentiDirigenti/DAGL/Convenzione%20Normattiva%20%206%20agosto%202019%20% 20con%20firma%20digitale.pdf

[15] http://documenti.camera.it/leg18/resoconti/commissioni/stenografici/html/59/audiz2/audizi one/2019/11/14/indice_stenografico.0004.html.

[16] Open Data initiative of European Commission: https://digital-strategy.ec.europa.eu/en/policies/ open-data

participation, rules of law certainty (e.g., UK[17], Chile; Cifuentes-Silva 2019, US[18], Luxembourg[19], France[20], Africa Laws association[21], Brazil Senate[22], FAO; Palmirani 2018, etc.) and also the Senate of Italy[23] produced a Linked Open Data corpus and a repository of all the bills in Akoma Ntoso reusable with open license (cc-by 3.0)[24]. Additionally, the European Institutions[25] and the United Nations[26] approved Akoma Ntoso as official XML standard for managing legal documents.

In August of 2020 a new portal of Normattiva has been emitted with a new adaptive interface, with a new semantic search engine based on CELI technologies[27]. This part of the project was included in the larger view of Lexdatafication action. The Akoma Ntoso conversion of Normattiva is part of the agreement between IPZS, but we believe that the outcome of Lexdatafication, approved by Digital Transformation Team as good results, should be reused and disseminated.

The University of Bologna has investigated with a scientific method (interview to many stakeholders and social media campaign for making survey) which barriers are now blocking this valuable outcome. Apparently, there are many reasons:

1. The policy-makers is not very interested to provide Legal Open Data information (e.g., Open data is not part of the current working plan with IPZS), so there is not a strong commitment on the political level concerning the task of converting Akoma Ntoso;
2. There is not a clear power endorsement that provides liability and accountability on this issue between the different committees and the authorities involved at government level in Normattiva management (e.g., Ministry of Innovation, Ministry of Justice, Ministry of Finance, DAGL, AGID, etc.);

---

[17] https://www.legislation.gov.uk/uksi/2020/1586/contents/made/data.akn.

[18] United States Code of Office of the Law Revision Counsel, https://uscode.house.gov/download/download.shtml.

[19] Luxembourg Official Gazette.http://data.legilux.public.lu/eli/etat/leg/loi/2021/01/09/a12/jo/fr/xml.

[20] LegiFrance, https://www.data.gouv.fr/fr/datasets/dispositifs-des-textes-monalisa-akoma-ntoso/; https://www.data.gouv.fr/fr/datasets/projets-de-loi-de-finances-redaction-de-1ere-lecture-au-senat-resultant-des-travaux-de-lassemblee-nationale/.

[21] Laws Africa Association, https://laws.africa/.

[22] LexML-Brazil, http://projeto.lexml.gov.br/documentacao/Parte-3-XML-Schema.pdf.

[23] Senate of Italy Open Data portal, https://dati.senato.it/sito/home.

[24] Senate of Italy bulk in AKN, https://github.com/SenatoDellaRepubblica/AkomaNtosoBulkData.

[25] AKN4EU https://op.europa.eu/it/web/eu-vocabularies/akn4cu.

[26] AKN4UN https://unsceb-hlcm.github.io/. In March 2017 the High-Level Committee on Management (HLCM) adopted the UN Semantic Interoperability Framework for normative and parliamentary documents (UNSIF) developed by the HLCM Working Group on Document Standards (WGDS) and based on the Akoma Ntoso OASIS standard, https://unsceb.org/sites/default/files/imported_files/CEB-2017-3-HLCM33-Summay%20of%20Conclusions-FINAL_0.pdf.

[27] CELI company that developed the new search engine of Normattiva, https://www.celi.it/blog/2018/05/lexdatafication_forumpa_agid_celi/.

3. AGID has not enough human resources and political power to push the tasks declared in the ordinance 2/2019 even if it is very convinced about the benefits that these actions could produce in whole Italian digital society;

4. IPZS, that is the technical entity in charge of Normattiva, declares that the technical results produced of Lexdatafication, that were developed in jointing with them, are not reusable because the software is not compatible with the internal architecture;

5. Private publisher market is not in favour to legal open data initiatives in Italy because they afraid to lose their relevant position in the market of the legal documentation.

At the contrary in the current pandemic situation there is the strong need to have open legal data reusable with open license, especially concerning the rules of COVID-19[28]. There is a strong movement arising in Canada, Australia, New Zealand called "Rules as Code"[29] for transforming the norms, written in natural language, into machine-consumable formats and to make possible that legal documents are really effective and useful for the automatic checking compliance (e.g., using artificial intelligence systems). The European Directive PSI enacted in 2019 pushes in this direction as well and Italy, one of the countries more affected by the attack of the COVID-19 but with a strong legal informatics tradition in this field, cannot lose this occasion that passes through the legal open data. We believe that the output of Lexdatafication project made by the University of Bologna must be released in open license to the research community, to any company that would like to improve the transparency, to any public administration that intend to foster the results for improving the open access to the legal information, the machine-computable approach, and the digital transformation of the public administration.

3   The Methodology

3.1   The Legal Theory Pillars

The methodology adopted in Lexdatafication project by the University of Bologna with the agreement of IPZS and of the Digital Transformation Team is based on the following principles:

1. **Authoritativeness**. We used the original sources produced for the Official Gazette. We would had made the natural language processing starting from the HTML published on the Web portal, but this methodology was prone to errors and not based on the authentic information originally provided by the official Gazette. For this reason, we used only the database of IPZS sources.
2. **Integrity**. We did not change the original sources. All the enrichment processes must not modify the authoritative sources even if errors are discovered (e.g., typos).

---

[28] https://www.datibenecomune.it/ it is a movement that is asking the release of open data concerning the COVID-19.

[29] https://www.oecd.org/innovation/cracking-the-code-3afe6ba5-en.htm.

3.2 The Technical Principles

During the process of conversion in NIR, we used the following design principles:

1. **Metadata separation**. We have implemented the separation of the metadata from the content that the original NormaInRete XML standard did not implement. We have annotated the modifications using *<activeModifications>*[30] and *<passiveModifications>* in Akoma Ntoso metadata part without modifying the official content.
2. **Enrichment**. We have added granularity of mark-up and accuracy in the structural mark-up and in the normative references.
3. **Correction of annotation**. We have corrected the wrong mark-up used in NIR like preamble, conclusion, modifications.
4. **Consistency check**. After the conversion we had made a consistency check on the metadata and the annotation in the text (e.g., *@eId, <heading>, <num>, @refersTo, @href*) in order to avoid that metadata contains different information with respect to the text due to different sources in the original databases.

3.3 The Legal Validation Approach

A bulk converter, managed at server side, is not sure at 100% and we needed to implement also a validation strategy. For this reason, we have defined a policy for validating the outcome:

1. **Quantitative statistics**. We have implemented a visual dashboard that could permit to the human expert to monitor the conversion on all the sample used for the conversion (2008–2018[31]). Quantitative statistics were produced and visualized (see the paragraph 6.1.).
2. **Qualitative assessment**. We have asked to a PhD student in Law to randomly take ten documents for each typology of law (normal law with chapter and titles, normal law with only articles; modificatory law, consolidated law, law of conversion of decree, budget law, ratification of treaties/agreement law, legislative decree, law with annexes, law with tables, law of conversion of decree with annexes) and to evaluate them under qualitative point of view (see the paragraph in Sect. 6.2).

4 The Technical Solutions

The technical solutions were made based on a pipeline of software developed in Python, Javascript, node.js server side and we investigated two different options.

---

[30] < activeModifications> are the metadata concerning the modifications made by a given act to other acts; < passiveModifications> are the metadata concerning the modifications undergone by an act.

[31] The interval of years was limited for technical reasons considering that the connection between University of Bologna and IPZS was done via dedicated VPN for security reasons and also to avoid overloading the IPZS servers.

4.1 First Option: Conversion from NIR to AKN

We considered using a converter from NIR XML to AKN XML, but the current NIR-XML is very poor and incorrect. To make refinement on the top of NIR-XML would have amounted to a violation of the principle to not modify the authorial source (Fig. 1).

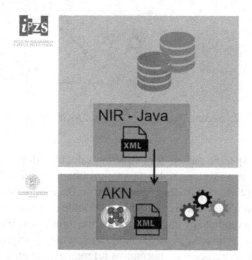

Fig. 1. First Hypothesis of NIR2AKN conversion.

Secondly to modify the Java component that now produces the NIR-XML serialization means to depend too much to the original expressiveness of NIR-XML standard and not to foster the benefits of the new AKN-XML standard that is more effectiveness and expressive (e.g., consolidation, modifications, metadata, ELI integration, etc.) than NIR.

4.2 Second Option: Conversion from Legacy Databases to AKN

We decided to use the original information spread in three different databases of IPZS that are used in the Java component for producing NIR-XML. Using the same original sources, but with the knowledge of AKN, we have produced an API RESTful capable to use Json file produced by IPZS from the original different databases. The Json file is at an intermediate level of modelling of the information intended to produce AKN as final output. The API component could take in input a single Json file or a zip file or a demon that each night could process the Gazette of the day and to produce the AKN serialization. In this manner the output of the serialization could be managed by Data Analytics Framework (DAF) or by Web portal (Fig. 2).

The pipeline[32] uses the existing data expressed in Json file and it is mapped in AKN elements and attributes when they are already tokenized for this task, otherwise we extract some legal knowledge from the text with a set of parsers developed in Python.

---

[32] http://sinatra.cirsfid.unibo.it/node/normattiva2akn/.

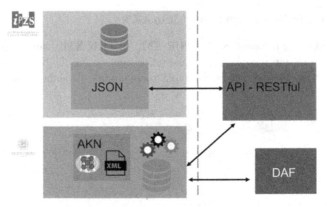

Fig. 2. Second Hypothesis Database2AKN conversion.

Additionally, we use specific parsers for extracting valuable information inside of existing fields for reaching a more granular annotation like location, roles, persons inside of signatures, or more *quotedStructure/quotedText* in the modificatory documents. We also refine the partition's *id* that are wrong (e.g., double assigning) from time to time, and we also refine incomplete or impartial normative references. We have in this manner produced a more precise AKN-XML mark-up reaching paragraphs and list levels (e.g., letters, points), more accurate normative references, detailed signatures, correct annotation in the preamble's sentences, and URI identifications using FRBR annotation and ELI naming convention. We also use preparatory documentation links, coming from Senate of Italy and Chamber of Deputies, to enrich the plurality of information. Normal Rate standard was not designed for parliamentary acts, so Akoma Ntoso allows to extend the interoperability also to this important category of legal documents and to use a unique uniform standard of representation. Additionally, Akoma Ntoso has a more sophisticated metadata elements for modelling convoluted situations related to the temporal modifications (e.g., suspension of interval of time, derogations and retroactive annulment).

5   Akoma Ntoso Modelling

5.1   AKN Serialization Process

The conversion pipeline was applied on only ten years of Official Gazette documentation (2008–2018) because the IPZS asked to not use all the database for avoiding the overload of their servers. The conversion produced Akoma Ntoso-XML at level 2.d according to the conformity rules defined at international standardization level by OASIS[33]. The AKN elements choice was synchronized with the choices already made by the Senate of Italy in its portal about the bills (Table 1).

---

[33] http://docs.oasis-open.org/legaldocml/akn-core/v1.0/akn-core-v1.0-part1-vocabulary.html.

Table 1. Converting elements and attributes mapping.

Category of mark-up	Element in AKN	Function
Document type	\<act\>	Constitution, law, legislative decree, ministrerial decree, regulation, real decree, treaty, agreement, etc.[*]
Document type	\<doc\>	Annexes, table, attachments
IRI	FRBRwork, FRBRalias, FRBRexpression, FRBRmanifestation	FRBRalias contains ELI e NIR existing URI for permitting to maintain the correlation with the other naming conventions
Macro structure	\<preamble\>, \<preface\>, \<body\>, \<conclusions\>	
\<preamble\> subdivision	\<formula\>, \<citations\>, \<recitals\>	
Inline elements	\<docNumber\>, \<docDate\>, \<docTitle\>	
Elements for the provisions	\<book\> \<part\> \<title\> \<chapter\> \<session\> \<division\> \<transitional\>	Grouping elements
Basic unit	\<article\>	
Structure of \<conclusions\>	\<location\> \<date\> \<role\> \<person\> \<signature\>	
Each partitions has sub-elements	\<num\> \<heading\>	For each partition we have sub-elements. In NIR the \<num\> was not marked-up
Sub-elements inside of the \<article\>	\<paragraph\> \<list\> \<point\> \<intro\> e \<wrap\>	
ids	@eId	Each partition has a unique id
Normative references	\<ref\> \<mref\> \<rref\>	
Semantic references	\<references\>	
Publication metadata	\<publication\>	Publication metadata from the Official Gazette
QuotedStructure	\<quotedTex\>	
QuotedText	\<quotedStructure\>	
modification	\<mod\>	

(continued)

Table 1. (*continued*)

Category of mark-up	Element in AKN	Function
Passive modification	<passiveModifications>	We used the existing formatting annotation (bold, italic, double brackets) for inferring the passive modifications and to model them in XML. The repeals of the textual fragment unfortunately are not inferable using the original data provided, but in the future we could access to other internal databases
Active modification	<activeModifications>	This information was not included in the Json file and we need to use parsers to deduct the action made by the modifications
Document lifecycle	<lifecycle>	We used the existing information for rebuilding the lifecycle of the document that did not exist in the previous NIR annotation
Annexes	<attachments>	The Json file was not always correct when the annexes were
Table	<table>	The original source includes the tables in ASCII format
Semantic annotation	Date, roles, persons	We used NER parsers base on RegEx and spaCy[**] to detect them
Preparatory parliamentary links	<component>	We have included a <component> in order to annotate the preparatory parliamentary links
Authorial Notes	<authorialNote>	We use this tag in place where the notes are authorial. Otherwise we use <notes> in the <metadata> part

[*]https://www.gazzettaufficiale.it/eli/tables/resource-type,
[**]https://spacy.io/

5.2 Enrichments

The NIR markup produced by Normattiva portal presents different problems:

1. where it exists a preamble, the first article of the document includes the preamble. With the AKN conversion we have managed this problem and we have relocated the preamble in the correct part of the XML node hierarchy. Additionally, we inferred also the *recitals* and the *citations* sentences;
2. the *conclusion* was included in the last article of the XML-node tree, now with AKN we have relocated the conclusion in the correct node of the XML hierarchy;
3. in the past NIR-XML serialization the *number* of the article and the *heading* were erroneously marked up (e.g., not bis, ter). Now with AKN each partition has *<num>* and *<heading>* in the correct location. All the Latin adverbs were in AKN conversion detected and properly marked-up and we have recalculated all the *ids* assigning unique identifiers of each partition (Table 2);

Table 2. New serialization in AKN after parsers.

Previous mark-up	−<articolo id="2"> <num>Art. 2 bis.</num> −<comma id="art2-com1"> <num>1</num> −<corpo> <h:p h:style="text-align: center;">Art. 2-bis </h:p> <h:br/> <h:br/> <h:p h:style="text-align: center;">(Ambito soggettivo di applicazione). </h:p> <h:br/>
AKN conversion	<article eId="art_2_bis"> <num>Art. 2 bis</num> <heading>(Ambito soggettivo di applicazione).</heading> <content><p>…

4. the normative references were not marked-up, and also the internal references were not included. With AKN we have marked-up all the normative references, external and internal;
5. the passive modifications were not marked-up as well as the active modifications. With AKN conversion we have detected the passive modifications using parsers and the same for the active modifications even if less accurately for scarcity of information in the original Json file;
6. we have detected the lifecycle of the document with the AKN metadata;
7. AKN conversion marked-up sub-paragraphs, letters, points, numbers, list that in the original version of NIR-XML was not present;
8. AKN conversion detected and marked-up quotedText inside of modificatory acts;
9. AKN conversion detected, isolated and marked-up the notes that were located after the text without distinction between official law content and editorial annotation;
10. in the original NIR-XML the annexes were annotated without a hierarchical nested order, using AKN we recognised the nested annexes.

We can conclude that our AKN conversion pipeline, using a hybrid architecture based on XML-hierarchical structure, patterns-oriented methodology (Di Iorio 2012, 2014) and parsers (e.g., NER using both RegEX and existing engine like Spacy – Sovrano 2020), has improved the legal knowledge modelling of the Normattiva XML output, without introducing errors but, on the contrary, resolving several structural mistakes of serialization in XML.

6   Validation

6.1   Quantitative Statistics

We have developed a dashboard portal[34] that provides statistics of validation and non-validation, also including the report of the errors, so we can have also the possibility to produce graphs concerning the main errors occurred like in these additional projects developed by the University of Bologna[35]. The validation error report was done based on categories of errors operating in the following parts of the XML: URI, eId, structures, heading, content, normative references, inline elements, modifications, lists, annexes, table, notes, components, preamble, preface, body, conclusions, active/passive modifications, metadata, *quotedText/quotedStructure*.

The statistics produced the following: 83% valid documents over 2055, with the best percentage in the law-act (95%) and the worst performance in the decree (58%). This is very useful for permitting a better refinement of the algorithm of conversion in the future and to customize the patterns for the different typologies of documents (Fig. 3).

We can also see the valid and invalid documents in a special dashboard in order to amend and refine them with special editors like LIME[36].

---

[34] http://bach.cirsfid.unibo.it/lexdatafication-dashboard (only FireFox or Chrome). The bulk with all the document is published here: https://gitlab.com/CIRSFID/lexdatafication/-/tree/master/normattiva2akn.

[35] http://bach.cirsfid.unibo.it/node/sofia-dashboard/; http://bach.cirsfid.unibo.it/node/ananas-dashboard/.

[36] http://sinatra.cirsfid.unibo.it/lime-cassazione/ - LIME editor developed with the support of the Court of Cassation.

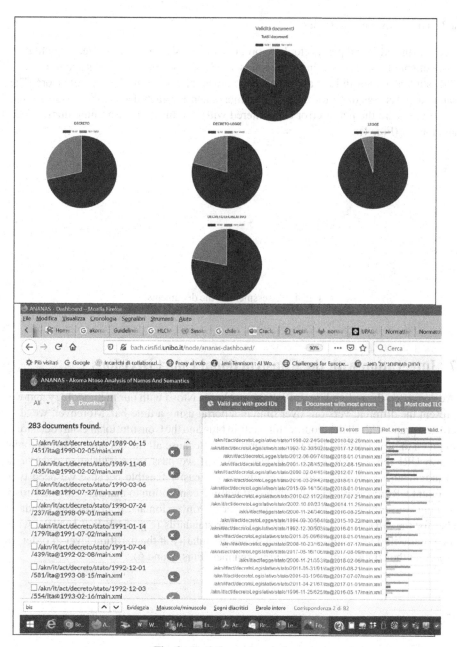

Fig. 3. Statistics and analytics.

6.2 Quantitative Assessment

Because the NLP and parsers tool are prone to errors, we have also started a qualitative assessment of the results using a randomised selection of the documents and to ask to a PhD student, expert in Law and XML, to validate them dividing for type of errors. The same analysis was done for type of documents and for each decade. We discover that the annexes are the major error encountered with also the quotedStrcuture/quotedText, and the list (Fig. 4).

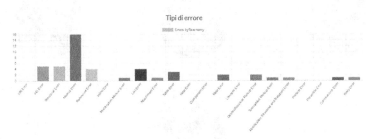

Fig. 4. Assessment of the documents.

7   Integration with the Constructional Court Decisions

We converted the Italian law n. 40/2004 in Akoma Ntoso with our methodology and we present the different versions over time in a portal using a time-bar. Moreover, we also create a graph with all the modificatory acts in blue and the Constitutional Court decisions in green to provide complete legal information. Because all the documents use Akoma Ntoso URI the navigation is guaranteed even if we pass from legislative document to judgments: using ELI and ECLI[37] this is not easily feasible because we need two different URI resolvers with two different naming convention to manage. Akoma Ntoso naming convention provides a common layer of conversion and interoperable standard also for the URI coming from different other standards (e.g., ELI, ECLI, NIR), with the possibility to harmonize the level of the granularity at the partition level in order to represent the citations in more accurate manner (Figs. 5 and 6).

---

[37] https://dati.cortecostituzionale.it/ECLI/ECLI.

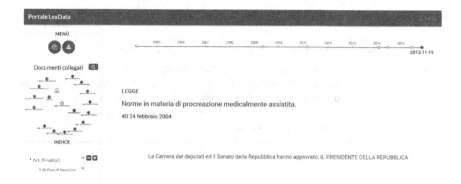

Fig. 5. Proof of concept of Akoma Ntoso converter.

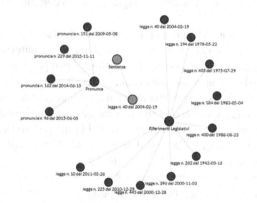

Fig. 6. Network graph of the references.

8   Conclusion and Future Work

The Lexdatafication project permits to demonstrate that we can convert old databases of Official Gazette coming from IPZS in Akoma Ntoso improving significantly the current NIR-XML serialization. We experimented a hybrid approach combining XML patterns-oriented approach (e.g., content model – Di Iorio 2012, 2014), parsers (e.g., NLP, NER), AKN hierarchical mark-up (e.g., XSD prescriptiveness). We have also developed a methodology for monitoring the validation of bulk conversion and for assessing the quality of the transformation, producing a dashboard where the experts could, in a future, to contribute to refine the Akoma Ntoso XML. We also have used the quantitative statistics and the qualitative evaluation for underlining the main problems in our legislation and to detect the anomalies. This statistical information will help us to refine the pipeline of the converter and to differentiate the rules of NLP according to the patterns detected in each historical period of Italian legislation. We have also detected anomalies that we can correct in the future version, considering also that some of them are due to the wrong characters of the origin format (e.g., hidden chars from old legacy databases) (Table 3).

Table 3. Future work of Akoma Ntoso converter of Italian Official Gazette in Normattiva portal.

Structural limitations	1. Annexes are very difficult to markup an also tables
Anomalies	2. the sub-paragraphs without numbering are not always detected
	3. *list* of *letters* with mixed numbering (e.g., 0a, aa, bb, etc.) are difficult to detect when algorithm of numbering is not codified;
	4. exceptions in the linguistic formulation of the preamble;
	5. *heading* with anomalies;
	6. modifications with incorrect quoted symbol opened/closed pairs

Finally, we have shown that using Akoma Ntoso, both in legislation and in Constitutional Court, we can provide better information to practitioners and citizens integrating in a unique portal all the legal knowledge navigable using a common naming convention (AKN-URI).

We believe that this scientific project has created valuable results in term of different kind of findings: a) in terms of legal methodology for monitoring the conversion from one format to another (dashboard); b) creating a mapping analysis of NIR into AKN for Italian legislation and also concerning Constitutional Court decisions; c) in term of assessment and evaluation of the legislative Italian corpus detecting anomalies and patterns differentiated for decades and legislatures; d) implementing a pipeline of software based on parsers and ANK patterns, available in open source; e) producing a sample of AKN serialization of Italian legislation (2008–2018) and of the Constitutional Court decisions (1956–2021); f) providing an integration of different heterogenous legal sources harmonized using a common standard AKN. For these reasons we believe that it is unfortunate that these outcomes, funded with public money, are not reused, also with the necessary modifications and improvements for covering the lacks and weaknesses discovered, as a starting point of the next task of the conversion of Normattiva corpus in Akoma Ntoso.

Acknowledgment. Lexdatafication project was supported and co-funded by the Digital Transformation Team of the Presidency of the Council of Ministries of Italy between 2018–2019 and co-funded by the University of Bologna using internal economical resources.

References

Di Iorio, A., Peroni, S., Poggi, F., Vitali, F.: A first approach to the automatic recognition of structural patterns in XML documents. In: Proceedings of the 2012 ACM Symposium on Document Engineering, pp. 85–94 (2012)

Di Iorio, A., Peroni, S., Poggi, F., Vitali, F.: Dealing with structural patterns of XML documents. JASIST **65**(9), 1884–1900 (2014)

Palmirani, M.: Legislative XML: principles and technical tools. ROMA, Aracne 2012, 161 (2012)

Palmirani, M.: Akoma Ntoso for making FAO resolutions accessible. In: Peruginelli, G., Faro, S. (eds.) Knowledge of the Law in the Big Data Age. Frontiers in Artificial Intelligence and Applications 2019, vol. 317, pp. 159–169. IOS Press, Amsterdam (2018). https://doi.org/10.3233/FAIA190018

Palmirani, M., Vitali, F.: Akoma-Ntoso for legal documents. In: Sartor, G., Palmirani, M., Francesconi, E., Biasiotti, M.A. (eds.) Legislative XML for the Semantic Web, pp. 75–100. Springer Netherlands, Dordrecht (2011). https://doi.org/10.1007/978-94-007-1887-6_6

Sovrano, F., Palmirani, M., Vitali, F.: Deep learning based multi-label text classification of UNGA resolutions. In: ICEGOV 2020: Proceedings of the 13th International Conference on Theory and Practice of Electronic Governance, New York, ACM, 2020, pp. 686–695 (2020)

Vitali, F.: A standard-based approach for the management of legislative documents. In: Sartor, G., Palmirani, M., Francesconi, E., Biasiotti, M.A. (eds.) Legislative XML for the Semantic Web, pp. 35–47. Springer Netherlands, Dordrecht (2011). https://doi.org/10.1007/978-94-007-1887-6_4

Vitali, F., Palmirani, M.: Akoma Ntoso: flexibility and customization to meet different legal traditions. Presented at Symposium on Markup Vocabulary Customization, Washington, DC, July 29, 2019. In: Proceedings of the Symposium on Markup Vocabulary Customization. Balisage Series on Markup Technologies, vol. 24 (2019)

A Critical Reflection on ODRL

Milen G. Kebede[1]([✉])[iD], Giovanni Sileno[1][iD], and Tom Van Engers[2]

[1] Informatics Institute, University of Amsterdam, Amsterdam, The Netherlands
{m.g.kebede,g.sileno}@uva.nl
[2] Leibniz Institute, University of Amsterdam/TNO, Amsterdam, The Netherlands
vanengers@uva.nl

Abstract. Rights expression languages (RELs) aim to express and govern legally binding behavior within technological environments. The Open Digital Rights Language (ODRL), used to represent statements about the usage of digital assets, is among the most known RELs today and has become a W3C recommendation to enhance the web's functionality and interoperability. This paper reflects on the representational power of ODRL from a practical perspective; utilizing use cases and examples, we discuss the challenges, issues, and limitations we came across while investigating the language as a potential solution for the regulation of data-sharing infrastructures.

Keywords: Policy expression languages · ODRL · Normative specification · Data sharing infrastructure

1 Introduction

Data usage control is one of the mechanisms that enable data owners to exercise their control, but, more generally, it concerns any party holding certain rights on data to exercise those rights. Data sharing agreements and licenses specify how, by whom, for what purposes, and under which conditions data may be used. In distributed data sharing infrastructures, policies and data-sharing agreements governing, e.g. the use of personal data, need to be expressed in a machine-readable knowledge representation language to support enforcement in all nodes; otherwise, policies can not be applied systematically, increasing the risk of non-compliance. Automating (at least partially) these policies fosters better transparency and eases the audibility of activities and inter-organizational transactions at the organizational level.

Rights expression languages (RELs) are originally proposed for representing policies and utilized for specifying *digital rights* in different domains of application [13]. The primary function of those rights is to manage and protect digital

This work is supported by the project Enabling Personalized Interventions (EPI, grant 628.011.028), funded by NWO in the Commit2Data—Data2Person program, and by the project Data Logistics for Logistics Data (DL4LD, grant 628.009.001), funded by NWO and TKI Dinalog in the Commit2Data initiative.

V. Rodríguez-Doncel et al. (Eds.): AICOL-XI 2018/AICOL-XII 2020/XAILA 2020, LNAI 13048, pp. 48–61, 2021.
https://doi.org/10.1007/978-3-030-89811-3_4

assets.[1] Several RELs exist, among which the Open Digital Rights Language (ODRL) [12], the Extensible Access Control Markup Language (XACML) [1], and the Enterprise Privacy Authorization Language (EPAL) [16].

In this paper, we focus on the Open Digital Rights Language (ODRL), a language that in recent years has gained popularity both in theoretical and practical settings, reaching the status of W3C recommendation [12]. The language is presented as being neutral to the technology used to implement usage control and is intended to be flexible enough to allow for the creation of new actions and constraints for data access policies. As our use cases focus on automating data-sharing agreements in the context of healthcare and logistics research, we found the language relevant to our research. Our goal is to utilize a language that supports the specification of normative constructs as those specified in regulations, agreements and consents. While extended versions of XACML support partial specification and enforcement of laws and regulations, it lacks, for instance, the support for "system obligations" [15], i.e. obligations the system has to perform on certain events such as notification of data breach. On the other hand, EPAL [4] is designed for writing enterprise privacy policies but it lacks e.g. reasoning support for conflicts or other relevant constructs [2].

Previous work investigated ODRL's suitability for different scenarios and from different perspectives, and also proposed various extensions [6,7,19]. The present contribution shares similar motivations, although our analysis focuses on the general modeling process and requirements taking the standpoint of a designer aiming to model a policy in ODRL. Additionally, we consider crucial institutional patterns that were only partially covered before, such as *delegation*. As an institutional construct, delegation is particularly relevant (and delicate), as it brings to the foreground the requirements of meeting the needs of stakeholders while maintaining accountability. The general aim of this paper is then to present the current challenges in using ODRL for specifying policies, elaborating on the experiences acquired on a data-sharing use case in the healthcare domain. Several limitations of the ODRL language are discussed, such as the lack of monotonicity in representing delegation scenarios, semantic ambiguity in the usage of "duty", granularity in identifying parties and transformational aspects of rules.

The paper is structured as follows. In the next section, we provide some references to related work. Section 2 gives an overview of the core model of the ODRL language, and in Sect. 3, we report our practical investigation of the language. We conclude with a discussion in Sect. 4.

1.1 Related Work

The ODRL language passed through several iterations, and the language maintainers and developers have shown openness to feedback from the community.

---

[1] In the past, rights expression languages and related technologies have been criticized for resulting in stronger restrictions than what generally granted by law. However, if RELs enable to create policy layers to integrate policies derived from legislation such as e.g. GDPR, they also make it possible to counterbalance excessive protections required by businesses.

Contributions in the literature range from suggesting extensions of the ODRL informational model [12], typically motivated by specific application domains, to introducing formal specifications, and to mapping of the language to other languages.

De Vos et al. [6] propose the application of an extended/revised ODRL model to capture the semantics of legal regulations such as the GDPR and organizational business policies. The proposed policy profile, the "regulatory compliance profile", can be used to model regulatory requirements and business policies via nested permissions, prohibitions, obligations, and dispensations. Shakeri et al. [19] consider the use of the ODRL in the context of digital data markets (DDMs). They extend the ODRL model by defining categories of assets and adding the *input* property. The first helps to solve the inconvenience of defining rules for every asset in the digital data market, while the second allows for defining the data used as input for data processing. Fornara et al. [7] extend the ODRL model in two directions: by inserting the notion of *activation event/action*, and by considering the temporal aspects of the deontic concepts (permission, obligation, and prohibition) as part of the application-independent model. The activation event/action notion is further expressed by events/actions as complex constructs having types and application-independent properties.

There are relatively few research efforts made towards the formalization of the semantics of ODRL. Garcia et al. [9] have formalized the implicit semantics of ODRL schemas and connected ODRL to the IPRonto ontology. They conclude that their approach can make semantic queries possible and enable specialized reasoners over licenses. Steyskal et al. [21] address ambiguities that might emerge based on explicit or implicit dependencies among actions. They propose an interpretation of ODRL policy expressions' formal semantics to enable rule-based reasoning over a set of policies. Arnab et al. [3] extended ODRL and XrML, a REL that allows content authors to set access control rights to their content. The extensions enable end-users to request the modification of current rights and allow rights-holders to grant or refuse the request. Steyskal et al. [20] demonstrate the ODRL's ability to express a large variety of access policies for linked data through different examples. These authors aim to mitigate issues with linked data regarding expressive access policies, introducing pricing models for online datasets, and providing a human and machine-readable form for metadata descriptions.

RELs are also used for governance in multimedia assets and intellectual property protected content. Rodriguez-Doncel et al. [17] present the MPEG-21 contract ontology (MCO), a part of the standard ISO/IEC 21000. MCO is an ontology that represents contracts that describe rights on multimedia assets and intellectual property protected content. It describes the contract model and key elements such as the parties in the contract and the relevant clauses conveying permissions, obligations and, prohibitions. Another work by Rodriguez-Doncel et al. [18] presents a dataset of licenses for software and data, expressed as RDF for use with resources on the web. They use ODRL 2.0 to describe rights and conditions present in licenses. It provides a double representation for humans and machines alike and can enable generalized machine-to-machine commerce if generally adopted.

There exist a few contributions aiming to model delegation policies (a central scenario in this paper) using the ODRL language. For instance, Grunwel et al. [10] focus on an information accountability framework that uses ODRL to model policies for delegation. In their work, they conclude that ODRL meets the requirement to model delegation policies, given that constraints and duties can be used to express the party to whom access is delegated, expiration of the access, and the types of actions.

At higher-level, the studies presented above approach ODRL taking into account a specific use case, in many cases extending the language based on the use-case requirements. Our approach differs in that we take into account a wider range of institutional constructs including duties, power, delegation and other relevant normative concepts to identify the challenges for future extensions of the language.

2   Modeling with ODRL

The Open Digital Rights Language (ODRL) is designed as a policy expression language, aiming to provide a flexible and interoperable information model, vocabulary, and encoding mechanism for representing normative statements concerning digital content and services [12]. It evolved through the years from a digital rights expression language for expressing simple licensing mechanisms for the use of digital assets to accommodating privacy policies [8]. The model is built using Linked Data principles; however, its semantics is described informally as no formal specification is provided. In the remainder of this section, we provide an overview of the ODRL information model, focusing on the main classes that are of interest for the institutional constructs under our attention (see Fig. 1).

Overview of Core ODRL Classes. An **Asset** is a digital resource that might be subject to a Rule. It has an *asset identifier* property and can be any form of identifiable resource. A **Party** refers to an entity such as a person, organization or collection of entities that undertake roles in a rule. It should have a party identifier. An **Action** class represents operations that can be exercised on assets; the association with the asset is specified via the *action* property in a rule. The **Constraint** class refines the specification of action or declares the conditions applicable to a rule by using an expression that compares two operands with an operator. When the comparison returns a match, it is considered satisfied. It has a *constraint* identifier, a *right-operand* property value data type of the right operand, a unit used in the *right-operand* and the status property generated from the *left-operand* action.

The **Rule** class is a super-class collecting the common characteristics of the three types of normative statements considered in ODRL: permission, prohibition, and duty. It concerns an action, which might be further refined. It must contain a *target* property (indicating the asset subjected to the rule), and might

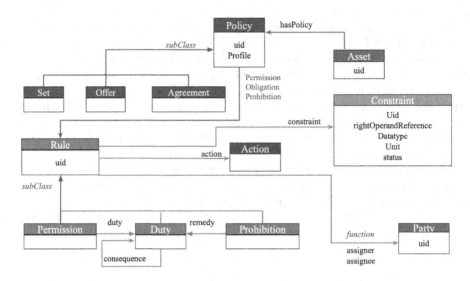

Fig. 1. Simplified view on the Information Model of ODRL 2.21

have an *assignee* and *assigner* properties (linking the rule to the associated parties). A **Permission** allows an action over an asset if all constraints are satisfied and if all duties are fulfilled. It may include one or more **duty** property values. A **Prohibition** disallows an action over an asset if all constraints are satisfied. The **remedy** property may be used when an action infringes the prohibition. A **Duty** is the obligation to exercise an action. It is fulfilled when all constraints and refinements are satisfied and have been exercised. It may have the **consequence** property, which is an additional duty that must be fulfilled in case of violation.

A **Policy** collects a group of rules (at least one) and can be qualified as Set, Offer and Agreement. It has a unique identifier, should have at least one rule, and a *profile* property to identify the ODRL profile the policy conforms to. An ODRL profile is defined to provide vocabulary terms that can be used in ODRL policies that require them, typically to be shared within a community of practice. A *set* supports expressing generic rules without further instantiating the parties involved. An *offer* supports 'offerings' of rules from assigner parties—it is used to make available policies to a wider audience but does not grant any rules. It specifies one party, the assigner, not the assignee. An *agreement* supports granting of rules from assigner to assignee parties and is typically used to grant the terms of the rules between the parties. Therefore, an *agreement* will specify both assigner and assignee parties. In the remainder of this section, we provide an overview of the ODRL information model, focusing on the main classes that are of interest to the use-case (see Fig. 1).

3 Criticalities of ODRL

The following section will report on our experience concerning the use of ODRL in modeling patterns relevant to data-sharing agreements, highlighting the issues that emerged in the exercise. We wrote the examples with respect to the ODRL documentations on the information model[2], informal semantics, use-cases and vocabulary of the language[3].

3.1 Illustrative Use Case: Delegation

Data-sharing scenarios, at times, require individuals to act on behalf of another. For example, a guardian may be required to act on behalf of a minor; or a carer on behalf of a person unable to grant or deny access to data. Similar patterns occur at the level of institutes. Additionally, research institutes might grant rights to be used by partner institutes under certain conditions to promote a shared research goal. This section will focus on institutional delegation scenarios. For instance, *suppose OrganizationX, an institution in the Netherlands maintaining a registry of patient data, forms a data-sharing agreement with OrganizationY, an institution in Belgium. The data-sharing agreement grants OrganizationY the permission to access the data and the possibility of delegating this permission to a third party, OrganizationZ. Consequently, the latter will be allowed to have access to the data if OrganizationY decides to delegate the permission received from OrganizationX.* Several contextual information might limit permissions and delegations, typically by means of constraints; these refinements will be neglected for now.

ODRL provides two main higher-level actions: `transfer` and `use`. According to the ODRL vocabulary `use` actions refers to any use of the asset (e.g. "play" music or "read" file), while the `transfer` actions explicitly refers to the transfer of ownership of the asset (lost by the agent, gained by the recipient) in its entirety (e.g. "sell" or "give"). This form of delegation (in the sense of transfer of rights) maps to a transfer action as shown in listing 1:[4]

```
"@type": "agreement",
"permission":
    "assigner": "OrganizationX", "assignee": "OrganizationY",
    "action": "transfer", "target": "datasetA"
```

Listing 1. Delegation as transfer.

The code above is an *agreement* (that is, in ODRL terms, there is an assigner and assignee) between *OrganizationX* and *OrganizationY*, for transferring

---

[2] https://www.w3.org/TR/odrl-model/.

[3] https://www.w3.org/TR/odrl-vocab/.

[4] The original JSON code is at https://grotius.uvalight.net/ODRL-policies. For space reasons, here we will omit accolades, use indenting for nested lists, empty lines to separate policies.

(ownership of) datasetA from *OrganizationX* to *OrganizationY*. Here, ownership is assumed to include the possibility of transferring the asset again to someone else (e.g., *OrganizationZ*). This model can be used as a specification for *non-monotonic* delegation, where the grantor loses the permission delegated. However, the same model can not be used to specify *monotonic* delegation scenarios where the grantor maintains the delegated permission.

This especially becomes problematic to capture the power relationship between parties; e.g., the party in power has to maintain ownership of the asset, or "veto" power to either constrain or revoke granted rights, as well as the power to transfer and/or lose ownership of the asset entirely. For these limitations, we consider the following alternative model:

```
"@type": "agreement",
"permission":
    "assigner": "OrganizationX", "assignee": "OrganizationY",
    "action": "grantUse", "target": "datasetA",
    "duty": [{ "action": "nextPolicy", "target": "ex:newPolicy" }]

"@type": "set",
"uid": "ex:newPolicy",
"permission": [{ "action": "read", "target": "datasetA" }]
```

Listing 2. Delegation as granting conditional usage.

In the code above, a combination of actions is used to restrict the permission to use the target asset *datasetA*. The action `grantUse` enables the assignee to create policies about the target asset (whose implicit owner is the assigner) for third parties (so it provides an implicit but limited form of institutional power) and is recommended in the ODRL vocabulary to be used with the `nextPolicy` action. The function of `nextPolicy` (which, to reiterate, is an action, not a policy) is to indicate the policy that applies to a third party for their use of the Asset (see e.g. [20]). In this way, however, usage rights are restricted only to a third party and not further. In some cases, delegated parties need to be allowed to delegate. A possible model (possibly abusing the intended use of `grantUse`) would be the one expressed below:

```
"@type": "agreement",
"permission":
    "assigner": "OrganizationX", "assignee": "OrganizationY",
    "action": "grantUse", "target": "datasetA",
    "duty": [{ "action": "nextPolicy", "target": "ex:newGrantPolicy" }]

"@type": "set",
"uid": "ex:newGrantPolicy",
"permission":
    "action": "grantUse", "target": "datasetA",
```

```
"duty": [{ "action": "nextPolicy", "target": "ex:newPolicy" }]

"@type": "set",
"uid": "ex:newPolicy",
"permission": [{ "action": "read", "target": "datasetA" }]
```

Listing 3. Delegation as nested granting of conditional usage.

This extension may enable us to form a hierarchical structure one step further than the previous example, yet it can not represent the full transfer of delegating power to a chain of delegators of unspecified length.

Other relevant aspects of delegation, e.g. the *revocation* of rights, also can not be specified within ODRL. While expressions in ODRL provide terms for specifying deadlines or expiration dates using the constraint class, updating activities are not considered. To conclude, the current ODRL model fits some delegation scenarios, but lacks expressiveness to accommodate others. Additionally, the intricate forms to specify these models make it difficult to identify the standard reusable components, and obscure the fact that we are dealing with a delegation pattern.

3.2 Additional Issues

In this section, we address additional limitations of ODRL that we have identified during our modeling experience.

Ambiguous Semantics for Duty. Duty in its common legal sense is an action that an agent is obliged to do; otherwise, there will be a violation (see, e.g., Hohfeld's framework of primitive legal concepts [11]). In principle, the duty class provides this concept, e.g., in Listing 3, with an obligation rule:

```
"@type": "agreement",
"obligation":
    "assigner": "OrganizationX", "assignee": "OrganizationZ",
    "action": "compensate",
    "refinement":
        "leftoperand":"payAmount", "operator": "eq",
        "rightOperand" {"@value": "2000.00", "@type": "xsd:decimal"},
        "unit": "http://dbpedia.org/resource/Euro"
```

Listing 4. Duty class in a policy with obligation rule

The policy above states that *OrganizationX* assigns to *OrganizationY* the duty of compensating the former with 2000 euro. However, with a non-intuitive terminological overlap, a permission rule (i.e., a rule containing a permission property) contains an inner **duty** property (2.6.5 of the ODRL Information Model)—linking to an instance of duty class—that in ODRL serves as a pre-condition for acquiring the permission:

```
"@type": "agreement",
"permission":
    "assigner": "OrganizationX", "assignee": "OrganizationY",
    "action": "use", "target": "datasetA",
    "duty":
        "action":"pay",
        "refinement":
            "leftoperand": "payAmount", "operator": "eq",
            "rightoperand": {"@value": "500.00", "@type": "xsd:decimal"},
            "unit": "http://dbpedia.org/resource/Euro"
```

Listing 5. Duty property in a policy with permission rule.

In the policy above, *OrganizationX* permits *OrganizationY* to use *datasetA*, conditionally to *OrganizationY* paying 500 euros. *OrganizationY* has a choice. The organization can choose not to pay and disregard access or pay and then acquire permission to use datasetA. Looking at Hohfeld's theory again [11], the position of *OrganizationY* is not a duty, but rather an *institutional power*: by performing the action described in the "duty" property, the assignee will enjoy the permission. Note that, for making the policy-relevant, an implicit assumption needs to be introduced here: that the use of the data is forbidden in general.

This also pinpoints another issue. If we are accepting the interpretation of this duty object as a precondition, it is not clear whether the *consequence* property (meant to trigger compensation measures to violation) can be used here: if the precondition is not satisfied, then the permission does not hold, so there cannot be a violation. We have found no specific constraint in the ODRL Information Model.

Lack of Granularity in Identifying Parties. The ODRL language considers only two functional roles for agents (assignor and assignee), a choice which raises several concerns. First, it is not clear if the assigner counts as the policy's creator and/or as the claim-holder (correlative of the duty-holder/assigner). Second, the roles relevant to norms and roles relevant to actions can be entirely disjoint: e.g., the party to which the duty is assigned can be different from the party that produces the performance removing a duty. For instance, a carer might have the duty to perform a particular check in due time. Indeed, some actions in the ODRL vocabulary allow refinements that enable specifying performer and recipient roles (e.g., `trackingparty`, `trackedparty` for the "track" action), but these are *ad-hoc* solutions, whereas a systematic approach, e.g., based on thematic roles of action, instead enhances readability and re-usability of patterns for different interactions.

Transformational Aspects. The activation or revocation of rules is a critical dimension in normative reasoning. Deontic relations are not fixed and change with interactions among parties. ODRL suggests to use the constraint class where temporal and contextual information can be specified to activate or terminate rules; it also provides the consequence and remedy class for enforcing actions against violations, but this is not always sufficient. For instance, regulations such as the GDPR place great importance on data subject rights; in data sharing scenarios, patients have the right to grant, change, or revoke their consent. Changes such as those consequent to patients withdrawing their consent (i.e., triggered by action) need to be captured to maintain lawful data processing. Furthermore, change also occurs at the level of parameters of policies. Suppose, OrganizationX has to pay 10% of a specific fee up to the end of 2020, and some action is possible that modifies the percentage to be paid. Based on our experience with ODRL, it is not possible to represent this mechanism, as it lacks a general approach to define in a machine-readable way the semantics of actions in terms of institutional or extra-institutional effects.

Handling Conflicts. ODRL provides a strategy to resolve conflicts that arise when merging policies due to policy inheritance [14]. It uses the `conflict` property which can take either the `perm`, `prohibit` alternatively, `invalid` values to decide which rule takes precedence over the other. For example, if the conflict property is set to "perm", then the permission will override the prohibition.

While this is one way to handle conflict between rules, for more complex scenarios, other factors such as attributes of the parties and contextual information can provide a richer input for setting the conflict property. The norm in Listing 6 states that data can not be shared outside of the EU, but if the recipient has a cross-border agreement and the purpose for sharing data is an emergency (e.g., an outbreak), then data may be shared.

```
"@type": "agreement",
"conflict": "perm",
"prohibition":
    "action": "share", "target": "datasetA",
    "constraint":
        "leftOperand": "spatial", "operator": "neq",
        "rightOperand": "https://www.wikidata.org/wiki/Q458"
"permission":
    "action": "share", "target": "datasetA",
    "refinement":
        "and": { "@list": [{"@id": "ex:c1"}, {"@id": "ex:c2"}] }

"@type": "constraint", "uid": "ex:c1",
"leftOperand": "purpose", "operator": "eq",
"rightOperand": {
    "@value":"emergency",
    "@type":"xsd:string"
}

"@type": "constraint", "uid": "ex:c2",
```

```
"leftOperand": "recipient", "operator": "eq",
"rightOperand": {
    "@value": "partOfcrossborderAgreement",
    "@type": "xsd:string"
}
```

Listing 6. Conflict property set to `Perm`, indicating that permission overrides prohibition.

The example above demonstrates that conflict resolution specifications are independent from the contextual information (i.e. constraints). Rather than a static, abstract `conflict` property, a more reasonable choice would be to take constraints into consideration and to implement principles as *lex specialis*. (Additional mechanisms are also required in principle for *lex posterior* and *lex superior*.)

Additional Limitations. So far, we discussed a focused selection of the considerations we drew over our interaction with ODRL, and acknowledged additional challenges, here reported only succinctly. Normative statements are about actions, while regulations, often, are about outcomes. For instance, a specific data processing can be licit (i.e., permitted) as performed on public sources, yet the output (e.g., discriminatory decision-making) might still be illicit. Second, there are instances where action might result in creating a new asset. For example, a rule might state that *"If an asset is copied, it must be attributed to a certain party"*. The rule on the original asset needs to be modified when it is copied. These changes in activity need to be reflected in the rules. Third, the higher-level distinction between use and transfer actions is simplistic, even if considering only digital assets. Looking at transfer only in terms of ownership does not allow us to consider, e.g., physical movement of data from one premise to another without changing the data rights-holder.

Finally, ODRL does not provide an exact model of the policy *life-cycle*, which has a potential application value as it enables capturing policy design patterns. Suppose a company makes an offer for the use of their dataset under a certain payment. If another company takes up the offer, then the policy should evolve to an agreement. It is not clear from the information model whether and how the ODRL will express these changes.

4   Discussion and Conclusion

The ODRL came a long way from its initial conception and its wide adoption can be attributed to its accessibility and expressiveness. In this paper, we addressed some significant limitations on the current version of the ODRL language: the lack of monotonicity in representing delegation scenarios, semantic ambiguity in the usage of "duty"'', granularity in identifying parties, and transformational aspects of rules.

Our findings are in contrast with those of Grunwell et al. [10] that conclude that the ODRL meets the requirements for representing access delegation policies. As stated by their work, the requirements for delegation policy are easy revocation, time dependency, and granularity. While this might hold for static policies, we have shown that this does not hold for dynamic policies as for instance the scenario in which a patient can revoke their consent at any given time. There is no mechanism to represent such events in ODRL. Fornara et al. [7] claim that the expressivity of their work is the same as the ODRL model in that it is possible to express deontic relations using both models. In our work, we found that the expressivity of the ODRL language in representing deontic relations is not enough in cases where more roles emerge, thus exceeding the limitation of two complementary roles such as assignee and assigner or creditor and debtor [7]. Additionally, these authors consider the lifecycle of the rules (i.e. their dynamics) while our work covers the lifecycle of a policy. Steyskal et al.[20] demonstrate that ODRL is suitable to express access policies for linked data by providing different scenarios. One of the examples covered is the introduction of payment duties. They illustrate that duty assignment can be easily defined but they also find the semantics of duty ambiguous and not explicit enough to express simple assignments. We confirm their analysis, and we additionally found that the semantics to specify the modification of actions is missing, and this is necessary, particularly in payment scenarios (e.g., for changing rates).

In next steps, we will focus on studying whether the policy specification language eFLINT [5] can overcome some of the issues covered here. An essential aspect of eFLINT is that it is an action-based language and derives normative positions of actors from the actions they perform (permission) or expected to perform (duties) at a given moment. This simplifies to perform e.g. compliance checking of scenarios or software implementations as they are inherently action-based. Furthermore, normative aspects of the language are based on the framework of normative concepts proposed by Hohfeld, supporting the use of a primitive for legal power—the ability to grant or remove permissions and duties assigned. These features should mitigate some of the limitations of ODRL discussed above, such as the representation of delegation and transformational aspects. The next step is to validate the expressiveness and tractability of the language with several use-cases from finance, healthcare, and other data marketplaces.

As future work, we plan to perform a systematic comparison between ODRL and eFLINT to extract common underlying models and test whether the interoperability of the two is feasible. Our vision with respect to an integration of ideas from ODRL and eFLINT is the development of a self-contained policy specification language that is as much as possible independent on the application, or the implementation framework.

References

1. Anderson, A., et al.: Extensible access control markup language (XACML) version 1.0. OASIS (2003)
2. Anderson, A.H.: A comparison of two privacy policy languages: EPAL and XACML. In: Proceedings of the 3rd ACM workshop on Secure web services, pp. 53–60 (2006)
3. Arnab, A., Hutchison, A.: Extending ODRL and XrML to enable bi-directional communication. University of Cape Town, Technical report (2004)
4. Ashley, P., Hada, S., Karjoth, G., Powers, C., Schunter, M.: Enterprise privacy authorization language (EPAL). IBM Res. **30**, 31 (2003)
5. van Binsbergen, L.T., Liu, L.C., van Doesburg, R., van Engers, T.: eFLINT: a domain-specific language for executable norm specifications. In: Proceedings of the 19th ACM SIGPLAN International Conference on Generative Programming: Concepts and Experiences, pp. 124–136 (2020)
6. De Vos, M., Kirrane, S., Padget, J., Satoh, K.: ODRL policy modelling and compliance checking. In: Fodor, P., Montali, M., Calvanese, D., Roman, D. (eds.) RuleML+RR 2019. LNCS, vol. 11784, pp. 36–51. Springer, Cham (2019). https://doi.org/10.1007/978-3-030-31095-0_3
7. Fornara, N., Colombetti, M.: Using semantic web technologies and production rules for reasoning on obligations, permissions, and prohibitions. AI Commun. **32**(4), 319–334 (2019)
8. Gajanayake, R., Iannella, R., Sahama, T.: An information accountability framework for shared ehealth policies. In: Data Usage Management on the Web: Proceedings of the WWW2012 Workshop, pp. 38–45. Technische Universitat Munchen-Institut fur Informatik (2012)
9. García, R., Gil, R., Gallego, I., Delgado, J.: Formalising ODRL semantics using web ontologies. In: Proceedings of the 2nd International Workshop on ODRL, pp. 1–10 (2005)
10. Grunwel, D., Sahama, T.: Delegation of access in an information accountability framework for eHealth. In: ACM International Conference Proceeding Series (2016)
11. Hohfeld, W.N.: Fundamental legal conceptions as applied in judicial reasoning. Yale Law J. **26**(8), 710–770 (1917)
12. Iannella, R., Villata, S.: ODRL Information Model 2.2. W3C Recommendation (2018)
13. Jamkhedkar, P.A., Heileman, G.L., Martínez-Ortiz, I.: The problem with rights expression languages. In: Proceedings of the ACM Conference on Computer and Communications Security, pp. 59–67 (2006)
14. Karafili, E., Lupu, E.C.: Enabling data sharing in contextual environments: policy representation and analysis. In: Proceedings of the 22Nd ACM on Symposium on Access Control Models and Technologies, pp. 231–238 (2017)
15. Leicht, J., Heisel, M.: A survey on privacy policy languages: Expressiveness concerning data protection regulations. In: 2019 12th CMI Conference on Cybersecurity and Privacy (CMI), pp. 1–6. IEEE (2019)
16. Pellegrini, T., et al.: A genealogy and classification of rights expression languages-preliminary results. In: Data Protection/LegalTech-Proceedings of the 21st International Legal Informatics Symposium IRIS, pp. 243–250 (2018)
17. Rodríguez-Doncel, V., Delgado, J., Llorente, S., Rodríguez, E., Boch, L.: Overview of the MPEG-21 media contract ontology. Semant. Web **7**(3), 311–332 (2016)

18. Rodriguez-Doncel, V., Villata, S., Gómez-Pérez, A.: A dataset of rdf licenses. In: JURIX, pp. 187–188 (2014)
19. Shakeri, S., et al.: Modeling and matching digital data marketplace policies. In: Proceedings of the IEEE 15th International Conference on eScience, eScience 2019, pp. 570–577 (2019)
20. Steyskal, S., Polleres, A.: Defining expressive access policies for linked data using the ODRL ontology 2.0. In: Proceedings of the 10th International Conference on Semantic Systems, pp. 20–23 (2014)
21. Steyskal, S., Polleres, A.: Towards formal semantics for ODRL policies. In: Bassiliades, N., Gottlob, G., Sadri, F., Paschke, A., Roman, D. (eds.) RuleML 2015. LNCS, vol. 9202, pp. 360–375. Springer, Cham (2015). https://doi.org/10.1007/978-3-319-21542-6_23

Automating Normative Control for Healthcare Research

Milen G. Kebede[✉] [iD]

University of Amsterdam, Science Park 904, Amsterdam, The Netherlands
m.g.kebede@uva.com

Abstract. There is an increasing need for norms to be embedded in technology as the widespread deployment of big data analysis applications increases. However, existing methodologies do not provide automated policy enforcement mechanisms especially for policies derived from legislation and contractual agreements. Consequently, data access is hindered and collaborations derailed due to fear data misuse and high non-compliance fees. This research aims to automate normative controls in healthcare, such as data sharing agreements, and ultimately, enforce these policies for compliant data usage and access which encourages collaboration and facilitates research outcomes while maintaining accountability. This paper outlines the PhD research questions, current approaches and preliminary results.

Keywords: Policy specification language · Ontology · Access control model

1 Problem Statement

In several domains of application, easier accessibility to data has the potential to produce a decisive positive impact [35]. This is particularly true in healthcare research. Current IT infrastructures used by organisations in the healthcare domain to run their business processes typically rely on specific access-control methods, such as the Role based access control model (RBAC), that employ static policies [30]. However, the introduction of legislation such as the General Data Protection Regulation (GDPR) [1] in to this systems creates more complexities due to the complexity and dynamic nature of such normative artefacts. This creates the need for patient data registry maintainers to develop data sharing infrastructures that enforces privacy policies derived from legislation and data sharing agreements, to ensure compliance and encourage collaborative research.

While data sharing encourages collaboration, improves treatment outcomes and maintain accountability, it can also create the opportunity for misuse of

This work is part of the Enabling Personalized Interventions (EPI) project and is supported by NWO in the Commit2Data – Data2Person program under contract 628.011.028.

V. Rodríguez-Doncel et al. (Eds.): AICOL-XI 2018/AICOL-XII 2020/XAILA 2020, LNAI 13048, pp. 62–72, 2021.
https://doi.org/10.1007/978-3-030-89811-3_5

data. Data sharing agreements are signed with the goal of preventing misuse and complying with regulations. These agreements regulate contracting parties on how they can share data with each other [21]. Enforcing this agreements is a challenging task considering the complexity of the legal documents. This results in a more cautious and conservative behaviours among data registry maintainers which forces data to stay in silos. Similarly, collaboration between different stakeholders is discouraged due to the challenges of maintaining trust in an environment where decisions can not be traced at system level. Being able to trace back the source of a problem is a necessary requirement for responsibility attribution; lack of this function is detrimental to social maintenance.

On the other hand, current IT infrastructures are not able to take into account that access and use of data is regulated at several levels, whose normative sources (users' consent, contractual agreements, laws) will change in time. Therefore, there is a need for new techniques to automatically enforce policies extracted from these agreements as well as abstract over the complexity of the documents and capture dynamic aspects of policies. This research addresses the challenges of automating privacy policies from legislation and contractual agreements and the automatic enforcement of these policies using an access control mechanism. Identifying the rules relevant to an access request can be challenging, given there are several normative dispositions that may be applicable. Additionally, when rules are taken from different sources, inconsistent policies result in conflict. In the following section, current work in legal ontologies, policy specification langauges and access control models will be presented.

2   Related Work

2.1   Legal Ontologies

To address the research goal, existing work on legal ontologies, policy specification languages and access control models is addressed.

Several ontologies are developed to model data-sharing agreements, some of which are designed to regulate data usage and privacy-aware data access [19], to specify contracts, to manage data-flows designed for linked open data environments [12] and to provide legal knowledge modelling of the GDPR core concepts [25]. In general, ontologies have gained momentum in recent years due to their potential as tools to conceptualize and specify shared knowledge as well as organize information, and to reduce the complexity of knowledge management and engineering. These ontologies are tailored to model general or specific kinds of legal knowledge. The LKIF core ontology is a library of ontologies relevant for the legal domain [13]. It can serve as a resources for legal inference, it facilitates knowledge acquisition, and can serve as a basis for semantic annotation of legal information sources.

The LegalRuleML aims to model the interpretation of a rule, the temporal evolution of norms and provides a classification of deontic operators [4]. It encourages the effective exchange and sharing of legal knowledge and reasoning between legal documents, business rules, and software applications. The work on [25] introduces

Pronto which is a legal ontology which provides legal knowledge modelling of the core concepts of General Data Protection Regulation (GDPR). It models deontic concepts and uses the LKIF core ontology to model actions and roles. The UFO-L core ontology represents rights and duty relations and aimes at making more explicit the elements of legal relations [11]. Ontologies are also used to support the application of data-sharing agreements (DSA) in a collaborative health research data sharing scenario by providing the appropriate vocabulary and structure to log privacy events in a linked data based audit log [19].

2.2   Policy Specification

A right expression language (REL) is a machine-readable language used typically in digital rights management systems for regulating usage and access control of digital assets. There are several applications to rights expression languages such as stating copyright and expression of contractual language. Some example of RELs are the Extensible Access Control Markup Language (XACML), Enterprise Privacy Authorisation Language (EPAL), and the Open Digital Rights Language (ODRL) [2,3,26]. The goal in this research is to utilize a language that supports specifying normative constructs as those specified in privacy regulations and agreements. While extended versions of XACML support partial specification and enforcement of laws and regulations, it lacks for the support for "system obligations" [18]. These are obligations the system has to perform on certain events such as notification of data breach. On the other hand, EPAL is designed for writing enterprise privacy policies but lacks reasoning support for conflicts or other relevant constructs.

RELs are also used for governance in multimedia assets and intellectual property protected content. The work on [28] present the MPEG-21 contract ontology (MCO), a part of the standard ISO/IEC 21000. MCO is an ontology that represents contracts that describe rights on multimedia assets and intellectual property protected content. It describes the contract model and key elements such as the parties in the contract and the relevant clauses conveying permissions, obligations and, prohibitions. Another work [29], presents a dataset of licenses for software and data, expressed as RDF for use with resources on the web. They use ODRL 2.0 to describe rights and conditions present in licenses. It provides a double representation for humans and machines alike and can enable generalized machine-to-machine commerce if generally adopted.

2.3   Access Control Models

Access control is the process of determining the permissiblity of any access request to perform a specific action on the system such as a read or a write on a data object that belongs to a data subject [5]. Typically, an access control model aims to protect the data object from unauthorized access based on specific access control policies. A number of access control models are proposed to control users' access to data and information resources. The early models presented in literature include the discretionary access control (DAC) [31], mandatory access control (MAC) [24], and role-based access control (RBAC) [32]. In RBAC, access to various resources is regulated on the basis of the role played by the data-consumer. These models fail to capture the dynamic nature of policies [35]. As a result, the need for flexible and dynamic access control systems has led to the emergence of the attribute-based access control (ABAC) and the usage control model (UCON) [15,38].

Attribute-based access control (ABAC), is regulated more generally on the basis of the value of attributes of the user while a usage control model (UCON) provides a means for fine grained control over access permissions though attributes. Even though ABAC allows for relationship among parties to be captured, the work in [7] states that ABAC might be lacking when the complexity and dynamically of systems grows thereby making it difficult to capture chains of interpersonal relationships. Other models such as the Relationship-Based Access control, are aimed towards community-centered systems [10]. Access decisions in this model are made based on the social relationships of the parties. This types of models allow for contextual information to be taken into account during access decision making. The gap identified here is the consideration of policies, within access control systems, from various sources of norms which raises the need for policy combination mechanisms as well as conflict resolution mechanism.

3 Research Questions

Given the problem statement and the relevance of this research, the main question this research aims to answer is **how can we develop solutions for the acquisition and application of contractual and other legal requirements for data processing in the healthcare domain, to enable embedded compliance in a distributed data sharing environment?**

Given that our research is restricted to a specific domain, healthcare, the normative artifacts that regulate processing of personal data in this domain need to be identified. After identifying the artefacts, relevant articles and clause that are associated with personal data processing will be extracted. Consequently, the first research question is:

RQ1. Which of the normative artefacts and articles that regulate data sharing systems in the healthcare domain are relevant to this research?

Data sharing systems need to comply with the regulations and data sharing agreements that regulate the parties involved. The policy specification languages utilised to specify such rules need to capture the dynamic nature and complexity of these documents. The policy specification language also should enable a complaint access control mechanism by allowing for the specification of expressive,fine grained and flexible policies. To develop a clear understanding of existing work and identify the relevant policy specification languages, the following research question is derived.

RQ2. What type of policy specification language can be developed or selected from existing languages to specify policies from applicable legislation and contractual agreements in healthcare?

Access control models should manage the complexity and dynamic nature of policies as well as enforcing these policies to ensure compliance. Data sharing systems involve different parties with whom agreements are made. In addition to privacy policies derived from legislation, each party will have their own authorisation requirements to the resources they own which will be specified in the policies. These policies will be composed into a single policy to determine how the asset is utilised. As a result, there is a need to combine these policies and to deal with any inconsistencies that may arise. To mitigate these issues, the third research question is formulated.

RQ3. How are policies from various sources of norms combined and inconsistencies handled during access decision making?

4   Proposed Approach

The goal of this research is to capture and enforce normative controls that regulate data sharing infrastructures within healthcare. This research is part of the Enabling Personalised Interventions project (EPI).

The EPI project aims to enable personalised diagnosis by developing real-time monitoring services and digital health twins. EPI aims to empower data subjects and providers through self-management, shared management and personalization across the full health spectrum. It will provide a platform based on secure and trustworthy distributed data infrastructure, that provides actionable and personalised insights for prevention, management and intervention to providers and patients. One of the use cases under EPI is the DIGP registry. Diffuse intrinsic pontine glioma (DIPG) is a rare pediatric brain cancer for which there is no curative treatment, despite decades of clinical trials [37]. In order to advance the progress and pace of DIPG research, the SIOPE DIPG Network and Registry was established. This cancer registry aims to overcome the current lack of clinical, imaging and biologic data and improve academic research on DIPG.

4.1   Identifying Regulatory and Organizational Requirements

The SIOPE DIPG Registry collects information on DIPG patients across Europe and a partner registry in North America, known as the International DIPG Registry, includes patient data from the USA, Canada, Australia and New Zealand. The DIPG network has provided us with different legal documents such as data sharing regulations, data sharing agreements and patient consent forms. Data sharing agreements is an agreement that regulates contracting parties on how they can share data with each other. Its purpose is to define what parties are required to do with respect to condition specified in the agreement [19].

The first stage of this research is to investigate and identify the relevant articles and clauses associated with processing of personal data. Data sharing agreements consist of terms about the data sharing agreement itself as well as terms concerning the data sharing process. From these documents, relevant articles that specify permissions, prohibitions and obligation will be extracted.

Legal documents make references to other legal documents, for example, the data sharing regulation makes several references to the GDPR. Therefore, relevant articles from the GDPR will be extracted. One of the challenges faced during this stage is the difficulty of representing norms accurately due to the complexity of regulatory documents such as the data sharing agreements.

4.2 Formalizing Policies from Regulatory Documents

The readability and usability of the policy specification language plays a central role for interoperability of policies. The goal in this stage is to develop a generic ontology that captures the concept and principles of the GPDR that apply to all context of personal data processing. Additionally, a specialised ontology that captures the concepts and principles of the data sharing agreements will be developed. This may impact the GDPR depending on the interpretation and application of different national and corporate policies. Several policy specification languages will be investigated to identify the ones that fit the use-case requirements. Policy specification languages such as the ODRL, eFlint and XACML are examples of policies investigated through examples from the DIPG use-case.

The policy specification language should also specify both higher and lower level policies. Higher level policies express general level requirements and rights that are specified in legislation , contractual agreements and regulatory requirements. Lower level policies describe how privacy requirements can be implemented in data sharing application such as access control policies. Such policies express what a subject is permitted or prohibited to do in relation to a particular asset e.g. a policy that states who can access a certain dataset [22].

4.3 Developing an Access Control Mechanism

The policies from the above ontology will be enforced through an access control mechanism in the data sharing infrastructure. Enforcing policies derived from various norms is not a trivial task. In collaborative data sharing environment, other than the data sharing agreements, parties can also create policies to protect their assets which results in various policies implicating one asset. Some of the existing solutions evaluate the policies of an asset individually, then apply strategies to combine decisions. While others, use an authoritative approach in which policies are combined in a predefined manner [8,20]. These type of approaches will be investigated to determine the policy composition algorithm to be developed.

When policies are derived from various norms, it is possible that we might end up a policy set granting and denying access for the same request to the same asset which creates conflicts. Existing conflict resolution strategies will be investigated. Recent work in this aspect have analysed conflict resolution from a game theoretic point of view and some graph-theoretic models [14,34,36]. We will investigate existing work and develop conflict resolution strategies.

5   Preliminary Results

This section presents our experiences with two policy specification languages in formalizing data sharing scenarios and policies.

5.1   Open Digital Rights Language

In recent years ODRL has gained popularity both in theoretical and practical settings. Our use cases focus on automating data-sharing agreements in the context of healthcare, we found ODRL to be of interest and relevant to our research.

The Open Digital Rights Language (ODRL) is designed as a policy expression language, aiming to provide a flexible and interoperable information model, vocabulary, and encoding mechanism for representing normative statements concerning digital content and services [16]. It evolved through the years from a digital rights expression language for expressing simple licensing mechanisms for the use of digital assets to accommodating privacy policies [17]. The W3C currently supports the ODRL Information Model 2.2 Recommendation. The model is developed using Linked Data principles; however, its semantics is described informally as no formal specification is provided.

Previous work investigated the language's suitability for different scenarios and from different perspectives, and some have proposed the extension of the language [9,23,33]. Our work shares similar motivations, although our analysis focuses on the general modeling process and requirements, as practitioners aiming to model a policy in ODRL. Additionally, vital institutional patterns that were only partially covered before, as delegation, were considered. Delegation is a particularly relevant (and delicate) institutional construct as it brings to the foreground the requirements of meeting the needs of stakeholders while maintaining accountability.

Using ODRL, patterns relevant to data-sharing agreements, highlighting the issues that emerge in the exercises were modelled. The examples were modelled with respect to the ODRL documentations on the information model, informal semantics, use-case and vocabulary of the language. We report our experiences concerning the limitations identified on the current version of the ODRL language. The main limitation identified are: the lack of monotonicity in representing delegation scenarios, semantic ambiguity in the usage of "duty," granularity in identifying parties, and transformational aspects of rules.

5.2   Data Sharing Policy Specification Using the eFlint Language

Our work describes how data sharing agreements specified using the eFlint language can be used as a means to disseminate certain types of usage and access control policies. In order to specify data sharing policies, we adopt a domain specific language, eFlint, developed to formalising different sources of norms [6]. The peculiarities of eFlint is that it is an action-based language and that the normative positions of actors are derived from the actions they can perform or are expected to perform. Compliance checking of scenarios and software implementations is simplified because scenarios and software implementation are action based.

It's theoretical foundations are found in transition systems and in Hohfeld's framework for legal fundamental conceptions. This means that eFlint is able to express normative positions, such as, the representation of 'duties' or 'power'. eFlint follows a legal case analysis method that involves interpretation, qualification and assessment of policies. It makes distinction between physical and institutional reality. This realities hold when actors interact with objects, each other and abstractions over physical reality. Additionally, eFlint allows for normative relations to change over time.

In this work, we formalize the semantic of terms, data usage policies and business rules from the data sharing regulation document. The early finding from this work demonstrates that eFlint specifications can be re-usable because types in eFlint can be redefined by subsequent type declaration. A generic interpretation can be used by several application by letting each application specialise certain types to the domain of the application. The concepts of the GDPR can be re-used across projects by utilising the references the DIPG regulatory document makes to the GDPR. Second, eFlint is flexible, i.e., the language can be used to specify different sources of norm such as the GDPR, data sharing regulatory documents and access control rules. Additionally, eFlint allows us to make a connection from higher level policies (GDPR) to lower-level policies (access control policies). We found eFlint to be expressive enough to specify granular policies, therefore our current formalisation match the granularity of the document.

6 Conclusion

There is an urgent need to share data among institutions that reside in the same continent as well as institutions across boarders. The motivation for this research is to contribute to one of the FAIR principles of "Accessibility" [1]. Data should be easily accessible especially in the healthcare. While there are several policy specification languages able to express and govern legally binding behaviour within technological environments, there are still some limitations such as the expressivity of the language in terms of capturing legal concepts [27]. One of the goals behind this research is to model a policy language that is able to represent legal fundamental concepts that can be expressive, granular and flexible enough to be used in a distributed environment.

Access control policies should enhance interoperability while being suitable for the underlying domain of application, in this case, healthcare. As such designing the right specification and enforcement mechanism for access control policies will have organizational benefits. Stakeholders should be enabled to define the structure of their policies in terms of applicable regulation and data sharing agreements to incorporate security, privacy and business requirements into policies. In future work, an evaluation method for the data sharing ontology as well as the access control mechanism will to measure performance overheads and efficiency of deploying access control mechanisms in the EPI distributed data sharing infrastructure.

Acknowledgment. This dissertation project is supervised by Prof. Tom van Engers, Dr. L. Thomas van Binsbegen and Dr. Dannis van Vuurden. This research is part of the EPI project and is supported by NWO in the Commit2Data – Data2Person program under contract 628.011.028.

References

1. 2018 reform of eu data protection rules. https://ec.europa.eu/commission/sites/beta-political/files/data-protection-factsheet-changes_en.pdf
2. Anderson, A., et al.: Extensible access control markup language (XACML) version 1.0. OASIS (2003)
3. Ashley, P., Hada, S., Karjoth, G., Powers, C., Schunter, M.: Enterprise privacy authorization language (EPAL). IBM Res. **30**, 31 (2003)
4. Athan, T., Boley, H., Governatori, G., Palmirani, M., Paschke, A., Wyner, A.: Oasis legalruleml. In: Proceedings of the Fourteenth International Conference on Artificial Intelligence and Law, pp. 3–12 (2013)
5. Bertino, E., Bettini, C., Ferrari, E., Samarati, P.: An access control model supporting periodicity constraints and temporal reasoning. ACM Trans. Database Syst. (TODS) **23**(3), 231–285 (1998)
6. van Binsbergen, L.T., Liu, L.C., van Doesburg, R., van Engers, T.: eFLINT: a domain-specific language for executable norm specifications. In: Proceedings of the 19th ACM SIGPLAN International Conference on Generative Programming: Concepts and Experiences, pp. 124–136 (2020)
7. Crampton, J., Sellwood, J.: Path conditions and principal matching: a new approach to access control. In: Proceedings of the 19th ACM Symposium on Access Control Models and Technologies, pp. 187–198 (2014)
8. Damen, S., den Hartog, J., Zannone, N.: CollAC: collaborative access control. In: 2014 International Conference on Collaboration Technologies and Systems (CTS), pp. 142–149. IEEE (2014)
9. De Vos, M., Kirrane, S., Padget, J., Satoh, K.: ODRL policy modelling and compliance checking. In: Fodor, P., Montali, M., Calvanese, D., Roman, D. (eds.) RuleML+RR 2019. LNCS, vol. 11784, pp. 36–51. Springer, Cham (2019). https://doi.org/10.1007/978-3-030-31095-0_3
10. Gates, C.: Access control requirements for web 2.0 security and privacy. IEEE Web **2**, 12–15 (2007)
11. Griffo, C., Almeida, J.P.A., Guizzardi, G.: A pattern for the representation of legal relations in a legal core ontology. In: JURIX, pp. 191–194 (2016)
12. Hadziselimovic, E., Fatema, K., Pandit, H.J., Lewis, D.: Linked data contracts to support data protection and data ethics in the sharing of scientific data. In: SemSci@ ISWC, pp. 55–62 (2017)
13. Hoekstra, R., Breuker, J., Di Bello, M., Boer, A., et al.: The LKIF core ontology of basic legal concepts. LOAIT **321**, 43–63 (2007)
14. Hu, H., Ahn, G.J., Zhao, Z., Yang, D.: Game theoretic analysis of multiparty access control in online social networks. In: Proceedings of the 19th ACM Symposium on Access Control Models and Technologies, pp. 93–102 (2014)

15. Hu, V.C., Kuhn, D.R., Ferraiolo, D.F., Voas, J.: Attribute-based access control. Computer **48**(2), 85–88 (2015)
16. Iannella, R., Villata, S.: ODRL information model 2.2. W3C Recommendation (2018)
17. Karafili, E., Lupu, E.C.: Enabling data sharing in contextual environments: policy representation and analysis. In: Proceedings of the 22nd ACM on Symposium on Access Control Models and Technologies, pp. 231–238 (2017)
18. Leicht, J., Heisel, M.: A survey on privacy policy languages: expressiveness concerning data protection regulations. In: 2019 12th CMI Conference on Cybersecurity and Privacy (CMI), pp. 1–6. IEEE (2019)
19. Li, M.: DSAP: data sharing agreement privacy ontology. Ph.D. thesis (2018)
20. Mahmudlu, R., den Hartog, J., Zannone, N.: Data governance and transparency for collaborative systems. In: Ranise, S., Swarup, V. (eds.) DBSec 2016. LNCS, vol. 9766, pp. 199–216. Springer, Cham (2016). https://doi.org/10.1007/978-3-319-41483-6_15
21. Matteucci, I., Petrocchi, M., Sbodio, M.L., Wiegand, L.: A design phase for data sharing agreements. In: Garcia-Alfaro, J., Navarro-Arribas, G., Cuppens-Boulahia, N., de Capitani di Vimercati, S. (eds.) DPM/SETOP-2011. LNCS, vol. 7122, pp. 25–41. Springer, Heidelberg (2011). https://doi.org/10.1007/978-3-642-28879-1_3
22. Casassa Mont, M., Pearson, S., Creese, S., Goldsmith, M., Papanikolaou, N.: A Conceptual Model for Privacy Policies with Consent and Revocation Requirements. In: Fischer-Hübner, S., Duquenoy, P., Hansen, M., Leenes, R., Zhang, G. (eds.) Privacy and Identity 2010. IAICT, vol. 352, pp. 258–270. Springer, Heidelberg (2010). https://doi.org/10.1007/978-3-642-20769-3_21
23. Fornara, N., Colombetti, M.: Operational semantics of an extension of ODRL able to express obligations. In: Belardinelli, F., Argente, E. (eds.) EUMAS/AT -2017. LNCS (LNAI), vol. 10767, pp. 172–186. Springer, Cham (2018). https://doi.org/10.1007/978-3-030-01713-2_13
24. Osborn, S.: Mandatory access control and role-based access control revisited. In: Proceedings of the Second ACM Workshop on Role-Based Access Control, pp. 31–40 (1997)
25. Palmirani, M., Martoni, M., Rossi, A., Bartolini, C., Robaldo, L.: Pronto: privacy ontology for legal compliance. In: Proceedings of the European Conference on e-Government, ECEG 2018, pp. 142–151 (2018)
26. Pellegrini, T., et al.: A genealogy and classification of rights expression languages - preliminary results. Jusletter IT, pp. 1–8 (2018)
27. Pellegrini, T., et al.: A genealogy and classification of rights expression languages-preliminary results. In: Data Protection/LegalTech-Proceedings of the 21st International Legal Informatics Symposium IRIS, pp. 243–250 (2018)
28. Rodríguez-Doncel, V., Delgado, J., Llorente, S., Rodríguez, E., Boch, L.: Overview of the mpeg-21 media contract ontology. Semant. Web **7**(3), 311–332 (2016)
29. Rodriguez-Doncel, V., Villata, S., Gómez-Pérez, A.: A dataset of rdf licenses. In: JURIX. pp. 187–188 (2014)
30. Rostad, L., Edsberg, O.: A study of access control requirements for healthcare systems based on audit trails from access logs. In: 2006 22nd Annual Computer Security Applications Conference (ACSAC 2006), pp. 175–186. IEEE (2006)
31. Sandhu, R., Munawer, Q.: How to do discretionary access control using roles. In: Proceedings of the Third ACM Workshop on Role-Based Access Control, pp. 47–54 (1998)
32. Sandhu, R.S., Coyne, E.J., Feinstein, H.L., Youman, C.E.: Role-based access control models. Computer **29**(2), 38–47 (1996)

33. Shakeri, S., et al.: Modeling and matching digital data marketplace policies. In: Proceedings of the IEEE 15th International Conference on eScience, eScience 2019, pp. 570–577 (2019)

34. Squicciarini, A.C., Shehab, M., Wede, J.: Privacy policies for shared content in social network sites. VLDB J. **19**(6), 777–796 (2010)

35. Wilkinson, M.D., et al.: The fair guiding principles for scientific data management and stewardship. Sci. Data **3**(1), 1–9 (2016)

36. Xiao, Q., Tan, K.L.: Peer-aware collaborative access control in social networks. In: 8th International Conference on Collaborative Computing: Networking, Applications and Work sharing (CollaborateCom), pp. 30–39. IEEE (2012)

37. van Zanten, S.E.V., et al.: Development of the siope dipg network, registry and imaging repository: a collaborative effort to optimize research into a rare and lethal disease. J. Neuro-Oncol. **132**(2), 255–266 (2017)

38. Zhang, X., Parisi-Presicce, F., Sandhu, R., Park, J.: Formal model and policy specification of usage control. ACM Trans. Inf. Syst. Secur. (TISSEC) **8**(4), 351–387 (2005)

Logic, Rules, and Reasoning

Principles and Semantics: Modelling Violations for Normative Reasoning

Silvano Colombo Tosatto[1]([envelope]) [iD], Guido Governatori[1] [iD],
and Antonino Rotolo[2] [iD]

[1] CSIRO, Data61, 41 Boggo Road, Dutton Park, QLD 4102, Australia
{silvano.colombotosatto,guido.governatori}@data61.csiro.au
[2] Dipartimento di Scienze Giuridiche, Università di Bologna,
Via Zamboni 33, 40126 Bologna, Italy
antonino.rotolo@unibo.it

Abstract. The present paper proposes a structural operational semantics and the related semantics for normative systems. The proposed approach focuses on explicitly representing in force obligations and violations as events in a temporal framework, determining the state of a normative system. In the paper we use a set of core principles, defining some of the properties required when reasoning about norms, to motivate the semantics of the approach. Finally, we show that the proposed approach is capable of reasoning about more complex legal scenarios.

Keywords: Normative reasoning · Violations · Contrary to duty · Compensations

1 Introduction

Norms and regulations are a critical part of our modern society, and are being used as a mechanism to direct the behaviour of the members belonging to such a group. With this in mind, it is natural to try to replicate such mechanisms in synthetic communities such as multi-agent systems.

The discipline studying how to reason about norms and related concepts is referred to as *Normative Reasoning*, and one of the most well known techniques used to reason about normative systems is deontic logic [5]. Despite deontic logic is known to have many limitations, as shown for instance by Parent and van der Torre [12] on the particular topic concerning the *pragmatic oddity*,[1] some of the basic concepts and insights developed are still key and survive in more modern approaches capturing the interaction between time and regulations, such as Lorrini [10], Ågotnes et al. [1], Marín and Sartor [11], Chesani et al. [4] and Alrawagfeh [2], have proposed extensions of the Event Calculus to model obligations and related notions; however, Hashmi et al. [8] point out several shortcomings of using such extensions, and proposed a novel extension of Event Calculus

---

[1] The issue arising from the co-existence of obligations, which, while logically consistent is semantically odd.

© Springer Nature Switzerland AG 2021
V. Rodríguez-Doncel et al. (Eds.): AICOL-XI 2018/AICOL-XII 2020/XAILA 2020, LNAI 13048, pp. 75–89, 2021.
https://doi.org/10.1007/978-3-030-89811-3_6

based on the complete (temporal) abstract classification of normative requirements presented in their following work [9]. An alternative popular approach, is instead based on commitments, in other words, the coordination between agents belonging to a community is governed by the commitment made from an agent to another concerning certain actions or obligations. One of the approaches based on commitments is for instance the one discussed by Prankaj et al. [14].

This paper introduces a normative reasoning approach focused on explicitly representing in force obligations and violations in the state of the system. The contributions of the paper are threefold: **I** An explicit illustration of some of the principles required to be followed when reasoning about obligations and eventual violations in a system. **II** A structural operational semantics[2] allowing to reason about norms and deontic concepts in temporal scenarios, meaning that it allows to reason about obligations and violations along a temporally ordered sequence of states of the world, while preserving the principles. **III** Finally we show the introduced semantics is not plagued by issues such as the *pragmatic oddity*.

The present paper is structured as follows: Sect. 2 introduces the core principles required to reason about norms in a temporal setting, Sect. 3 introduces the informal semantics and illustrates how each concept can be used to represent the behaviour of some real world case. Section 4 shows how the semantics follows the principles, and how it avoids the *pragmatic oddity*. Finally Sect. 5 concludes the paper.

2   Normative Principles

The following principles, listed in Principles 1, are required to hold at every stage of a normative system to ensure that its behaviour does not lead to undesirable conclusions. The principles are derived from the ones introduced informally by Governatori [6].

Principles 1

1. *There cannot be a violation without a violated in force obligation in the current or the previous temporal frames.*
2. *There cannot be a compensatory obligation in force without a violation to be compensated.*
3. *A compensated violation is no longer maintained in the following temporal frames.*

The first principle covers the requirement that a violation requires the existence of an obligation that is *in force*, which means that a situation in which such an obligation should be the case happened. Moreover, the principle also specifies that a violation can outlive the in force obligation associated to it, as in some cases, it is desirable to terminate a violated in force obligation, but to maintain track of the violation.

---

[2] A structural operational semantics focuses on describing the computational steps, opposed to natural semantics which focus on describing the outcomes.

The second principle illustrates the correlation between compensatory obligations and the violations they can potentially compensate. In normative systems, both in reality and in synthetic environments, is not uncommon that a violation occurs, hence one of the priorities of normative systems is to deal with such scenarios, which usually involves setting in force additional obligations that once fulfilled would nullify the occurred violation. The principle specifies that for such kind of compensatory obligations to be in force, a violation that they are supposed to compensate must exist.

The third and final principle also deals with compensatory obligations, however it focuses on the fact that a violation should not be maintained in a normative system when the associate compensatory obligation is satisfied. This is related to the *double jeopardy doctrine*, stating that one should not be punished twice for a single violation. Thus the third principle specifies that a violation decades when compensated.

Finally, notice that the first and the third principles are related to the lifecycle of compensatory obligations in relation to their triggering violation, and describe the desired behaviour of a violation after its compensatory obligation is being fulfilled.

3   Semantics

In this section we introduce the informal semantics and the operational semantics that an agent within a multi-agent community can use to reason about its obligations related to its own behaviour and governed by the overlaying normative system. Moreover we show how the approach allows to reason about the normative components identified by Hashmi et al. [9], and we show that Principles 1 are preserved.

Before proceeding to introduce the semantics relative to our approach, we want to highlight that the approach is designed to deal with temporal settings, which, for simplicity, we represent as sequences of separate temporal instants of appropriate duration and associated to a label and a representation of the state of the world. For instance, the following temporal instant: (t_i, φ_i), is represented by the identifying label t_i and a state $\varphi_i{}^3$.

3.1   Obligations

One of the core elements of a normative system is represented by the norms governing it. In the present paper we focus on obligations, defining the requirements of the norms. Independently of the types of an obligation, it is described as a set of three propositional formulae, in a similar fashion as done by Governatori and Rotolo [7]. The three propositional formulae are used to describe the properties of an obligation, namely its *trigger*, its *requirement*, and its *deadline*.

---

[3] For the sake of simplicity, we assume these states to be represented as propositional formulae.

The *trigger* of an obligation represents the condition required in the context of a temporal frame to set the obligation in force, in other words when the triggering condition[4] that makes the obligation relevant in that context and is then required to be fulfilled. The *requirement* of an obligation represents what the obligation requires when in force to be fulfilled. In addition to the requirement, the *type* of an obligation also determines whether a sequence of temporal frames fulfils an obligation when it is in force. Finally, the *deadline* of an obligation determines when an obligation in force stops to be in force in accordance to the temporal frames' contexts. For certain types of obligation, its fulfilment or violation is determined when it stops to be in force.

Definition 1 (Obligation). *Let α, β, and γ be propositional formulae. An obligation is defined as follows:*

$$\mathcal{O} = O_{(p|y)}^{(a|m)}(\alpha, \beta, \gamma)$$

where:

$(a|m)$ *determines whether the obligation is of type* achievement *or* maintenance *respectively,*
$(p|y)$ *determines whether the obligation is* perdurant *or* yielding,
α *represents the* trigger,
β *represents the* requirement,
and γ *represents the* deadline.

The trigger of an obligation determines the required condition for the obligation to be in force, while the deadline determines when the obligation stops to be in force. The requirement describes what is necessary for the obligation to be fulfilled once in force, depending also on whether it is of type achievement or maintenance. Finally, whether an obligation is perdurant or yielding, determines whether an in force obligation persists after being. violated or not.

Obligation in Force. When an obligation is inactive, it does not influence the state of a normative system. In other words it cannot be violated. Differently, when an obligation is *in force*, it can be either violated or satisfied, hence affecting the compliance state. Informally, we consider the state of a normative system to be compliant, when no violations are present.

In force obligations are treated as events and are associated with the temporal frame label triggering them, as shown in Definition 2.

Definition 2 (Obligation in Force). *Given an obligation $\mathcal{O} = O_{(p|y)}^{(a|m)}(\alpha, \beta, \gamma)$, and a temporal frame t_i associated to a context, represented by a propositional formula ϕ. $\mathcal{O}$ becomes in force in t_i if $\phi \models \alpha$ and is represented as follows:*

$$\mathcal{F}(\mathcal{O}, t_i)$$

---

[4] Given a state φ_i associated to a temporal instant and trigger condition of an obligation represented by the propositional formula α, the state triggers the obligation, also referred to as setting the obligation in force, if and only if $\varphi_i \models \alpha$.

When an obligation is in force and a temporal frame associated to a context with a state φ_j, such that $\varphi_j \models \gamma$, then the obligation is not longer in force.

Encountering a state satisfying the *deadline* condition of an obligation is just a way to terminate an in force instance of such obligation. Further ways to terminate an in force instance of an obligation depend on the type of the obligation and on other features, and we introduce them in Definition 4, Definition 5, and Definition 6.[5]

Violations. Similarly as for obligations in force, we handle violations as events, and their presence in a normative system determines its compliance status. Moreover, as inactive obligations cannot be violated, a violated event is associated to both the in force obligation event being violated, as well as to the label of the temporal frame whose context caused the violation. This way of representing violations brings two main advantages. The first advantage is that it allows a single in force obligation to be potentially violated multiple times in different temporal frames. The second advantage, following from the first, is that each violation can be independently tracked and be dealt with. We show how violation events are represented in Definition 3.

Definition 3 (Violation). *Given an obligation in force $\mathcal{F}(\mathcal{O}, t_i)$ and a temporal frame t_j where $\mathcal{O}$ is violated. A violation of $\mathcal{F}(\mathcal{O}, t_i)$ in t_j is represented as follows:*

$$\mathcal{V}(\mathcal{F}(\mathcal{O}, t_i), t_j)$$

Achievement Obligation. An achievement obligation requires that its requirement is fulfilled in at least one of the temporal frames when the obligation is in force, and before the obligation's deadline is satisfied by one of the temporal frames' context. Notice that for the following examples of the obligation types (achievement and maintenance), we assume them to be *yielding*. For now, it is sufficient to know that a yielding obligation ceases to be in force when it is violated. We provide a proper description of the semantics of yielding and perdurant obligations in Definition 6.

Definition 4 (Achievement Obligation). *An achievement obligation, either persistent or yielding, is represented as follows:*

$$O^a_{(p|y)}(\alpha, \beta, \gamma)$$

fulfilment *exists a temporal frame within the interval, composed of a sequence of temporal frames, when the obligation is in force, whose associated state fulfils the obligation's condition. Once an in force instance of an achievement obligation is fulfilled, it is terminated.*

---

[5] Notice that Definition 6 introduces an exception to the in force instance termination, rather than defining additional ways of terminating an instance.

violation *every temporal frame in the interval, composed of a sequence of temporal frames, when the obligation is in force does not have an associated state fulfilling the obligation's requirement.*

We show in Example 1 how we represent a violation of an in force instance of a yielding[6] achievement obligation through a list of temporal instants.

Example 1 (Violated in Force Obligation). Given an obligation $\mathcal{O} = O_y^a(\alpha, \beta, \gamma)$, and a sequence of time instants t_1, ..., t_4 with their associated states, represented by a propositional formula, below the respective time instant label. The obligation $\mathcal{O}$ is violated as shown in Table 1, and we show the states in which the in force instance holds in the third line of the table and the states in which the violation holds in the fourth line.

Table 1. In force and violation.

t_1	t_2	t_3	t_4
$\emptyset$	α	$\alpha \wedge \gamma$	$\alpha \wedge \gamma$
	$\mathcal{F}(\mathcal{O}, t_2)$		
		$\mathcal{V}(\mathcal{F}(\mathcal{O}, t_2), t_3)$	$\mathcal{V}(\mathcal{F}(\mathcal{O}, t_2), t_3)$

The in force instance originating in t_2 is violated in t_3 as the deadline of the obligation is satisfied in the state and no of the states within the interval where the in force instance holds contain the obligation's requirement.

Maintenance Obligation. A maintenance obligation requires that each temporal frame's context satisfies the obligation's requirement when in force. An obligation is in force from the temporal frame following the temporal frame whose context satisfies the obligation's trigger, and until the temporal frame whose context satisfies the obligation's deadline, included.

Definition 5 (Maintenance Obligation). *A maintenance obligation, either perdurant or yielding, is represented as follows:*

$$O_{(p|y)}^m(\alpha, \beta, \gamma)$$

fulfilment *every temporal frame in the interval, composed of a sequence of temporal frames, when the obligation is in force has an associated state fulfilling the obligation's requirement.*
violation *exists a temporal frame within the interval, composed of a sequence of temporal frames, when the obligation is in force, whose associated state does not fulfil the obligation's requirement. Once an in force instance of a maintenance obligation is violated, it is* terminated.

---

[6] An obligation is yielding if its in force instances do not persist when violated. The difference between yielding and perdurant is discussed in Sect. 3.1.

Perdurant and Yielding Obligations. In a normative system, it is crucial to analyse the behaviour of an in force obligation affected by a violation. Independently than the type of obligation considered, two possible behaviours are whether the in force obligation would stay in force or not after a violation is being detected.

For achievement obligations, it can be the case that an obligation should stay in force after the deadline has been reached and a violation has been detected. This allows the possibility to fulfil the requirement of the obligation in future temporal frames.

For maintenance obligations, violations can be identified while the obligation is in force instead of only when the deadline is reached, usually terminating it. Sometimes it is desirable to keep the obligation in force despite the violation. This allows the detection of further deviations from the desired requirement.

Independently than the type of obligation, we refer to obligations whose in force state would survive violations as *perdurant*, while those whose in force event terminates when violated are referred to as *yielding*. Notice that due to their behaviour, *perdurant* obligations allow multiple violations to be associated to a single in force event, and each of these violations happens in a different temporal frame. Finally, notice also that for *perdurant achievement* obligations, a violation is identified when a temporal frame satisfying the deadline condition is reached, hence we can assume that for these kind of obligations the in force event is extended to the next temporal frame whose context satisfies the deadline condition.

Definition 6 (Perdurant and Yielding Obligations). *Given a persistent obligation $\mathcal{O} = O_p^{(a|m)}(\alpha, \beta, \gamma)$, when an instance of $\mathcal{O}$ is in force, identified as the event $\mathcal{F}(\mathcal{O}, t_i)$, its in force status does not terminate when violated. Meaning that, depending on its type, when a violation is identified, the following is the case:*

achievement *the in force event is maintained in the normative system state, and its in force interval is extended to the next temporal frame satisfying the deadline condition.*
maintenance *the in force event is maintained in the normative system.*

Differently, a yielding obligation terminates its in force status when it is violated. Depending on its type, when a violation is identified, the following is the case:

achievement *the in force event is terminated.*
maintenance *the in force event is terminated and at the temporal frame the violation happens, effectively reducing the original in force interval.*

We show in Example 2 how a perdurant obligation is represented in the semantics introduced.

Example 2 (Speed Limit). Considering a scenario involving traffic regulations, in particular the speed limit imposed while driving within a urban area, as defined by the following obligation: $\mathcal{O} = O_p^m(\alpha, \beta, \neg\alpha)$ where:

α the vehicle is in a urban area

β the speed is 50 km/h or lower

We show in Table 2 and in Table 3 the differences of the behaviour of obligations in the scenario described above. In Table 2 we show the behaviour of $\mathcal{O}$ as a maintenance perdurant obligation.

Table 2. Example 2: perdurant obligation

t_1	t_2	t_3	t_4
$\neg\alpha$	$\alpha \wedge \beta$	$\alpha \wedge \neg\beta$	$\neg\alpha$
	$\mathcal{F}(\mathcal{O}, t_2)$	$\mathcal{F}(\mathcal{O}, t_2)$	
		$\mathcal{V}(\mathcal{F}(\mathcal{O}, t_2), t_3)$	$\mathcal{V}(\mathcal{F}(\mathcal{O}, t_2), t_3)$

In Table 3 we show the behaviour of $\mathcal{O}$ as a maintenance yielding obligation: $\mathcal{O} = O_y^m(\alpha, \beta, \neg\alpha)$.

Table 3. Example 2: Yielding obligation

t_1	t_2	t_3	t_4
$\neg\alpha$	$\alpha \wedge \beta$	$\alpha \wedge \neg\beta$	$\neg\alpha$
	$\mathcal{F}(\mathcal{O}, t_2)$		
		$\mathcal{V}(\mathcal{F}(\mathcal{O}, t_2), t_3)$	$\mathcal{V}(\mathcal{F}(\mathcal{O}, t_2), t_3)$

Notice that the behavioural difference is in t_3, where the perdurant obligation is still in force despite being violated, while its yielding counterpart is not.

3.2 Compliance

The goal of a normative system is to guide the behaviours of involved entities towards acceptable states. By following the norms governing such system, the system state is kept *compliant* with the regulations governing it. Differently failing to follow the regulations leads to a *non compliant* state, which as we have shown previously, not fulfilling the requirements of an obligation deriving from the normative system's regulation results in a violation. Informally, we consider a normative system to be in a compliant state if it does not contain violations.

Compensations. While compliance as the absence of violations appears to be rigid, as whenever a normative system transitions from its compliant state to a non compliant one, then it is not possible to return to a compliant state. As the violation of a norm constituting a normative system represents befalling into a non compliant state. In such a situation, it is often the case that it is preferable to provide a way of allowing to return to a compliant state. Such ways are represented in a normative system by *compensatory obligations*, obligations which in force trigger is represented by the appearance of violations in the state of the system, and whose task is to repair to such violations though their fulfilment, practically erasing them, hence allowing the system to be again compliant.

The behaviour of compensations and their relation with their triggering violation is described in Definition 7.

Definition 7 (Compensations). *Let $\mathcal{O}$ be and $\mathcal{O}'$ be two obligations, $\mathcal{O}'$ is the compensatory obligation of $\mathcal{O}$, and their relation is expressed as follows:*

$$\mathcal{O} \otimes \mathcal{O}'$$

The trigger of $\mathcal{O}'$ can be considered the following:

$$\mathcal{V}(\mathcal{F}(\mathcal{O}, t_i), t_j)$$

And the effects of fulfilling, or violating $\mathcal{O}'$ are the following:

fulfilment *the violation $\mathcal{V}(\mathcal{F}(\mathcal{O}, t_i), t_j)$ is removed from the system state.*
violation *a violation for $\mathcal{O}'$ is introduced in the system state.*

Additionally, the in force *instance of a compensatory obligation is terminated when the associated violation disappears from the system state.*

Notice that Definition 7 includes a condition that allows to terminate an in force instance of a compensatory obligation when the related violation event is no longer part of the system state. Intuitively, a compensatory obligation without a violation to compensate does not offer any practical utility in a normative system, hence when the associated violation event disappears, then the same happens to the in force instance of the compensatory obligation.

We show in Example 3 how the semantics introduced can be used to represent and reason about the behaviour of a compensatory obligation.

Example 3. Considering the following legal text fragment from YAWL[7] Deed of Assignment, Clause 5.2[8]:

"Each Contributor indemnifies and will defend the Foundation against any claim, liability, loss, damages, cost and expenses suffered or incurred by the Foundation as a result of any breach of the warranties given by the Contributor under clause 5.1."

The two following obligations can be formalised from the fragment:

$$\mathcal{O} = O_y^m(\alpha_s, \beta, \perp)$$

Where:

---

[7] Yet Another Workflow Language.
[8] http://www.yawlfoundation.org/files/YAWLDeedOfAssignmentTemplate.pdf, retrieved on March 28, 2013.

α is a contributor

β not to breach of the warranties given by the Contributor under clause 5.1.

$$\mathcal{O}' = O_y^m(\mathcal{V}(\mathcal{F}(\mathcal{O}, t_i), t_j), \alpha', \top)$$

Where:

α' indemnifies and will defend the Foundation against any claim, liability, loss, damages, cost and expenses suffered or incurred by the Foundation

For the relation between the two obligations $\mathcal{O}$ and $\mathcal{O}'$, we borrow the syntax used by Calardo et al. [3] and is represented as follows:

$$\mathcal{O} \otimes \mathcal{O}'$$

We show in Table 4, and Table 5 how the pair of obligation and related compensatory obligation behave in the following cases, respectively when the obligation is violated, and when the obligation is violated but the violation is compensated.

Table 4. Example 3: Compensation: violation

t_1	t_2	t_3	t_4
$\emptyset$	α	$\alpha \wedge \neg\beta$	$\alpha \wedge \neg\beta$
	$\mathcal{F}(\mathcal{O}, t_2)$		
		$\mathcal{V}(\mathcal{F}(\mathcal{O}, t_2), t_3)$	$\mathcal{V}(\mathcal{F}(\mathcal{O}, t_2), t_3)$
		$\mathcal{F}(\mathcal{O}', t_3)$	$\mathcal{F}(\mathcal{O}', t_3)$

Table 5. Example 3: Compensation: compensated violation

t_1	t_2	t_3	t_4
$\emptyset$	α	$\alpha \wedge \neg\beta$	$\alpha \wedge \neg\beta \wedge \alpha'$
	$\mathcal{F}(\mathcal{O}, t_2)$		
		$\mathcal{V}(\mathcal{F}(\mathcal{O}, t_2), t_3)$	
		$\mathcal{F}(\mathcal{O}', t_3)$	

4   Discussion

We show in this section how the principles are followed by the operational semantics proposed. We also show how such semantics allows to reason about particular normative scenarios, namely the pragmatic oddity.

4.1   Evaluating the Principles

We evaluate whether the semantics proposed follows the underlying principles identified.

First Principle. To show that the principle is correctly followed, we consider two cases: the first case involves the scenario where a violation coexists in the same state as its related in force obligation. For this case, notice that the semantics defined in Definition 3 allow to spawn a violation event only when the related obligation is in force. Meaning that a violation cannot exist without the related obligation being in force. Considering now the case when a violation exists in a state without its related in force obligation, we know that according to Definition 2, an in force obligation generally terminates when the deadline is verified. Moreover, the same is the case for the in force termination semantics depending on the obligation type or properties, as shown in Definition 4, Definition 5, and Definition 6. Notice that the semantics concerning the termination of an in force obligation does not influence violations, hence an existing violation related to an in force obligation can survive the in force obligation being terminated. Moreover, notice that according to Definition 6, a *yielding* obligation terminates its in force event when violated, while the newly spawned violation event survives. Therefore, the principle is followed by the semantics.

Second Principle. The second principle is concerned with the relation between a violation and an compensatory obligation in force related to the violation. Following the semantics described in Definition 7, it can be noticed that in order for a compensatory obligation to be set in force, it requires the primary obligation to be violated. Which is also shown in the *trigger* representation of the compensatory obligation, whose condition is represented by the violation of the primary obligation (or the previous obligation in the chain in case of concatenated compensations). Moreover, the semantics described in Definition 7, also covers the situation where a violation event is removed from the system state, and in that case every associated compensatory obligation in force instances are also terminated. Which follows the second principle.

Third Principle. The third and last principle is concerned with the persistence of a violation in the state. In particular it states that when a compensatory obligation is fulfilled, which is in force if and only if there is a violation event to be compensated (as stated by the second principle), then the associated violation ceases to be part of the state. Intuitively, a single instance of a violation should not be compensated multiple times. Following the semantics described in Definition 7, we can see that when a compensatory obligation is fulfilled, then the associated violation is removed from the system state, which follows what is dictated from the third principle.

4.2   Dealing with the Pragmatic Oddity

The pragmatic oddity is a well known paradox in normative reasoning, which set of regulations leads to a contradictory conclusion.

   The structure behind the oddity is the following: a given obligation in force is violated, leading to its compensatory obligation being set in force, which, in turn leads to both the secondary obligation and the primary obligation being in force together. This as been shown in the past to lead to scenarios where the co-existence of the primary and secondary obligations appears odd. Prakken and Sergot [13] illustrate the problem brought by the pragmatic oddity using the following scenario:

Example 4 (Pragmatic Oddity)

D1 There ought to be no dog.
D2 If there is a dog, there ought to be a warning sign.
D3 There is a dog.

   Prakken and Sergot observe that, when these rules are used in a situation where a dog is present, then the *odd* conclusion is the one requiring an *ideal* scenario where there should not be a dog, as required by rule **D1** in Example 4, and where there is a warning sign, as is derivable from the rules **D2** and **D3** in Example 4.

Example 5 (Reasoning about the Pragmatic Oddity). The regulations from Example 4 can be formalised as follows:

$$\mathcal{O}_{\mathbf{D1}} \otimes \mathcal{O}_{\mathbf{D2}}$$

$$\mathcal{O}_{\mathbf{D1}} = O_y^m(\alpha_s, \neg\beta, \bot)$$

Where:

α represents an initial state setting the prohibition of having a dog in force.
β having a dog

$$\mathcal{O}_{\mathbf{D2}} = O_y^a(\mathcal{V}(\mathcal{F}(\mathcal{O}_{\mathbf{D1}}, t_i), t_j), \gamma, \delta)$$

Where:

γ there is a warning sign.
δ represents the deadline before which it is required to fulfil the compensatory obligation.

   We illustrate in Table 6, how reasoning about the pragmatic oddity using the informal semantics introduced in the previous section does not achieve a contradictory conclusion.

Table 6. Pragmatic oddity

t_1	t_2	t_3	t_4
α	β	$\beta \wedge \gamma$	$\beta \wedge \gamma \wedge \delta$
$\mathcal{F}(\mathcal{O}_{\mathbf{D1}}, t_1)$			
	$\mathcal{V}(\mathcal{F}(\mathcal{O}_{\mathbf{D1}}, t_1), t_2)$		
	$\mathcal{F}(\mathcal{O}_{\mathbf{D2}}, t_2)$		

Notice that **D3** is represented by the context of the temporal frames t_2, t_3 and t_4, where β is true.

Table 7. Pragmatic oddity not compensated

t_1	t_2	t_3	t_4
α	β	β	$\beta \wedge \delta$
$\mathcal{F}(\mathcal{O}_{\mathbf{D1}}, t_1)$			
	$\mathcal{V}(\mathcal{F}(\mathcal{O}_{\mathbf{D1}}, t_1), t_2)$	$\mathcal{V}(\mathcal{F}(\mathcal{O}_{\mathbf{D1}}, t_1), t_2)$	$\mathcal{V}(\mathcal{F}(\mathcal{O}_{\mathbf{D1}}, t_1), t_2)$
	$\mathcal{F}(\mathcal{O}_{\mathbf{D2}}, t_2)$	$\mathcal{F}(\mathcal{O}_{\mathbf{D2}}, t_2)$	
			$\mathcal{V}(\mathcal{F}(\mathcal{O}_{\mathbf{D2}}, t_2), t_4)$

Moreover, if we consider Table 7, where it is illustrated a scenario where a warning sign is not being set up, we can see that the state remains non compliant as the violation of having a dog is not compensated, and an additional violation of not having set up the warning sign is introduced in t_4.

Finally, notice that in both scenarios, the odd conclusion of having the prohibition of having a dog ($\mathcal{O}_{\mathbf{D1}}$ in force) and having the obligation of having a warning sign for the dog ($\mathcal{O}_{\mathbf{D2}}$ in force), is not the case, as these obligations are never in force at the same time.

5   Conclusion

The present paper introduces a structural operational semantics for normative reasoning in terms of their obligation and violations. We initially introduce some core principles aimed at guiding the design of formalisms to reason about normative systems. These principles are an evolution of the ones earlier discussed by Governatori [6], and extends these concepts by taking into consideration how temporal aspects influence the interaction between the various normative elements involved in a normative system, such as for instance obligations and violations.

The structural operational semantics proposed in the paper handles in force obligations and violations as events, allowing to differentiate different instances of an obligation in force, from the abstract obligation itself considered as a normative concept. Moreover, the proposed approach also allows to precisely define a connection between the violations occurring in a normative system, to the responsible instances of an obligation. The relations defined between obligations, their in force intervals and eventual violations captured by the operational syntax and semantics, represent the main contribution of the paper. While the set of core principles appears initially simple, we strongly believe that following them when dealing with normative reasoning in temporal settings, is of paramount importance.

While structural description of the semantics naturally lends itself as a temporal solution for normative reasoning, we are aware that these kinds of approaches are often criticised. While we don't argue against the critiques towards temporal normative solutions, we highlight that our solutions focuses on explicitly representing in force instances of obligations and their violations as additional normative concepts. The explicit separation of such concepts from the obligations, allow to reason about them independently from the original obligation description. Moreover, such separation allows to handle situations where the life-cycles of these concepts become independent, for instance, considering *abrogation*,[9] which would not be able to be differently handled from *annulment*[10] if such separation would not be the case.

We conclude the paper with a brief discussion on the computational complexity of the proposed approach. Evaluating the state of an obligation over a timeline composed by a sequence of states is directly related to the computational complexity of the logic used to evaluate whether a state in the timeline satisfy the elements composing the obligation. This computational cost is applied for each state of the timeline and for each obligation being evaluated.

References

1. Ågotnes, T., van der Hoek, W., Rodríguez-Aguilar, J.A., Sierra, C., Wooldridge, M.: A temporal logic of normative systems. In: Makinson, D., Malinowski, J., Wansing, H. (eds.) Towards Mathematical Philosophy. TL, vol. 28, pp. 69–106. Springer, Dordrecht (2009). https://doi.org/10.1007/978-1-4020-9084-4_5

2. Alrawagfeh, W.: Norm representation and reasoning: a formalization in event calculus. In: Boella, G., Elkind, E., Savarimuthu, B.T.R., Dignum, F., Purvis, M.K. (eds.) PRIMA 2013. LNCS (LNAI), vol. 8291, pp. 5–20. Springer, Heidelberg (2013). https://doi.org/10.1007/978-3-642-44927-7_2

3. Calardo, E., Governatori, G., Rotolo, A.: A preference-based semantics for CTD reasoning. In: Cariani, F., Grossi, D., Meheus, J., Parent, X. (eds.) DEON 2014. LNCS (LNAI), vol. 8554, pp. 49–64. Springer, Cham (2014). https://doi.org/10.1007/978-3-319-08615-6_5

---

[9] The removal of obligations from a normative system without influencing eventual past violations, and scenarios where the obligation applies.

[10] The removal of obligations from a normative system, while removing also its past effects.

4. Chesani, F., Mello, P., Montali, M., Torroni, P.: Representing and monitoring social commitments using the event calculus. Auton. Agents Multi-Agent Syst. **27**(1), 85–130 (2013)
5. Gabbay, D., Horty, J., van der Meyden, R., Parent, X., van der Torre, L. (eds.) Handbook of Deontic Logic and Normative Systems, 1 edn. College Publications (2013)
6. Governatori, G.: Representing business contracts in RuleML. Int. J. Coop. Inf. Syst. **14**(02n03), 181–216 (2005)
7. Governatori, G., Rotolo, A.: A conceptually rich model of business process compliance. In: Proceedings of the Seventh Asia-Pacific Conference on Conceptual Modelling - Volume 110, APCCM 2010, Darlinghurst, Australia, Australia, pp. 3–12. Australian Computer Society Inc. (2010)
8. Hashmi, M., Governatori, G., Wynn, M.T.: Modeling obligations with event-calculus. In: Bikakis, A., Fodor, P., Roman, D. (eds.) RuleML 2014. LNCS, vol. 8620, pp. 296–310. Springer, Cham (2014). https://doi.org/10.1007/978-3-319-09870-8_22
9. Hashmi, M., Governatori, G., Wynn, M.T.: Normative requirements for regulatory compliance: an abstract formal framework. Inf. Syst. Front. **18**(3), 429–455 (2016)
10. Lorini, E.: Temporal logic and its application to normative reasoning. J. Appl. Non-Class. Log. **23**(4), 372–399 (2013)
11. Marín, R.H., Sartor, G.: Time and norms: a formalisation in the event-calculus. In: ICAIL, pp. 90–99 (1999)
12. Parent, X., van der Torre, L.: The pragmatic oddity in a norm-based semantics. In: Proceedings of the 16th International Conference on Artificial Intelligence and Law (ICAIL-2017) (2017)
13. Prakken, H., Sergot, M.: Contrary-to-duty obligations. Studia Logica Int. J. Symb. Log. **57**(1), 91–115 (1996)
14. Telang, P., Singh, M.P., Yorke-Smith, N.: A coupled operational semantics for goals and commitments. J. Artif. Intell. Res. **65**, 31–85 (2019)

Towards a Formal Framework for Partial Compliance of Business Processes

Ho-Pun Lam[1](✉), Mustafa Hashmi[1,2,3], and Akhil Kumar[4]

[1] Data61, CSIRO, Canberra, Australia
{brian.lam,mustafa.hashmi}@data61.csiro.au
[2] Federation University, Brisbane, QLD 4000, Australia
[3] La Trobe Law School, La Trobe University, Melbourne, Australia
[4] Smeal College of Business, Penn State University, University Park, PA 16802, USA
akhilkumar@psu.edu

Abstract. Binary "YES-NO" notions of process compliance are not very helpful to managers for assessing the operational performance of their company because a large number of cases fall in the grey area of partial compliance. Hence, it is necessary to have ways to quantify partial compliance in terms of metrics and be able to classify actual cases by assigning a numeric value of compliance to them. In this paper, we formulate an evaluation framework to quantify the level of compliance of business processes across different levels of abstraction (such as task, trace and process level) and across multiple dimensions of each task (such as temporal, monetary, role-, data-, and quality-related) to provide managers more useful information about their operations and to help them improve their decision making processes. Our approach can also add social value by making social services provided by local, state and federal governments more flexible and improving the lives of citizens.

Keywords: Partial compliance · Business process modelling · Compliance measures · Process compliance

1 Introduction

When designing business processes (BPs), practitioners always assume that the business model will be executed as planned. However, this is impractical in many situations. For example, cost fluctuations, equipment and resource availability, time constraints, and human errors can cause disruptions. In response to this, it is crucial for the practitioners to have a complete picture of the status of their running business processes—for taking strategic decisions on identifying, forecasting, obtaining and allocating required resources, and to be notified if any non-compliance issues are identified during execution.

Let us illustrate this idea by examining the payment process model as shown in Fig. 1, which consists of a sequence of tasks to be performed. Accordingly, a customer is required to make the payment within 15 days upon receiving the invoice; if not, the invoice must be paid with 3% per day interest in addition

© Springer Nature Switzerland AG 2021
V. Rodríguez-Doncel et al. (Eds.): AICOL-XI 2018/AICOL-XII 2020/XAILA 2020, LNAI 13048, pp. 90–105, 2021.
https://doi.org/10.1007/978-3-030-89811-3_7

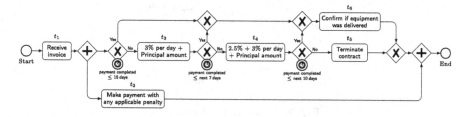

Fig. 1. Fragment of payment-making process (adopted from: [4])

to principal amount within the next 7 days. For any subsequent days hereafter within the next 10 days, an additional 2.5% interest will be added to the total payment as penalty, which will be calculated based on the principal amount. The contract will be terminated automatically upon 3 consecutive defaults.

Now, consider two compliant executions performed by two business customers of the company, customers A and B. Customer A strictly follows the *normal* sequence and makes the payment within 15 days after receiving the invoice. Customer B instead delayed the payment and paid the bill (together with interest and penalty) 3 weeks later. If we ascribe value to this process depending on the billing company's revenue, both executions positively contribute to it, as both customers did make their payments after receiving the invoices. However, the deferred payment of company B may affect the cash flow of the service provider company. Moreover, both these scenarios represent examples of partial compliance because there was a violation on the temporal dimension. Other violations may occur along other compliance dimensions such as: *money*, when monetary payments are not made according to agreements; *roles*, when individuals who perform certain tasks like approvals, etc., are not in the normal or authorized, or delegated, role; *data*, when the complete data required to perform a task is not available; and *quality*, when the quality of the work performed by a task is sub-standard. For each dimension, there are prescribed ranges of values or performance indicators in which a task is considered to be compliant on that dimension. If the indicator values within a narrow range are outside this normal range, then the task is said to be in partial compliance on that dimension. Finally, if the indicator does not fall into either of these two ranges then it is said to be non-compliant. A dimension can also be related to an attribute value. Thus, *payInDays* attribute represents the number of days within which payment is made after the invoice is sent to the customer. This attribute corresponds to the temporal dimension and can be used interchangeably with it.

Existing systems and compliance management frameworks (such as Declare [11], SeaFlows [10], COMPAS [14], etc.) only provide an *all-or-nothing* type of binary answer, i.e., *YES* if the BP is *fully* compliant; and *NO* if any non-compliant behavior has been detected at some point during execution, which is not *informative* and raises a simple yet significant question of whether the *whole* process is not compliant or *only* a part of it, and whether corrective actions should be performed

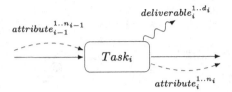

Fig. 2. Annotations of a task

from the point where the non-compliant behavior was detected, or from an earlier point.

Recently, some efforts coining the notion of *partial compliance* have been reported. For example, the approach in [9] returns the status of a BP as *ideal, sub-ideal, non-compliant* and *irrelevant*. Based on the notion of decision lattices [6], Morrison *et al.* [12] categorizes the compliance status as *Good, Ok,* and *Bad*. However, the issues remain similar as: *to what extent the process is compliant and how much (or what kind of) additional resources are required to resolve any detected non-compliance issues?*

To answer this question, in this paper, we present a formal framework for evaluating the levels of compliance of a BP at different levels of abstraction during execution and auditing phases, aiming to provide more clear and useful information to users concerned in facilitating their decision making process when any non-compliance issues arise.

The rest of the paper is organized as follows. Next, in Sect. 2, we provide necessary background information and terminologies following which we introduce our proposed framework in Sect. 3. Examples illustrating how the proposed framework works in practice are presented in Sect. 4. Related work is discussed in Sect. 5 before the paper is concluded with final remarks and directions for future work in Sect. 6.

2 Background and Problem Statement

In this section, we first introduce the necessary background and terminologies for the understanding of our proposed framework, and subsequently derive the problem statement.

Structure of a Business Process

A BP is represented as a temporally and logically ordered, directed graph in which the nodes represent tasks of the process that are executed to achieve a specific goal. It describes what needs to be done and when (*control-flows* and *time*), who is involved (*resources*), and what it is working on (*data*) [3]. Essentially, a BP is composed of various elements which provide building blocks for aggregating loosely-coupled (atomic) *tasks* as a sequence in a process aligned with the business goals.

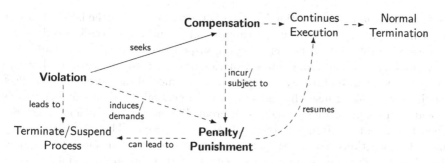

Fig. 3. Violation-compensation relationships of a partially compliant business process

Each task is an atomic unit of work with its own set of *input attributes*, which can be (partially) aggregated from *preceeding* task(s) or acquired from other sources, describing the *prerequisites* or *requirements* that the task has to comply with for its (full) execution, and *output attributes* that it has to produce upon execution and for propagation to *succeeding* tasks (as input), as shown in Fig. 2. Note that, in Fig. 2, the term *deliverable* is used to describe an output document or *artefact* that is produced by the task after execution and is not propagated to the next task. Technically, the values of attributes (both input and output) can have multiple dimensions, which may include information about time (or *temporal*), monetary, data, role, quality of service, or any combination of these. As an example, the value of payment and the payment due date of t_2 in Fig. 1 are from temporal and monetary dimensions, respectively.

A sequence representing the execution order of tasks of a BP in a given case is called a trace (a.k.a. occurrence sequence). Typically, a BP can be executed in a number of ways. For instance, below is the set of traces that can be generated from the business model, from *start* to the *end*, as shown in Fig. 1.

$$\mathfrak{T}^+ = \{\mathfrak{T}_1 = \langle t_1, t_2, t_3, t_4, t_5 \rangle, \quad \mathfrak{T}_2 = \langle t_1, t_3, t_2, t_6 \rangle,$$
$$\mathfrak{T}_3 = \langle t_1, t_2, t_3, t_4, t_6 \rangle, \quad \mathfrak{T}_4 = \langle t_1, t_3, t_4, t_2, t_5 \rangle,$$
$$\mathfrak{T}_5 = \langle t_1, t_2, t_3, t_6 \rangle, \quad \mathfrak{T}_6 = \langle t_1, t_3, t_4, t_2, t_6 \rangle,$$
$$\mathfrak{T}_7 = \langle t_1, t_2, t_6 \rangle, \quad \mathfrak{T}_8 = \langle t_1, t_3, t_4, t_5, t_2 \rangle,$$
$$\mathfrak{T}_9 = \langle t_1, t_3, t_2, t_4, t_5 \rangle, \quad \mathfrak{T}_{10} = \langle t_1, t_3, t_4, t_6, t_2 \rangle,$$
$$\mathfrak{T}_{11} = \langle t_1, t_3, t_2, t_4, t_6 \rangle, \quad \mathfrak{T}_{12} = \langle t_1, t_3, t_6, t_2 \rangle,$$
$$\mathfrak{T}_{13} = \langle t_1, t_6, t_2 \rangle\}$$

While it is always desirable that a BP behave strictly in accordance with the prescribed conditions, this may not always be the case in practice. A BP may deviate from its desired behavior in unforeseen circumstances and violate some (or all) of the conditions attached to it during execution.

Figure 3 illustrates what can happen when a violation occurs in a BP. The divergent behavior may cause a temporary suspension or (in some cases) termination of the process, and may also induce penalties. A *penalty* is a punitive measure (e.g. monetary or in some other form) enforced by company policy or a rule for the performance of an action/act that is proscribed, or for the failure to carry out some required acts. However, a violation of (mandatory) conditions does not necessarily imply automatic termination/suspension of a BP that would prevent any further execution. Certain violations can be compensated for [5], where compensation can be broadly understood as a remedial measure taken to offset the damage or loss caused by the violation. In general, legal acts and contracts provide clauses prescribing penalties and remedial provisions which are triggered when the deviations from the contractual clauses occur. These provisions may prescribe conditions that are subject to some penalties or punishments. As mentioned in the previous section, an execution with compensated violations (as in Fig. 1) leads to a sub-ideal situation [13], and is deemed partially-compliant. Nonetheless, the process can continue execution and complete normally once the compensatory actions are performed.

Next, we develop our framework in a formal manner.

3   Partial Compliance Framework

In this section we develop a partial compliance framework. The framework is based on the following principles or axioms underlying partial compliance:

Axiom 1. Compliance should not be binary 0/1 but should cover a spectrum of scenarios between 0 and 1.

Axiom 2. Partial compliance should be recognized and treated fairly.

Axiom 3. Partial compliance can be rectified by compensation mechanisms such as imposition of penalties, or sanctions that increase monotonically with the extent of the violation.

Axiom 4. The level of partial compliance decreases monotonically as the magnitude of the violation increases.

Throughout this section, we use the following notations: T is the set of unique task identifiers of tasks that appear in an instance of a trace $\mathfrak{T}$; and, A_n denote the set of attribute names of task t. Each attribute is mapped to a value v from a suitable numeric or categorical domain in a running instance. Thus, (a, v) is a attribute-value pair or tuple for an attribute in a task.

We introduce a partial compliance function ψ on task t to define partial compliance values for different attribute values under various compliance dimensions. Thus, $\psi_t(a, v, d)$ denotes the degree of partial compliance of attribute a of task t where the value of attribute a is equal to v, on compliance dimension d. This function maps attribute values of a task to a real-value in the $[0, 1]$ range that represents the degree of partial compliance, where 1 corresponds to full compliance. Thus, to formally describe the partial compliance for the running example in Fig. 1, we can write, $\psi_{t_2}(payInDays, 10 \text{ days}, Time) = 1$. This means that the partial compliance of task T_2 in dimension $Time$ is equal to 1 if the $payInDays$ attribute has a value of 10 days, which also represents compliance of task T_2 on the temporal dimension.

Given a task t, we denote metric $\mathcal{D}_t = \{\mathcal{D}_t^1, \ldots, \mathcal{D}_t^n\}$ as the set of compliance dimensions that relate to t attributes and denote its size as $|\mathcal{D}|$. Thus, for the running example of Fig. 1, $\mathcal{D}_{t_2} = \{\text{Monetary, Time, Percent}\}$ and $|\mathcal{D}| = 3$.

Hence, given a set of attribute names $\mathcal{A}_n$, it is necessary to determine which attributes relate to compliance and aggregate their individual compliance into a single metric of compliance. Thus, one can decide if a task is fully-, partially- or non-compliant. Accordingly, we introduce the following definitions:

Definition 1 (Aggregate attribute compliance metric). *Given a task $t \in \mathcal{T}$, $\mathcal{D}_t = \{\mathcal{D}_t^1, \ldots, \mathcal{D}_t^n\}$ as the set of n compliance dimension(s) of attributes of t, and an attribute aggregation operator $\odot$, then we define a compliance metric for attribute a of task t on dimension i be:*

$$M_t^d = \bigodot_{(a,v)|a \in \mathcal{A}_n} \psi_t(a, v, d)$$

be the aggregate compliance value across all task attributes for which $d \in \mathcal{D}_t$ is the dimension relevant to an attribute a in task t, v is the value of the attribute a in task t. In addition, we denote $\psi_t(a, v, d) = $ null if dimension d does not apply to attribute a in task t. Thus, any compliance dimension with a null value will be simply ignored from the aggregation. Finally, $M_t^i \in [0, 1]$.

Definition 2 (D-Compliant). *Given a task $t \in \mathcal{T}$, $\mathcal{D}_t = \{\mathcal{D}_t^1, \ldots, \mathcal{D}_t^n\}$ as the set of compliance dimensions that relate to its attributes, M_t^i as the aggregate attribute compliance metric per Definition 1, and $S^i, \Delta^i \in \mathbf{R}$, then,*

- *t is non-compliant on dimension $\mathcal{D}_t^i$ iff $M_t^i < S^i$;*
- *t is partially compliant on dimension $\mathcal{D}_t^i$ iff $S^i \leq M_t^i < S^i + \Delta^i$; and*
- *t is fully compliant on dimension $\mathcal{D}_t^i$ iff $M_t^i \geq S^i + \Delta^i$*

where S_t^i and Δ_t^i are the standard and threshold values for full and partial compliance, respectively. Note that the threshold represents a range or window around the standard value in which partial compliance is possible. These values are generally numeric constants provided by the domain experts to the analysts.

Definitions 1 and 2 define how the attribute metric value should be calculated and conditions for different levels of compliance, respectively. This means that a task is fully compliant if it is executing under some *ideal* situation; while a task is partially compliant if its attributes in $\mathcal{D}_t$ are to a large extent in accordance with the requirements specified but a few of them have been violated and remedial actions have been performed to repair/compensate the situation such that all violations identified have either been resolved or compensated; or a task is non-compliant otherwise.

The D-compliance score on dimension $\mathcal{D}_t^i$ is given by M_t^i and is a real value in $[0, 1]$. For a non-numeric value, the attribute dimension metric may be recorded on a qualitative scale such as a 3-point scale of (low, medium, high) or on a 5- or 7-point Likert scale. In this case, the points on the scale can be mapped uniformly to the 0–1 scale. Thus, by default, high would correspond to 1, medium to 0.67 and low to 0.33. Alternatively, a user-defined mapping function may be employed for this purpose. In general, rules can also be applied to determine a user-defined mapping function for nominal compliance values. Thus, given a task with an attribute a and a 3-point scale of (low, medium, high) in dimension d, a set of rules can be written as follows using three reasonable cut-off values:

$$\psi(a, v, d) = \begin{cases} 0.25 & \text{if } v = \text{``low''} \\ 0.50 & \text{if } v = \text{``medium''} \\ 0.90 & \text{if } v = \text{``high''} \end{cases}$$

Once the individual attribute value has been evaluated, they can be combined in different ways to obtain a dimension compliance score M_t^i.

Below are some alternative methods to compute the aggregation operator $\odot$ for an attribute.

1. *Average method.* Take a (weighted) average of attribute dimension metric values. This will give an average across the individual scores across all the applicable dimensions. For three dimensions with scores of 0.7, 0.9 and 1, the average would be 0.867. It is also possible to assign different weights to each dimension based on its importance.
2. *Product method.* Take the product of all attribute dimension metric values. In this case, we would multiply across all the $\psi(a, v, d)$'s. Thus, in the above example we would obtain 0.63. In general the product approach would lead to a lower value than the average approach.
3. *Rule-based method.* Apply a more general rule-based method to combine the individual metrics. Thus, a rule could be expressed as:

If $(\psi(a_1, v_1, d_1) < 0.5)$ AND $(\psi(a_2, v_2, d_2) < 0.5)$ then $M_t^i = 0$.

which states that if the partial compliance on metrics 1 and 2 is less than 0.5 then the task is non-compliant even though it is partially compliant on individual metrics, perhaps because these two metrics are very important.

The simplest implementation of M_t^i is to set $\odot$ to the (weighted) average of all non-null compliance values after evaluations, i.e., $M_t^i = \frac{1}{|\mathcal{D}_t^i|}\Sigma\psi_t(a)$. However, we should be cautious when selecting which function to use in computing M_t^i as setting $\odot = max$ would mean that whenever an attribute in a dimension is fully compliant, then the task will also be fully compliant in this particular dimension and similar will apply when we set $\odot = min$, which may not be something that we intended.

Example 1. A review loan application task has $S^i = 3$ days. $\Delta^i = 2$ days. If the task takes 4 days, it is partially compliant on the dimension $\mathcal{D}^{T1}$. But if it takes 6 days, it is non-compliant.

---

[1] From now on, we will use $\mathcal{D}^M$, $\mathcal{D}^T$, $\mathcal{D}^R$, $\mathcal{D}^D$, and $\mathcal{D}^Q$ to denote the monetary, time, role, data, and quality dimensions of a task, respectively.

Based on the definitions above, the level of compliance of tasks can be defined in terms of a metric outside a permitted range for one or more related dimensions, such as money, time, role, data, quality, etc. Thus, a task in the process may be required to be performed by a worker in a role using certain data inputs or documents. There is also a time limit for the completion of a task and a quality requirement. Finally, some tasks may also require the monetary payment of a fee (e.g. an application fee for admission to a school, processing fee for issuance of a passport or permit, etc.).

Definition 3 (T-compliance). *Given a task $t \in \mathcal{T}$, and $\mathcal{D}_t = \{\mathcal{D}_t^1, \ldots, \mathcal{D}_t^n\}$ as the compliance dimensions that correspond to its various attributes, then we define:*

- *t is* non-compliant *iff $\exists \mathcal{D}_t^i \in \mathcal{D}_t$, $\mathcal{D}$ is* non-compliant;
- *t is* fully compliant *iff $\forall \mathcal{D}_t^i \in \mathcal{D}_t$, $\mathcal{D}$ is* fully compliant;
- *otherwise, t is said to be* partially-compliant *meaning that some attributes are operating under* sub-ideal *conditions.*

Definition 4 ($\mathcal{P}_t$-Measure). *Given a task $t \in \mathcal{T}$, $\mathcal{D}_t = \{\mathcal{D}_t^1, \ldots, \mathcal{D}_t^n\}$; M_t^i the set of its attribute dimensions and dimension metrics as defined in Definition 1; and a dimension aggregation operator $\oplus$, then we define:*

$$\mathcal{P}_t = \bigoplus_{i \in [1,n]} M_t^i$$

as the task compliance measure, or $\mathcal{P}_t$-Measure, of task t.

The dimension aggregation operator $\oplus$ here works much like the attribute aggregation operator $\odot$ in Definition 2. It aggregates dimension metrics that were calculated for each dimension and returns a single value that represents the overall level of task compliance. However, as discussed above, the aggregation function should be chosen with care.

Example 2. A loan application process consists of 5 activities from submit application to receive final decision. The standard amount of time for it is 15 days. If the threshold Δ is 5 days and it takes 18 days to finish the loan application process, then it is partially compliant, showing that even when some activity(ies) in the process instance may be non-compliant, the instance itself can be compliant.

Consequently, given an instance of trace $\mathfrak{T}$ of a BP, one can simply calculate the level of compliance of $\mathfrak{T}$ by directly aggregating/averaging the $\mathcal{P}_t$-Measure value of each task. However, this may have some drawbacks as the aggregated value may not necessarily reflect the real situation of the *whole* trace. This is due to the fact that the changes made after any non-compliance issues might introduce new attributes (and/or values), and changes to the task. Besides, during execution other tasks may also impact the value of the attribute, averaging these values might not give correct performance of the attribute, hence it would not make sense.

To overcome these issues, we define trace compliance and a trace compliance measure based on the attribute dimension metrics, as follow.

Definition 5 ($\mathfrak{T}$-compliance). *Given an instance of trace $\mathfrak{T}$ of a BP, we define:*

- *$\mathfrak{T}$ is non-compliant iff $\exists t \in \mathcal{T}$, t is non-compliant;*
- *$\mathfrak{T}$ is fully compliant iff $\forall t \in \mathcal{T}$, t is fully compliant;*
- *otherwise, $\mathfrak{T}$ is said to be partially-compliant meaning that $\mathfrak{T}$ has been executed under some sub-ideal (or sub-optimal) conditions.*

Definition 6 ($\mathcal{P}_{\mathfrak{T}}$-Measure). *Given an instance of trace $\mathfrak{T}$ of a BP; $\mathcal{T}$ the set of unique task identifiers of tasks; $\mathcal{D}_{\mathfrak{T}} = \{\mathcal{D}_{\mathfrak{T}}^1, \ldots, \mathcal{D}_{\mathfrak{T}}^n\}$; M_t^i the set of attribute compliance dimensions that appear in $\mathfrak{T}$; the aggregate compliance metric of task t as in Definition 1; and $\otimes$ the task dimension aggregation operator, then we define the trace partial compliance measure as:*

$$\mathcal{P}_{\mathfrak{T}} = \bigotimes_{\mathcal{D}_{\mathfrak{T}}^i | \mathcal{D}_{\mathfrak{T}}^i \in \mathcal{D}_{\mathfrak{T}}} \underset{t | t \in \mathcal{T} \cap \mathcal{D}_{\mathfrak{T}}^i \in \mathcal{D}_t}{\arg\min} (M_t^i > 0)$$

As execution progresses, the aggregate compliance metrics of each task will be updated accordingly. Hence, to reflect this situation, the compliance measure of a trace is defined by the aggregated dimension metrics (across all dimensions). Naturally, if all metrics of a particular dimension are 0 for a task, then a zero value will be returned. Note here that an instance of trace can be D-compliant on multiple dimensions, yet it does not mean that it will automatically be $\mathfrak{T}$-compliant at the end.

Lastly, we give the following definition for the overall compliance of a process log consisting of multiple traces to conclude our framework.

Definition 7 ($\mathcal{P}_P$-Measure). *Given a BP P; T_P the set of log trace instances obtained after executing P; and $|\mathrm{T}_P|$, its size, then the compliance measure for the process P is given by:*

$$\mathcal{P}_P = \frac{1}{|\mathrm{T}_P|} \Sigma_{\mathfrak{T} \in \mathrm{T}_P} \mathcal{P}_{\mathfrak{T}}$$

where $\mathcal{P}_{\mathfrak{T}}$ is the $\mathcal{P}_{\mathfrak{T}}$-Measure of the trace instance $\mathfrak{T}$.

Here, the compliance measure of a BP, $\mathcal{P}_P$-Measure, is defined as the average value of the $\mathcal{P}_{\mathfrak{T}}$-Measure across all traces since each trace represents an independent execution of P and will not affect other ones.

It is important to note that we have defined our metrics at three levels of aggregation in a hierarchical manner, i.e., at the task, trace and process log levels. Depending upon the user application and requirements, metrics at one or more levels can used in conjunction with each other to gain multiple perspectives. Besides, it is possible that a metric may be violated at one level but may still be satisfied at another or a higher level, or vice-versa. Moreover, some metrics along some dimensions like time may be more meaningful at the instance level as in Example 2 since the total instance duration is more important for the customer than the duration of individual tasks. Other metrics may be more relevant at

the task level, such as the monetary amounts involved, etc. The process log level
metrics can give insights into the overall compliance level for the entire log over
a period of time, such as a week, month or quarter. Comparing such metrics
across several successive periods can provide managerial insights into overall
compliance trends.

4   Composite Measure Computations

Next, we discuss some scenarios in the context of a real-world example to illustrate how the proposed framework can be applied in practice to compute different levels of compliance by employing the *averaging method* discussed in the
previous section. For this purpose, we consider the invoice payment example
from Fig. 1, and provide some notation for our computations. We consider the
attribute aggregation operator $\odot$ to be the average of all attribute values pro-
jected in the dimension, i.e., $M_t^i = \frac{1}{|\mathcal{D}_t^i|} \Sigma \psi_t(a)$. Moreover, $\oplus$ is the compliance
dimension aggregation operator averaging the dimension index values for each
dimension in the task i.e., $M_t^i = \frac{1}{|\mathcal{D}_t^i|} \Sigma \psi_t(a)$, and $\otimes$ is the minimum of all values
for each dimension.

Table 1 illustrates the attributes and their possible values in the context of
Fig. 1. Attributes such as *description, invoiceValue* and *invoiceDate* are meta
information of the invoice and do not contribute to the compliance metrics. The
attributes *equipmentDeliveryDays* and *payInDays* denote the number of days
required to deliver the equipment(s) to the purchaser and the number of days
within which *full* payment must be made after the invoice is issued, respectively.
As shown, different values for these parameters are mapped to compliance levels based on the ψ projection function. Moreover, in this scenario, the partial
compliance cut-off value S and the threshold Δ are set to 0.3 and 0.4, respectively. Thus, a compliance value between 0.3 and 0.7 $(0.3 + 0.4)$ is considered
as *partially compliant*, and any value below 0.3 as *non-compliant*. Similarly, the
attribute *paymentReceived* is the amount paid by the customer, which includes
the principal plus any applicable interest and penalty. Notice from the table
that a payment of less than half of the amount due is deemed as non-compliant,
while other values of payment are considered as partially compliant. The two
attributes, *interest* and *penalty* are meta information that will be used to calculate the penalty when violations occur.

Full Compliance: Consider a scenario where *equipmentDeliveryDays* =
2 days, *payInDays* = 10 days and a payment of $500 has been received
from the purchaser, i.e., the equipment has been delivered and full payment has been received within the prescribed time frame. Hence, as an
example, consider the trace $\mathfrak{T}_4 = \langle t_1, t_2, t_6 \rangle$ which contains the attributes
$\langle paymentReceived, payInDays, equipmentDeliveryDays \rangle$, as illustrated in
Table 1. To compute the aggregate metric across the compliance dimensions
for an attribute, we first compute the individual compliance values along each
dimension and then aggregate them. The compliance metric for the monetary
(M) and temporal (T) dimensions for task t_2 are first computed as:

Table 1. Attributes metric of business process in Fig. 1

Attribute name (a)	Dimension	$\mathcal{A}_n$	$\psi(a)$	Cut-off (S)	Threshold (Δ)
invoiceValue (P)	Monetary	\$500	–	–	–
invoiceDate	Temporal	2019-04-01	–	–	–
equipmentDeliveryDays	Temporal	$\leq$3 days	1	0.3	0.4
		$\leq$7 days	0.5		
		$>$7 days	0		
payInDays	Temporal	$\leq$15 days	1		
		$\leq$22 days	0.6		
		$\leq$32 days	0.3		
		$>$32 days	0		
interest (Int)	Percentage	0%	–	–	–
		3%			
penalty (Pen)	Percentage	0	–	–	–
		2.5 %			
paymentReceived	Monetary	$<$50% $\times$ R	0	0.3	0.7
($R = P + Int + Pen$)		$<$75% $\times$ R	0.3		
		$<$80% $\times$ R	0.5		
		$<$R	0.9		
		$\geq$R	1		

$$M_{t_2}^M = \frac{1}{|\mathcal{D}_{t_2}^M|}\Sigma_{a\in\mathcal{D}_{t_2}^M}\psi_{t_2}(a) \qquad\qquad M_{t_2}^T = \frac{1}{|\mathcal{D}_{t_2}^T|}\Sigma_{a\in\mathcal{D}_{t_2}^T}\psi_{t_2}(a)$$
$$= \frac{1}{|\mathcal{D}_{t_2}^M|}(\psi_{t_2}(paymentReceived)) \qquad = \frac{1}{|\mathcal{D}_{t_2}^T|}(\psi_{t_2}(payInDays))$$
$$= \tfrac{1}{1}(1) \qquad\qquad\qquad\qquad\qquad = \tfrac{1}{1}(1)$$
$$= 1 \qquad\qquad\qquad\qquad\qquad\qquad = 1$$

Further, by Definition 4, we have:
$$\mathcal{P}_{t_2}\text{-Measure} = \frac{1}{|\mathcal{D}_{t_2}|}\Sigma_{i\in\mathcal{D}_{t_2}}M_{t_2}^i$$
$$= \frac{1}{|\mathcal{D}_{t_2}|}(M_{t_2}^T + M_{t_2}^M) = \tfrac{1}{2}(1+1)$$
$$= 1$$

Similarly, for task t_6, we have: $M_{t_6}^T = 1$ and $\mathcal{P}_{t_6}$-Measure $= 1$.
Hence, we have: $\mathcal{D}_{\mathfrak{T}_1} = \{\mathcal{D}_{\mathfrak{T}_1}^M, \mathcal{D}_{\mathfrak{T}_1}^T\}$, and $\mathrm{argmin}(\mathcal{D}_{\mathfrak{T}_1}^M) = 1$, and $\mathrm{argmin}(\mathcal{D}_{\mathfrak{T}_1}^T) = 1$.
Consequently, it follows that: $\mathcal{P}_{\mathfrak{T}_1}$-Measure $= \tfrac{1}{2}(1+1) = 1$.

Partial Compliance: Let us now turn to consider a different scenario where:
equipmentDeliveryDays = 2 days,
payInDays = 20 days
paymentReceived = \$575
In this scenario, the payable amount is now \$500 + \$75 = \$575, and has been fully paid by the customer. Thus, we can calculate the aggregate partial compliance measures for the tasks as follows:

$$interest = (20 - 15) \times 0.03\% \times \$500$$
$$= \$75$$

and $penalty = 0$.

Accordingly, we have the following:

Dimensions	Attributes	t_2	t_3	t_6
Temporal	$equipmentDeliveryDays$	–	–	1
	$payInDays$	0	0.6	–
Monetary	$paymentReceived$	0	1	–
	$\mathcal{P}_t$-Measure	0	0.8	1

Hence, $\arg\min(\mathcal{D}_{\mathfrak{T}}^{M}) = 0.6$, and $\arg\min(\mathcal{D}_{\mathfrak{T}}^{T}) = 1$.
Therefore, $\mathcal{P}_{\mathfrak{T}}$-Measure $= \frac{1}{2}(0.6 + 1) = 0.8$.

Non-Compliance. Lastly, consider the situation where no payment has been received after 32 days, i.e., the conditions of the contract have been violated and cannot be repaired. Thus, the contract will be deemed as terminated.

Dimensions	Attributes	t_2	t_3	t_4	t_6
Temporal	$equipmentDeliveryDays$	–	–	–	0
	$payInDays$	0	0	0	–
Monetary	$paymentReceived$	0	0	0	–
	$\mathcal{P}_t$-Measure	0	0	0	0

$\mathcal{P}_{\mathfrak{T}}$-Measure $= \frac{1}{2}(0 + 0) = 0$.

The partial complianc functions can also be introduced as mappings from a numeric domain of attribute values corresponding to the threshold window around the standard value for an attribute to the $[0, 1]$ range. These functions typically take a linear, concave or convex form depending upon how rapidly the distance from the standard value affects compliance. The shapes of typical functions have to be determined through empirical studies and this is out of the scope of the current work.

5 Discussion and Related Work

There are different ways to overcome various partial compliance scenarios as shown in Fig. 4. For each kind of deviation or case of partial compliance, one or more compensatory mechanisms may be provided for the task to resume execution.

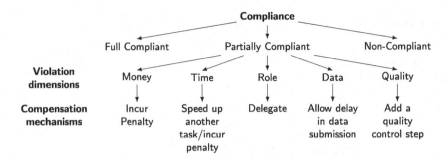

Fig. 4. Compliance dimensions and compensation mechanisms for partial compliance

For example, if a task is delayed it may be made up by speeding up a later task so that the customer of a service does not notice any increase in the total time for a process instance. A role violation occurs when an employee in the designated role is not available to perform a task. In such a case, a possible compensation is to assign the task to a delegate of the person who would normally perform it. For the data dimension, to process a passport application a user may be required to provide social security card or ID card and birth certificate, etc. If the user does not provide the birth certificate (or, say, one of three required documents), it may still be possible to process the application provided the missing document is submitted within one week of the application. Thus, the application may still be processed despite a minor violation. In the absence of such a mechanism, the application would have to be rejected, and then have to be resubmitted thus increasing the overall cost of processing it both for the citizen and for the governmental agency involved.

On the other hand, if the process instance itself takes longer than the standard time, then the customer has to be compensated by the service provider as per their agreement. With regard to the process model redo, restart, undo or abort the tasks can be other possible ways to overcome partial compliance issues. As they have their own complexities and impact on the execution of the individual tasks and the process on the whole, these are the topics of our further investigation.

Our framework determines partial compliance in a bottom up manner from the individual task level, to the trace level, and then a compliance score can thus be assigned to each trace. It is also possible to consider partial compliance at a still higher level of all process instances or cases in a day, week or month. The notion here would be to determine how many traces fall within a certain partial compliance level, say X% of instances have a compliance level more than 0.8 during a month. Moreover, a similar analysis may be done at the individual task level to determine what percentage of payment tasks had a compliance level of more than, say, 0.9 in a month. Such information can be very helpful to the management of a company. As we noted above, binary notions of compliance are not very useful from a management perspective for understanding the operational performance of a company. By introducing partial compliance in this way management can gain deeper insights into their operations. The values in Table 1 can be derived through empirical studies and from analysis of logs from previous executions of the business process model.

The problem of partial compliance has been studied widely in different domains. Gerber and Green [2] proposed the use of regression analysis scalable protocols to resolve some of the partial compliance issues that appear in field experiments. Jin and Rubin [7], on the other hand, proposed the use of principal stratification to handle partial compliance issues when analysing drug trials and educational testing. However, only a limited amount of work has targeted the area of improving business process compliance (BPC) management or measuring the level of compliance of a BP in a quantitative way. In the followings, we present some pertinent studies and discuss their strengths and limitations for the measurement of partial compliance.

In [13], Sadiq *et al.* introduced the notion of compliance distance as a quantitative measure of how much a process model may have to change in response to a set of rules (compliance objectives) at design time; or by counting the number of recoverable violations, how much an instance deviates from its expected behavior at runtime. This approach is extended in [9] to effectively measure the distance between compliance rules and organization's processes. To this purpose, the authors have divided the control objectives into four distinct classes of ideal semantics, namely: (i) *ideal*, (ii) *sub-ideal*, (iii) *non-compliant*, and (iv) *irrelevant*, and compute the degree to which a BP supports the compliance rules. Although their method provides computationally efficient means to analyze the relationships between the compliance rules and BPs, the heavily formalized rules have increased the complexity of the modelling process which is a potential obstacle to non-technical users.

Shamsaei [15] proposed a goal-oriented, model-based framework for measuring the level of compliance of a BP against regulations. In the paper, the author decomposed the regulations into different control rule levels, and then defined a set of key performance indicators (KPIs) and attributes for each rule to measure their level of compliance. The value of the KPIs can be provided either manually or from external data sources, and the satisfaction level of each rule is evaluated on a scale between 0 and 100 by considering the values of target, threshold, and worst, so that analysts can prioritize compliance issues to address suitably given the limited resources at their disposal.

Morrison *et al.* [12] proposed a generic compliance framework to measure the levels of compliance using *constraint semiring* (c-semiring) [1]. In their approach, imprecise or non-crisp compliance requirements will first be quantified by means of *decision lattices* (through the notion of *lattice chain*), which provides a formal setting to represent concept hierarchies and values preferences [6], such as {*Good, Ok, Bad*} or {*Good, Fair, Bad*}. These values will then be combined and utilized as a decision-making tool by c-semirings to rank the level of compliance of the

BPs. Essentially, the advantage of their framework lies in the ability in combining compliance assessments on various dimensions. Although the proposed approach is general enough to provide an abstract valuation at policy (business process in this case) level, the information about compliance at lower levels of abstraction is missing.

Kumar and Barton [8] discussed an approach for checking temporal compliance. They used a mathematical optimization model to check for violations. After a violation occurs it can also check whether the remaining process instance can be completed without further violations and determine the best way to do so. In this way, the level of compliance along the time dimensions can be managed.

6   Conclusions and Future Work

Compliance to policies, rules and regulations is usually treated in a rigid manner in business, government, and other kinds of organizations. Compliance pertains to matters that affect employees, customers, and just ordinary citizens. Rigid compliance means that either there is strict adherence to a rule or policy by an entity in which case the entity is compliant, else the entity is treated as being non-compliant or in violation of the rule or policy. In the real-world, however, such a binary approach is not very efficient because violations related to processing of applications, permits, invoices, fines, taxes, etc. may occur along a continuous spectrum and even minor violations may lead to cancellation of transactions or processes. Hence, it is important to recognize the extent of the violation and also allow for remediation or compensation mechanisms for them that are commensurate with the degree of the violation. This would enhance overall social value by reducing inefficiencies and cutting down wasteful work performed in the system.

In this paper, we propose notions of full-, partial- and non-compliance to describe the compliance levels of a business process during its execution, and, based on the information available on different compliance dimensions for the attributes of a task, we have proposed a metrics-based framework that can be used to measure the level of compliance and provide more information on the state of a BP instance during execution and auditing phases. The framework was developed from basic principles of partial compliance.

To realize the effectiveness of the proposed framework, from an implementation perspective, we are planning to implement it as a ProM Plugin[2] such that, given a process log, the application can automatically perform a compliance evaluation and analysis on the log, and generate a full report that shows compliance at multiple levels of aggregation. We would also like to test, validate and fine tune this tool by applying it to real-world logs, and seeking feedback from the domain experts about the perceived value they obtain from such an analysis. Further, it would be useful to extend the notions of compensation more formally.

---

[2] ProM Tools: http://www.promtools.org/doku.php.

References

1. Bistarelli, S.: Semirings for Soft Constraint Solving and Programming. Springer, Heidelberg (2004). https://doi.org/10.1007/978-3-540-25925-1
2. Gerber, A.S., Green, D.P.: Field Experiments: Design, Analysis, and Interpretation. W. W. Norton & Company (2012)
3. Governatori, G.: Business Process Compliance: An Abstract Normative Framework. IT Inf. Technol. 55(6), 231–238 (2013)
4. Hashmi, M.: A methodology for extracting legal norms from regulatory documents. In: 2015 IEEE 19th International Enterprise Distributed Object Computing Workshop, Adelaide, SA, Australia, pp. 41–50. EDOCW 2015, IEEE, September 2015
5. Hashmi, M., Governatori, G., Wynn, M.T.: Normative requirements for regulatory compliance: an abstract formal framework. Inf. Syst. Front. 18(3), 429–455 (2016)
6. Huchard, M., Hacene, M.R., Roume, C., Valtchev, P.: Relational concept discovery in structured datasets. Ann. Math. Artif. Intell. 49(1), 39–76 (2007)
7. Jin, H., Rubin, D.B.: Principal stratification for causal inference with extended partial compliance. J. Am. Stat. Assoc. 103(481), 101–111 (2008). https://doi.org/10.1198/016214507000000347
8. Kumar, A., Barton, R.R.: Controlled violation of temporal process constraints - models, algorithms and results. Inf. Syst. 64, 410–424 (2017)
9. Lu, R., Sadiq, S., Governatori, G.: Measurement of compliance distance in business processes. Inf. Syst. Manage. 25(4), 344–355 (2008)
10. Ly, L.T., Rinderle-Ma, S., Göser, K., Dadam, P.: On enabling integrated process compliance with semantic constraints in process management systems. Inf. Syst. Front. 14(2), 195–219 (2012)
11. Maggi, F.M., Montali, M., Westergaard, M., van der Aalst, W.M.P.: Monitoring business constraints with linear temporal logic: an approach based on colored automata. In: Rinderle-Ma, S., Toumani, F., Wolf, K. (eds.) BPM 2011. LNCS, vol. 6896, pp. 132–147. Springer, Heidelberg (2011). https://doi.org/10.1007/978-3-642-23059-2_13
12. Morrison, E., Ghose, A., Koliadis, G.: Dealing with imprecise compliance requirements. In: Proceedings of the 13th Enterprise Distributed Object Computing Conference Workshops, Auckland, New Zealand, pp. 6–14. IEEE, September 2009
13. Sadiq, S., Governatori, G., Namiri, K.: Modeling control objectives for business process compliance. In: Alonso, G., Dadam, P., Rosemann, M. (eds.) BPM 2007. LNCS, vol. 4714, pp. 149–164. Springer, Heidelberg (2007). https://doi.org/10.1007/978-3-540-75183-0_12
14. Schumm, D., Turetken, O., Kokash, N., Elgammal, A., Leymann, F., van den Heuvel, W.-J.: Business process compliance through reusable units of compliant processes. In: Daniel, F., Facca, F.M. (eds.) ICWE 2010. LNCS, vol. 6385, pp. 325–337. Springer, Heidelberg (2010). https://doi.org/10.1007/978-3-642-16985-4_29
15. Shamsaei, A.: Indicator-based policy compliance of business processes. In: Boldyreff, C., Islam, S., Leonard, M., Thalheim, B. (eds.) Proceedings of the CAiSE Doctoral Consortium 2011, London, UK, pp. 3–14, June 2011

A Case Study Integrating Knowledge Graphs and Intuitionistic Logic

Bernardo Alkmim[(✉)], Edward Haeusler, and Daniel Schwabe

Departamento de Informática, Pontifícia Universidade Católica do Rio de Janeiro,
Rio de Janeiro, Brazil
{balkmim,hermann,dschwabe}@inf.puc-rio.br

Abstract. In this work, we present a way to model and reason over Knowledge Graphs via an Intuitionistic Description Logic called iALC. We also introduce a Natural Deduction System for iALC to reason over our modelling of the information of the Knowledge Graphs. Furthermore, we apply this modelling to a case study in a context that aims to support the definition of concepts of Trust, Privacy, and Transparency, and the solution of apparent conflicts between them without the need for additional strategies, using only terms of the logic itself.

Keywords: Description logic · Intuitionistic logic · iALC · Knowledge graphs · Natural deduction

1 Introduction

There have been several different ways to represent knowledge throughout the years, either mimicking human cognitive behaviour or other approaches. Knowledge-Based Systems (KBSs) [17] are one of such methods, derived from Expert Systems, explained in [5] and [15].

One way to represent KBSs is via Knowledge Graphs (KGs), which can be seen in [19] and [24]. With their graph-like structure, KGs offer a versatile way to represent data that postpones the necessity for a schema, which adapts well to different domains. The fact that they are graphs is, as well, another positive point since this mathematical structure has an extensive literature and theory [4] behind it. This is especially useful when one utilises Machine Learning since KGs can connect it to Knowledge Representation, thus imbuing it with tools to solve specific questions in the data, such as Data Insufficiency or Zero-shot Learning.

Another way to represent knowledge is via Description Logics (DL) [2]. DLs consist in the representation of entities called *concepts*, *roles*, and *individuals*, as well as their relationships, based around *axioms*. This type of logic comes in different *flavours*, depending on which properties one wishes to utilise. One of the attractive points of DLs is the possibility to make syntactic reasoning (due to being a Logic) over the associated KB via whichever deduction system is fit.

The authors gratefully acknowledge financial support from CNPq.

V. Rodríguez-Doncel et al. (Eds.): AICOL-XI 2018/AICOL-XII 2020/XAILA 2020, LNAI 13048, pp. 106–124, 2021.
https://doi.org/10.1007/978-3-030-89811-3_8

For the domain of Law, in particular, there is the option of utilising Deontic Logic. However, it is criticised (as in [25], for instance) for not making explicit the models and theories for making legal KBSs, among other reasons. By choosing Description Logic, one may utilise a calculus for deductions, similarly to Deontic Logic but without facing the same modelling issues.

We then end up with two means to represent data: KGs, which are versatile and can represent data more accurately, and DLs, over which one can reason while extracting more direct *meaning* from the reasoning itself, at the same time. Due to their equivalence in representation, we can exploit the good parts of both these methods over the same KBS.

In this article, we present our approach for translating a representation of a KBS that focuses on the concepts of Trust, Privacy, and Transparency of information from a KG to a DL, namely, iALC, in order to reason over it.

iALC is a Logic that represents and reasons over the domain of Law, and its structure is established on Kelsenian Jurisprudence. Thus, legal individuals can be Valid Legal Statements (VLSs) in iALC, and the notion of precedence between Laws (as per Theory of Legal Order in [3]) can be directly represented in it as a relation between worlds in a Kripke Model. The choice to utilise a Description Logic instead of a type of the more usual Deontic Logic comes from the existence of some conceptual issues of Deontic Logic, especially when one considers the notion of validity of a normative sentence.

In Sect. 2, we introduce the concept of KGs and the framework that allows them to better deal with Trust, Privacy, and Transparency of information, motivating why we chose another way to deal with the same concepts. In Sect. 3, we explain the main concepts of the DL we choose to reason with, iALC. In Sect. 4, we contextualise the KBS and give a case study of formalisations in iALC and the reasonings made and discuss our decision-making throughout this process. In Sect. 5, we conclude the article and point to future directions of this work. In the Appendix A, we present the Natural Deduction system rules created for iALC to make the deductions themselves.

2   The KG Usage Framework and Motivation for iALC

There are multiple definitions for KGs in the literature and varied implementations (all discussed thoroughly in [11]). However, they all share a central idea: they utilise graphs to represent data and knowledge. The concept comes from the 70s [22], but Google popularised the term itself in 2012,[1] when they implemented the concept to create a system to support moving from search to question answering.

In general, the nodes in a KG represent entities of the natural world and the edges, relations between them. This is, however, a rather simplistic approach that is not much different from a labelled graph. In order to talk about knowledge, it is necessary to have a supplementary *ontology*, or a set of *rules* of the domain to be represented. One may use quantification of variables and separate one large KG into smaller ones to isolate certain parts of the original graph semantically.

In [23], the authors present a framework called KG Usage, which allows KGs to support Trust, Transparency, and Privacy concerns in its modelling. The author uses it

---

[1] https://www.blog.google/products/search/introducing-knowledge-graph-things-not/.

in applications for whom decisions based on security or disclosure of information or knowledge is essential. In the article, the authors define these three terms in regards to access to information: whether a specific piece of information can be *trusted*, or whether it has to remain *private* from others, or made *transparent*. In the context of Semantic Web, *trust* relates directly to the trust an agent has in a particular source of information.

This article will utilise the same example presented in [23], further explained in the third section of that article. To summarise, in a fictional context, a disaster occurs and a person, Ed, wishes to donate to a foundation to give aid to the people affected by it. He, however, wants to be sure that his money will be well spent, so he needs to *trust* the organisation to which he intends to give his money. There is another person, George, who is an officer in a non-governmental organisation (NGO) called ReliefOrg, whom Ed does not trust. Assuming that there is a law that requires NGOs to disclose publicly (*transparency*) their officers, Ed must have access to this information from ReliefOrg if he is to donate to it. On the other hand, George may not want to disclose his personal information to others (right to *privacy*), creating a conflict that must be resolved. In this case, since it is in the law that this kind of information must be disclosed, the transparency law has priority.

The approach there postulates the existence of a *trusted graph* containing all the statements an agent will rely upon to decide to take some action. Rules define when a statement should be included in this graph (*trust*), as well as when some action should be allowed (*transparency*) or denied (*privacy*). In the example, there are also other scenarios in this same context that relate the concepts of Trust, Privacy, and Transparency in similar ways.

It is worth noting that this hierarchy of normative sentences, i.e. rules, laws, was the starting point to formalise this information in iALC since the logic was created to deal with laws from the cradle.

One of the main issues presented in the KG Usage Framework is how to solve conflicts between applicable rules, especially when considering opposing privacy and transparency rules over the same piece of information. These kinds of conflicts are solved in the framework via six categories of strategies, identified in [9], implemented as algorithms, most of which require user involvement in run-time.

Using iALC, we can solve conflicts during the modelling since the precedence of norms is primitive in iALC, as seen in the next section. Then, the implementation of calculus for applying reasoning in the logic would not need to apply multiple secondary algorithms for each formula to solve possible conflicts; the conflicts would not exist.

The case study in Sect. 4 shows an example of a situation in which iALC solves an apparent conflict between privacy and transparency, more specifically in Sect. 4.6.

3 IALC

In [6], the authors propose Description Logics (DL) for semantic analysis of natural language utterances. In [8] the DL iALC (first introduced in [12] and further elaborated on in [7], [13] and [14]) was used as the basis of a solver/reasoner for the questions in the OAB Exam.[2] iALC is based on the canonical ALC, and was by definition made

---

[2] The Bar exam for candidate lawyers in Brazil.

to deal with law systems. In it, valid legal statements (VLS) are not propositions, but individuals in a legal ontology, i. e. one law cannot be true or false; it just either exists or not in a concept. The propositions we consider are some of the *concepts* of the ontology (i.e. formulas of the form $a : C$ or $C \sqsubseteq D$, for a nominal[3] and C, D concepts), representing the legal systems that can hold different kinds of laws. They also have a precedence relationship, derived by the hierarchy of individual laws of Kelsenian Jurisprudence [18]. As an example, one can imagine Brazilian Law or other similar systems. It has the Constitution, comprised of some general principles that must precede every other (usually more specific) law based on or created after it. With this notion, one can create different tiers of laws to not generate legal antinomies with more basic, fundamental laws (*ground norms*, as defined by Kelsen), with the fundamental laws preceding the lesser ones.

While in [1], iALC models and proves the correctness of the right choice of questions from the OAB Exam. In this article, the characteristics of this logic are more thoroughly explained, especially in how they make it well suited to deal with normative sentences, especially laws.

In general, in iALC, legal individuals are specific combinations of laws representing an artificial legal being, which can validate or not certain concepts. However, they do not represent individuals entirely outside of the legal expressiveness of the logic; they are mere legal projections of real-life concepts. For example, *john : Attorney* states that the legal individual *john*, which represents a legal document that proves that John is an attorney, is part of the concept *Attorney*, which represents the set of laws relating to attorneys. In general, to simplify, individuals like *john* are assumed to always be, from the start, in a nominal. Notice that everything in this example is in the ontology about laws.

Let us extend this example. Suppose that *John Doe* has passed the OAB Exam and now is an attorney. So, there is a legal document which says exactly that, and we can represent this as $j_a : Attorney$ (j_a representing our artificial abstraction over the fact that *John Doe* is a lawyer). However, as a Brazilian citizen, he has a birth certificate as well. Supposing we have a concept representing every legal statement valid for Brazilians, say, *BrCitizen*, we can also conclude that we can have $j_b : BrCitizen$ (j_b being a **different** legal statement, representing the fact that he is Brazilian). Now the precedence rules for legal individuals of iALC start to appear explicitly. Since, according to [14], from the intuitionistic aspect of the logic, a Kripke model in it is a Heyting algebra, every pair of worlds has a finite *meet* (say, j_c), related by *precedence* of laws to the others, in which are valid both the concepts *Attorney* and *BrCitizen*. In other words, given j_a and j_b, there is always a world j_c such that $j_c \preceq j_a$ and $j_c \preceq j_b$.

$$j_a \models Attorney \qquad j_b \models BrCitizen$$

$$j_c \models Attorney, BrCitizen$$

---

[3] A nominal is a concept defined extensionally, consisting of only one individual, by which it is referenced. A nominal for a is $\{a\}$. Throughout this article, we will omit this notation, referring to the nominals without the brackets.

Note that there is no world (legal individual) representing the physical individual *John Doe* fully since all that we deal with are laws and legal statements. From this, we can conclude that, conceptually, we never deal with people nor objects from the real world, only with legal statements over them.

These characteristics from iALC stem from it being an *intuitionistic* description logic. Intuitionistic logic is different from classical logic and is constructive by nature [16]. Intuitionistically, $\neg A$ being true can be understood as there being no way to construct a proof of A (or that, by assuming A to be true, one reaches a contradiction). With this, we have, for instance, that the Principle of the Excluded Middle, i.e. $A \vee \neg A$, is **not** valid in intuitionistic logic, even though it is in classical logic. In fact, intuitionistic logic is weaker than classical (see [16]), but it allows for non-conflicting existence of thought-to-be contradictory logical formulas *with fewer workarounds in the modelling* than Description Logics that have a classical notion of validity, such as ALC and other canonical DLs. In the domain of law, this allows, for instance, for the existence of a model representing two distinct legal systems, one in which the death penalty is not allowed and the other in which it is (referencing the same concept of death penalty). In a regular (i.e. Classical) DL, this cannot be represented as directly as in its intuitionistic counterpart when modelling since a Kripke model for a Classical Logic collapses all worlds (the legal individuals) to the same one, causing a contradiction by allowing and prohibiting the death penalty at the same time. Representing in a classical DL would force us to find a more conflated way to insert Legal Individuals into the logic at the risk of losing legibility and, even worse, soundness. In the case of death penalty, one would need a concept *DeathPenaltyBrazil* and another *DeathPenaltyTexas* as well as one or more structural rules connecting both concepts in one manner or another (and would have to find another way to model VLSs, instead of directly having them being worlds in the model). By having this intuitionistic perception, we better translate the foundations of Private International Law into the modelling, connecting VLSs who are part of different legal systems via the same concept, for example, *DeathPenalty*, without having to resort to many workarounds. Legal precedence is, as well, something defined constructively in the logic through the VLSs, and does not need additional formulations in the model.

This non-conflicting existence of different sets of laws (i.e. normative sentences, rules) relates directly to the existence of different KGs in Sect. 4, each with their own set of rules, which may contradict those of other KGs. iALC not only allows for those to coexist, but it also has means to solve those apparent contradictions, in an analogous way to how it deals with laws themselves. IALC can mix laws and KG rules seamlessly by treating them the same way, as shown in the following examples, but more specifically in their conclusion, Sect. 4.6.

Another characteristic that iALC can import from KGs in a simple and almost seamless way is that each KG from each agent can be represented directly as a VLS. With this, complex models with many different graphs can be represented more straightforwardly than with classical DLs.

4 Case Study

We will now show the modelling of essential information and a case study of rules[4] of the scenario presented in Sect. 2, based on the same rules from the original article, that modelled them in a KG divided into several Knowledge Items, each one being a nanopublication [10].

Since all of the following examples are in the same context, we advise the reader to follow them sequentially. As different features of the logic appear throughout, they will be explained, culminating in the primary motivation for using iALC in the final rule. The naming conventions of each rule are the same from [23].

To avoid further confusion, from here on out, the *rules* of the ND calculus will be referred to as *deduction steps*, *proof steps*, or just *steps*.

4.1 Base Knowledge

All agents share part of the information, which is the base knowledge for the scenario, namely: ReliefOrg is a non-profit organisation, AuditInc is a certification agency, George is an officer of ReliefOrg and AuditInc.

For the base graph knowledge, we have the following formulas:

$$relief : NonProfit^L$$
$$audit : CertAgency^L$$

$$NonProfit \sqsubseteq Organisation^L$$
$$CertAgency \sqsubseteq Organisation^L$$

$$george \quad officerInOrg \quad relief$$
$$george \quad officerInOrg \quad audit$$

In the formulas above, we can see that the individuals are represented by nominals in iALC, as would be expected. We utilise a *concept* to represent a category or set. In the first formula, we have that the individual *relief*, an individual representing ReliefOrg, is part of the concept *NonProfit*, a concept for non-profit organisations. One concept can be a subset of another, as in the third formula, which states that non-profit organisations are a subset of organisations in general. When one establishes a relationship between two individuals, it is translated into a *role*. In the fifth formula, for example, the role *officerInOrg* relates the individuals *george* and *relief*.

To model the deduction steps in iALC, we needed to make them somewhat general to chain formulas with one another. So, we used generic labels when modelling e.g. *relief* : *NonProfit^L*, where L is any list of labels for the concept *NonProfit*.

---

[4] The term *rule*, here, is not to be confused with the rules (of deduction steps, proof steps) from the ND calculus. When referencing *the* rule of each section, we are referring to the rules in boxes, which represent the title - and the main goal - of each section.

Different logics can have different kinds of calculi in order to reason over them and make actual deductions. For iALC, we developed a Natural Deduction (ND)[5] system (whose deduction steps are all in Appendix A), which has labels to facilitate the chaining of existential restrictions. Our ND system needs these chainings to connect in the same reasoning individuals related indirectly (via transitivity) in a fully *TBOX-esque* fashion. For instance, when an *agent* intends to do an *action* to an *organisation* that has a certain *officer*, we need to encapsulate all these different relations between distinct agents in one single formula in order to make more direct and structurally simple reasonings without losing meaning. Luckily, it is possible in DLs via applying restrictions sequentially on concepts. The labels on concepts also aid this chaining by giving the context at all times, thus maintaining the soundness of the system. This chaining of formulas is explained more concretely in Sect. 4.2.

The proof steps utilised in this document are, mostly, $\exists - intro$, $\exists - elim$, $\sqsubseteq - elim$, and $\sqcap - intro$. The steps used have to do with our modelling choices: most of the modelling involved existential restrictions ($\exists$), subjunctions of concepts ($\sqsubseteq$), and conjunctions of concepts ($\sqcap$). Also, since the cases are always instantiated, all the steps involved nominals to reflect the information as faithfully as possible, so the $-n$ indicator will not be shown due to limited space in the proofs.

4.2   Rule Ed1 (TRUST)

We begin with the first rule of knowledge graph for the agent Ed: Ed wants to donate to a certified not-for-profit organisation. More precisely, if an organisation *org* has (some) officer *of*, financial record *fin*, *fin* is audited by organisation *aud*, *aud* is certified by organisation *cert*, *cert* is a Certification Agency and the provenance graph for *cert* includes the fact that it is an accredited agency, then Ed's trusted graph includes the fact that *org* is a non-profit organisation. Some of the formulas are simple:

$$cert : CertAgency^L$$
$$cert : Accredited^L$$
$$Accredited \sqsubseteq CertAgency^L$$

While other sentences can be represented with role assertions, creating the related roles:

$$
\begin{array}{lll}
fin & auditedBy & aud \\
aud & certifiedBy & cert \\
org & hasFinRecord & fin \\
of & officerInOrg & org
\end{array}
$$

These formulas, however, do not fit well into our Natural Deduction system, since they involve information pertinent to the ABOX (assertion) part of iALC, and the ND

---

[5] Gentzen first introduced this deduction system in the 1930s. A complete and formal reference for its main concepts is [20]. ND is a deduction system with many similarities to how a person makes an actual deduction, which relates closely to a previous usage of iALC: answering questions to a Bar Exam. It is, in a way, a more readable Sequent Calculus. Another reason in its favour is the lack of support in general of inferential tools to Intuitionistic Logics.

system reasons over the TBOX (terminological) part of it. Luckily, they are equivalent to a more TBOX-friendly format, in the form of Restrictions of Concepts (in our case, existential restrictions).[6] We have not yet seen the case in which a universal restriction was necessary, but their modelling would be similar and would chain just as naturally as the existential ones (and iALC supports it if needed). This results in:

$$fin : \exists auditedBy.Organisation^L$$
$$aud : \exists certifiedBy.CertAgency^L$$
$$org : \exists hasFinantialRecord.FinRecord^L$$
$$of : \exists officerInOrg.Organisation^L$$

To the formulas, we give generic labels (namely L, representing a list of labels) to give them flexibility when used in proofs because, in those, they need to represent any context in which they appear. Once we give context, we substitute L with an adequate list of labels. This generic list of labels can be interpreted as the formula being valid regardless of the context in which it is inserted. The supposition that an organisation audits financial record fin, i.e. $\exists auditedBy.Organisation$, can be used in the situation presented or in any other. If we use fixed labels for this context, the isolated sentence may lose meaning in a different context (maybe one without a certification agency involved, for instance). After the proof, we will explain these concrete contexts more clearly. By chaining some of those in order to relate these different individuals, we have:

$$fin : \exists auditedBy.\exists certifiedBy.CertAgency^L$$

The formula above represents a refinement in $fin : \exists auditedBy.Organisation^L$ via $aud : \exists certifiedBy.CertAgency^L$, which is an existential restriction on $CertAgency$, a subconcept of $Organisation$.

$$org : \exists hasFinRecord.\exists auditedBy.\exists certifiedBy.Organisation^L$$

This last formula can be read as org has a (certain) financial record which is audited by an organisation certified by a certification agency, which is itself an organisation. Representing the entire rule, we have:

$$\exists hasFinRecord.\exists auditedBy.\exists certifiedBy.Organisation \sqsubseteq NonProfit^L$$

Showing that, for Ed's graph, a company that fits into these criteria must be a non-profit organisation.

In order to shorten the names of formulas so that the proofs fit in the page, the following changes will be made:

---

[6] The choice here is between universal and existential restrictions, but universal restrictions would imply that, for instance, the organisation org would have every financial record (in the world) related to it, which is very rarely the case.

$$Accredited \sqsubseteq CertAgency^L$$
$$Acc \sqsubseteq CA^L$$

$$aud : \exists certifiedBy.CertAgency^L$$
$$aud : \exists c.CA^L$$

$$fin : \exists auditedBy.\exists certifiedBy.Organisation^L$$
$$fin : \exists a.\exists c.O^L$$

$$org : \exists hasFinRecord.\exists auditedBy.\exists certifiedBy.Organisation^L$$
$$org : \exists h.a.\exists c.O^L$$

$$\exists hasFinRecord.\exists auditedBy.\exists certifiedBy.Organisation \sqsubseteq NonProfit^L$$
$$\exists h.\exists a.\exists c.O \sqsubseteq NP$$

To better illustrate this, here is a small (partial) model to illustrate these relations:

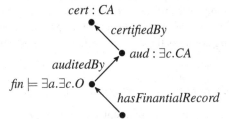

$$org \models \exists hasFinRecord.\exists auditedBy.\exists certifiedBy.Organisation$$

We make similar changes to subsequent proofs. So, here is the proof for the first rule in our ND system with the names shortened:

$$\cfrac{\cfrac{\cfrac{\cfrac{\cfrac{\cfrac{cert : Acc \quad Acc \sqsubseteq CA}{cert : CA}\sqsubseteq-e \quad CA \sqsubseteq O^{\exists h,\exists a,\exists c}}{cert : O^{\exists h,\exists a,\exists c}}\sqsubseteq-e}{aud : \exists c.O^{\exists h,\exists a}}\exists-i}{fin : \exists a.\exists c.O^{\exists h}}\exists-i}{org : \exists h.\exists a.\exists c.O}\exists-i \quad \exists h.\exists a.\exists c.O \sqsubseteq NP}{org : NP}\sqsubseteq-e$$

In this proof, the generic label L is substituted by the concrete labels, giving the context needed to the formulas. For instance, $fin : \exists a.\exists c.O^{\exists h}$ has the label $\exists h$, indicating that the restricted concept $\exists a.\exists c.O$ can be even further restricted via the role h (*hasFinantialRecord*), which connects *fin* to another individual, namely, *org*. Whereas $org : \exists h.\exists a.\exists c.O$ has no labels because they are not necessary. In the case of this formula, everything is explicit in the restrictions themselves.

A pre-proof step that was made was changing the formula $aud : \exists c.CA^L$ to $aud : \exists c.O^L$, due to the formula $CertAgency \sqsubseteq Organisation^L$, in a similar fashion to how it was done with $cert : CA^L$ in the proof via the $\sqsubseteq -e$ step. This could have been done with additional uses of $\sqsubseteq -elim - n$ in the proof itself, but it would become too convoluted to show, specially as a first example.

It is worth noting that, in our situation, of can be replaced by $george$, org by $relief$, and $cert$, by $audit$, showing that ed is aware that $relief$ is a non-profit organisation.

4.3 Rule Ed2 (Ed Does Not Trust George)

To capture the fact that Ed does not trust George (in the sense of donating to an organisation where George is an officer), we include the following rule. If Ed's trusted graph includes the fact that org is a non-profit organisation, George is officer of org, and Ed intends to take action a of type $Donate$ to org, then a is $Denied$.

We can set the formulas to:

$$george : \exists officerInOrg.Organisation^L$$
$$org : NonProfit^L$$

$$act : Donate^L$$
$$act : \exists directedTo.Organisation^L$$

$$ed : \exists intends.Donate^L$$
$$\text{(alternatively) } ed : \exists intends.\exists directedTo.Organisation^L$$

We want to get to $act : Denied^L$.

However, before going on to the proof, we need a formalisation of the fact that Ed does not trust George as an officer in an organisation, and that it nullifies his action: $ed \ \ notTrust \ \ george$, which becomes $ed : \exists notTrust.Person^L$, and, after another transformation:

$$ed : \exists notTrust.\exists officerInOrg.Organisation^L$$

We have just inserted a new concept, $Person$, not necessarily present in the context, which was necessary to categorise of what concept $george$ is a part. An analogous concept would be $Officer$, but the name would be solely to give semantic information to the reader. To the situation itself, we give the semantics of the concepts extensionally.

We also need to account for the fact that the action is directed towards a non-profit organisation:

$$(\exists intends.\exists directedTo.NonProfit \sqcap \exists noTrust.\exists officerIn.Organisation) \sqsubseteq$$
$$\exists intendsTo.Denied^L$$

Then, we have the following proof:

$$\cfrac{\cfrac{\cfrac{org : NP^{\exists i, \exists d}}{act : \exists d.NP^{\exists i}} \; \exists - i}{ed : \exists i.\exists d.NP} \; \exists - i \qquad ed : \exists nT.\exists o.O}{ed : (\exists i.\exists d.NP \sqcap \exists nT.\exists o.O)} \; \sqcap - i \qquad (\exists i.\exists d.NP \sqcap \exists nT.\exists o.O) \sqsubseteq \exists i.D^L$$

$$\cfrac{ed : \exists i.D}{act : D^{\exists i}} \; \exists - e$$

$$\sqsubseteq - e$$

4.4 Rule George1 (PRIVACY)

Now, let us focus on the graph of George, the officer. He has a rule that does not allow divulging (reading) any information about him, including his association with an organisation. This rule, which represents his right to *privacy*, and the rule from Sect. 4.5, which represents the *transparency* of a non-profit organisation are apparently in conflict. In the next sections, we explain how this conflict is solved. The proofs for the rules of his graph will be similar to those in the graph of Ed.

If *george* is an officer of organisation *org*, action *a* is of type *read*, and Agent *ag* intends *a* Then *a* is *Denied*.

$$org : Organisation^L$$
$$george \quad officerIn \quad org \implies george : \exists officerIn.Organisation^L$$

$$a : Read^L$$
$$ag \quad intends \quad a \implies ag : \exists intends.Read^L$$

$$a \quad directedTo \quad george \implies a : \exists directedTo.\exists officerIn.Organisation^L$$

We also have, by the privacy rule:

$$(\exists intends.\exists directedTo.\exists officerIn.Organisation \sqcap \exists intends.Read) \sqsubseteq$$
$$\exists intends.Denied$$

We want to arrive at $a : Denied^L$. The proof is as follows:

$$\cfrac{\cfrac{\cfrac{\cfrac{org : O^{\exists in, \exists diTo, \exists ofIn}}{g : \exists ofIn.O^{\exists in, \exists diTo}} \; \exists - i}{a : \exists diTo.\exists ofIn.O^{\exists in}} \; \exists - i}{ag : \exists in.\exists diTo.\exists ofIn.O} \; \exists - i \qquad \cfrac{a : R^{\exists in}}{ag : \exists in.R} \; \exists - i}{ag : (\exists in.\exists diTo.\exists ofIn.O \sqcap \exists in.R)} \; \sqcap - i \qquad (\exists in.\exists diTo.\exists ofIn.O \sqcap \exists in.R) \sqsubseteq \exists in.D}{\cfrac{ag : \exists int.D}{a : D^{\exists int}} \; \exists - e} \; \sqsubseteq - e$$

To contextualise, *org* is the organisation *relief*, and the agent who intends to read information on ReliefOrg, *ag*, is *ed*.

4.5   Rule Transp (Transparency)

The rule for the Transparency ruling over our agent George's privacy rule is: if an organisation *org* is of type *NonProfit* and has officer *of*, action *a* is of type *read* and an agent *ag* intends *a*, then A is *Allowed*.

There are also two additional pieces of information to the graph: the primary source to the transparency rule is Non-profit Act, and Non-profit Act is a Law.

$$org : NonProfit^L$$
$$off \quad officerIn \quad org \implies off : \exists officerIn.NonProfit^L$$

$$act : Read^L$$
$$act \quad directedTo \quad org \implies act : \exists directedTo.NonProfit^L$$

$$ag \quad intends \quad act \implies ag : \exists intends.Read^L \implies$$
$$ag : \exists intends.\exists directedTo.NonProfit^L$$

By transparency, $(\exists intends.\exists directedTo.\exists officerIn.NonP \sqcap \exists intends.Read) \sqsubseteq \exists intends.Allowed$. We also may represent $l : NonProfitAct$ as the legal individual for the Non-profit Act. With that, we should have $act : Allowed^L$.

This will not be used in the proof, but it is interesting to state the disjunction between the concepts *Denied* and *Allowed*: $Denied \sqsubseteq \neg Allowed$.

$$
\cfrac{
\cfrac{
\cfrac{
\cfrac{org : NP^{\exists in, \exists diTo, \exists ofIn}}{ofr : \exists ofIn.NP^{\exists in, \exists diTo}} \exists-i
}{a : \exists diTo.\exists ofIn.NP^{\exists in}} \exists-i
}{
\cfrac{ag : \exists in.\exists diTo.\exists ofIn.NP \quad \cfrac{a : R^{\exists in}}{ag : \exists in.R} \exists-i}{ag : (\exists in.\exists diTo.\exists ofIn.NP \sqcap \exists in.R)} \sqcap-i
\qquad \cfrac{(\exists in.\exists diTo.\exists ofIn.NP \sqcap \exists in.R) \sqsubseteq \exists in.A}{}
}{
ag : \exists int.A
} \sqsubseteq-e
}{a : A^{\exists int}} \exists-e
$$

Also, we have a secondary proof to show that the transparency rule (represented by the individual *rtr*) is preceded by the law *l*.

$$\cfrac{rtr : (\exists in.\exists diTo.\exists ofIn.NP \sqcap \exists in.R) \sqsubseteq \exists in.A \quad l \preceq rtr}{l : (\exists in.\exists diTo.\exists ofIn.NP \sqcap \exists in.R) \sqsubseteq \exists in.A} \preceq-elim$$

This ND step, $\preceq -elim$, is a way to introduce the precedence mechanism between individual laws directly into the Natural Deduction system. The smaller premise, $l \preceq rtr$, indicates the precedence among the individuals concerned. The bigger premise taking to the conclusion tells us that, since a concept is valid for an individual who is preceded by another, the one who precedes (in our case, *l*) must also validate such concept, due to the intuitionistic nature of the logic, that requires the construction of the validity of a concept to be passed through (in this case, to have been passed through) to the next individuals hierarchically.

At this point, we still need to relate the rules from Sects. 4.4 and 4.5, which could pose a problem since they apparently contradict each other, one stating that the action should be denied and the other stating that it should be allowed. The following rule, then, comes to the rescue.

4.6 MetaRule1 (Precedence)

This rule states that Laws precede personal privacy rules from the graphs of agents, such as the rule in Sect. 4.4.

This is easily solved by iALC with the formula *LegalRule* $\sqsubseteq$ *PersonalPrivRule*^{L}.

We can, then, conclude that the rule in Sect. 4.5 has precedence over the one in Sect. 4.4 since the law ruling over the transparency of non-profit organisations has precedence over George's right to privacy (at least when regarding his work at such an organisation).

This final rule shows the primary motivation for using iALC in this context of KGs to solve the apparent conflict between **transparency** and **privacy**. The previous examples did not show any specific motivation for using the logic itself; they were there to give context and show that the logic can model them without issues. This one, however, shows how natural it is to model the hierarchy of normative sentences in iALC without resorting to manually asserting that a *LegalRule* must have a kind of *blocker* of contradictory *PersonalPrivRule*s. This assertion is part of the construction itself.

5   Final Remarks

In this article, we presented the concepts of KGs and showed how to use them in contexts that involve the concepts of Trust, Privacy, and Transparency, as well as an alternate formalisation for them in iALC, an Intuitionistic Description Logic, to show one way one can reason over them, in a way that does not need additional strategies to deal with any apparent conflict between rules, such as those needed in the KG Usage Framework. This modelling and reasoning were proven successful in the context shown and conveyed how we can translate the complexities of KGs via the intuitionistic notion of truth that iALC.

For future work, we intend to extend and generalise the representation in iALC in this context of Trust, Privacy, and Transparency, especially since a lot of the decision making in terms of reasoning comes from the relations between laws that seem conflicting at first glance, but that are solved by mechanisms iALC is equipped to model.

In regards to the more theoretical side of the work, we aim to give a more formal introduction to the Natural Deduction system for iALC via showing its properties of soundness and completeness, for instance, which are out of the scope of this article, and possibly implement an automated reasoner to it for usage in a larger scale.

As an extension for iALC, we also intend to check the availability of adding a non-monotonic[7] type of reasoning to it, via the usage of Default Logic (first described by

---

[7] Usually, Logic has *monotonic* reasoning, meaning that given a set of formulas Γ (our premises) and a conclusion α, $\forall \gamma$ formula, $\Gamma \vdash \alpha \implies \Gamma, \gamma \vdash \alpha$. That is, no matter what other formula we add to the premises, we can still conclude α. In non-monotonic reasoning, this is not always the case.

Reiter in [21]). Non-monotonic reasoning can better represent the changes made to codes of laws as time goes by since certain laws can change previous ones.

A Natural Deduction System for iALC

In this appendix, we present the rules[8] of the ND calculus for iALC. In this notation, the premises of each rule are present over a bar (which represents the rule itself), and its conclusion lies below. We obtain proofs in ND a tree-like form by connecting these deduction steps.

In ND, rules come in two kinds: introduction and elimination rules. Introduction rules take some premises and produce a formula in conclusion with the related logical operator. For instance, the $\sqcap - intro$ generates a formula with the $\sqcap$ operator (meaning conjunction) in conclusion. Its meaning is as such: if a concept α is valid and a concept β is valid, then we can conclude that the concept $\alpha \sqcap \beta$, meaning the conjunction of α and β, is also valid. On the other hand, elimination rules, as the name suggests, take one of the premises with the related operator (called the *main premise* - other premises, if present, are *secondary* or *auxiliary premises*) and produce a different formula, thus *eliminating* the operator. For example, the $\sqsubseteq -elim$ rule takes the main premise, containing the $\sqsubseteq$ operator and an auxiliary premise, resulting in a formula without the operator involved in the main premise. An intuitive approach to this rule is that, given that α is valid and that α is a sub-concept of β, we can conclude that β is valid as well (this rule is, in fact, the *modus ponens* of iALC). Depending on the logic, the ND calculus may have other kinds of rules, called *structural*. Such is the case for the rules *Gen* and *Gen* − *n*.

An important feature present in ND is the mechanism of hypothesis *discharge*, present in some rules. When one discharges a hypothesis, it means that the deduction step is *self-contained*, i.e., the discharged hypothesis is not open in the proof. A closed hypothesis transfers the logical consequence up to the conclusion of the rule that discharged it. For instance, in the $\sqsubseteq -intro$ rule, if we somehow have that α is valid somewhere and that we end up at β being valid by applying possibly many ND rules, we can then conclude that $\alpha \sqsubseteq \beta$, and discharge the α above β. With this, there is no need to assume that α is valid every time one wants to show that β is valid since it was already proven, and one is allowed to utilise $\alpha \sqsubseteq \beta$ for the rest of the proof. Essentially, to discharge a hypothesis *closes* the respective branch. The discharge mechanism reduces the number of hypotheses needed. A proof that has open hypotheses draws the truth of its conclusion from the truth of these hypotheses. We can prove the hypotheses using a set of axioms or, as in this article, the base knowledge of the KGs.

Let α and β be concepts, x and y nominals, δ a formula, and R a role. L represents a list of labels (possibly empty). Labels represent either negation of concepts, or universal or existential restrictions on concepts, made *implicit*. They indicate a sort of *context* to the concept to which they are attached. $L^{\forall}$ is a list that restricts all labels in itself to $\forall R$ of some kind, and $L^{\exists}$ restricts all to $\exists R$. The $\perp$ concept is not valid in any world. We can see it representing falsehood.

---

[8] In this appendix, the world of KG is not in scope, so the term *rule* means deduction step.

Rules that utilise nominals and involve x and y assume a role connecting them, i.e. xRy. The n in the names of some rules indicate that those are the versions that utilise nominals. The *Gen* rules function as a *lift* to concepts, adding a universal restriction to the **end** of the list.

The *ex falso quodlibet* rules, *efq* and *efq* − *n*, represent the Principle of Explosion: from falsehood, anything follows.

The *reductio ad absurdum* rules, namely *raa* and *raa* − *n*, are not part of the ND calculus for iALC but are there to show that the calculus can be expanded from iALC to its classical counterpart, ALC, by having both of them. With their addition to the ruleset, nominals will have completely different meanings from the VLSs of iALC. Consequently, the models will lose their refinement obtained by the intuitionistic aspect of the logic. However, the deductions themselves will have the same structure, with rules that elevate the calculus to a classical framework.

$$
\cfrac{\begin{array}{c}[\alpha^{L1}]\\ \vdots\\ \beta^{L2}\end{array}}{\alpha^{L1} \sqsubseteq \beta^{L2}}\ \sqsubseteq-intro
\qquad\qquad
\cfrac{\alpha^{L1} \quad \alpha^{L1} \sqsubseteq \beta^{L2}}{\beta^{L2}}\ \sqsubseteq-elim
$$

$$
\cfrac{\begin{array}{c}[\alpha^{L}]\\ \vdots\\ \bot\end{array}}{(\neg\alpha)^{\neg L}}\ \neg-intro
\qquad\qquad
\cfrac{\alpha^{L} \quad (\neg\alpha)^{\neg L}}{\bot}\ \bot-intro
$$

$$
\cfrac{\alpha^{L^\forall} \quad \beta^{L^\forall}}{(\alpha\sqcap\beta)^{L^\forall}}\ \sqcap-intro
\qquad\qquad
\cfrac{(\alpha\sqcap\beta)^{L^\forall}}{\alpha^{L^\forall}}\ \sqcap-elim_1
$$

$$
\cfrac{(\alpha\sqcap\beta)^{L^\forall}}{\beta^{L^\forall}}\ \sqcap-elim_2
$$

$$
\cfrac{\alpha^{L^\exists}}{(\alpha\sqcup\beta)^{L^\exists}}\ \sqcup-intro_1
\qquad\qquad
\cfrac{(\alpha\sqcup\beta)^{L^\exists} \quad \begin{array}{c}[\alpha^{L^\exists}]\\ \vdots\\ \delta\end{array} \quad \begin{array}{c}[\beta^{L^\exists}]\\ \vdots\\ \delta\end{array}}{\delta}\ \sqcup-elim
$$

$$\frac{\beta^{L^\exists}}{(\alpha \sqcup \beta)^{L^\exists}} \sqcup - intro_2$$

$$\frac{y : \alpha^{\exists R, L}}{x : (\exists R.\alpha)^L} \exists - intro \qquad\qquad \frac{x : (\exists R.\alpha)^L}{y : \alpha^{\exists R, L}} \exists - elim$$

$$\frac{y : \alpha^{\forall R, L}}{x : (\forall R.\alpha)^L} \forall - intro \qquad\qquad \frac{x : (\forall R.\alpha)^L}{y : \alpha^{\forall R, L}} \forall - elim$$

$$\frac{}{\delta}\bot \; efq \qquad\qquad\qquad \begin{array}{c} [(\neg\alpha)^{\neg L}] \\ \vdots \\ \dfrac{\bot}{\alpha^L} \; raa(ALC) \end{array}$$

$$\begin{array}{c} [x : \alpha^{L1}] \\ \vdots \\ \dfrac{x : \beta^{L2}}{x : (\alpha^{L1} \sqsubseteq \beta^{L2})} \sqsubseteq -intro-n \end{array} \qquad \frac{x : \alpha^{L1} \quad x : (\alpha^{L1} \sqsubseteq \beta^{L2})}{x : \beta^{L2}} \sqsubseteq -elim-n$$

$$\begin{array}{c} [x : \alpha^L] \\ \vdots \\ \dfrac{x : \bot}{x : (\neg\alpha)^{\neg L}} \neg - intro - n \end{array} \qquad \frac{x : \alpha^L \quad x : (\neg\alpha)^{\neg L}}{x : \bot} \bot - intro - n$$

$$\frac{x : \alpha^{L^\forall} \quad x : \beta^{L^\forall}}{x : (\alpha \sqcap \beta)^{L^\forall}} \sqcap - intro - n \qquad \frac{x : (\alpha \sqcap \beta)^{L^\forall}}{x : \alpha^{L^\forall}} \sqcap - elim_1 - n$$

$$\frac{x : (\alpha \sqcap \beta)^{L^\forall}}{x : \beta^{L^\forall}} \sqcap - elim_2 - n$$

$$\frac{x : \alpha^{L^{\exists}}}{x : (\alpha \sqcup \beta)^{L^{\exists}}} \; \sqcup - intro_1 - n$$

$$\frac{x : (\alpha \sqcup \beta)^{L^{\exists}} \qquad \begin{array}{c}[x : \alpha^{L^{\exists}}]\\ \vdots \\ \delta \end{array} \qquad \begin{array}{c}[x : \beta^{L^{\exists}}]\\ \vdots \\ \delta \end{array}}{\delta} \; \sqcup - elim - n$$

$$\frac{x : \beta^{L^{\exists}}}{x : (\alpha \sqcup \beta)^{L^{\exists}}} \; \sqcup - intro_2 - n$$

$$\frac{x : \bot}{\delta} \; efq - n$$

$$\frac{\begin{array}{c}[x : (\neg \alpha)^{\neg L}]\\ \vdots \\ x : \bot \end{array}}{x : \alpha^L} \; raa - n(ALC)$$

$$\frac{\alpha^L}{\alpha^{L,\forall R}} \; Gen$$

$$\frac{x : \alpha^L}{y : \alpha^{L,\forall R}} \; Gen - n$$

$$\frac{\begin{array}{c}[x : \alpha^{\exists R, L1}]\\ \vdots \\ x : \beta^{\forall R, L2} \end{array}}{x : (\alpha \sqsubseteq \beta)^{\forall R, L1, L2}} \; \forall - \sqsubseteq - mix - i$$

$$\frac{x : \alpha^L \quad y \preceq x}{y : \alpha^L} \; \preceq - elim$$

References

1. Alkmim, B., Haeusler, E.H., Rademaker, A.: Utilizing iALC to Formalize the Brazilian OAB Exam. In: Nalepa, G.J., Atzmueller, M., Araszkiewicz, M., Novais, P. (eds.) Proceedings of the EXplainable AI in Law Workshop Co-located with the 31st International Conference on Legal Knowledge and Information Systems, (XAILA@JURIX 2018), Groningen, The Netherlands, December 12, 2018. CEUR Workshop Proceedings, vol. 2381, pp. 42–50. CEUR-WS.org (2018). http://ceur-ws.org/Vol-2381/xaila2018_paper_2.pdf
2. Baader, F., Calvanese, D., Mcguinness, D., Nardi, D., Patel-Schneider, P.: The Description Logic Handbook: Theory, Implementation, and Applications, 2nd edn, Cambridge University Press, Cambridge, January 2007. ISBN: 978-0-521-15011-8
3. Bobbio, N., dos Santos, M.: Theory of Legal Order, Ed Universidade de Brasília. (1995). ISBN: 9788523002763
4. Bondy, J., Murty, U.: Graph Theory. 1st edn, Springer Publishing Company, Incorporated (2008). https://dl.acm.org/doi/10.5555/1481153

5. Buchanan, B.G., Feigenbaum, E.A.: Dendral and meta-dendral: their applications dimension. Artif. Intell. **11**(1), 5–24 (1978). https://doi.org/10.1016/0004-3702(78)90010-3

6. Condoravdi, C., Crouch, D., de Paiva, V., Stolle, R., Bobrow, D.G.: Entailment, intensionality and text understanding. In: Proceedings of the HLT-NAACL 2003 Workshop on Text Meaning (HLT-NAACL-TEXTMEANING 2003) - Vol. 9. pp. 38–45. Association for Computational Linguistics, Stroudsburg, PA, USA (2003), https://doi.org/10.3115/1119239.1119245

7. de Paiva, V., Haeusler, E.H., Rademaker, A.: Constructive Description Logics Hybrid-Style. Electr. Notes Theor. Comput. Sci. **273**, 21–31 (2011). https://doi.org/10.1016/j.entcs.2011.06.010, International Workshop on Hybrid Logic and Applications 2010 (HyLo 2010)

8. Delfino, P., Cuconato, B., Haeusler, E.H., Rademaker, A.: Passing the Brazilian OAB exam: data preparation and some experiments. In: Wyner, A.Z., Casini, G. (eds.) Legal Knowledge and Information Systems - JURIX 2017: The Thirtieth Annual Conference, Luxembourg, 13–15 December 2017. Frontiers in Artificial Intelligence and Applications, vol. 302, pp. 89–94. IOS Press (2017). https://doi.org/10.3233/978-1-61499-838-9-89

9. Falcone, R., Castelfranchi, C.: Social trust: a cognitive approach. In: Castelfranchi, C., Tan, Y.H. (eds.) Trust and deception in virtual societies, pp. 55–90. Springer, Netherlands, Dordrecht (2001). https://doi.org/10.1007/978-94-017-3614-5_3

10. Groth, P., Gibson, A., Velterop, J.: The Anatomy of a Nano-publication. Information Services and Use, vol. 30, January 2010. https://doi.org/10.3233/ISU-2010-0613

11. Gutierrez, C., Sequeda, J.F.: Knowledge Graphs. Commun. ACM **64**(3), 96–104 (2021). https://doi.org/10.1145/3418294

12. Haeusler, E., De Paiva, V., Rademaker, A.: Intuitionistic Logic and Legal Ontologies. In: Legal Knowledge and Information Systems - JURIX 2010, Frontiers in Artificial Intelligence and Applications, vol. 223, pp. 155–158. IOS Press (2010). https://doi.org/10.3233/978-1-60750-682-9-155

13. Haeusler, E.H., Paiva, V.d., Rademaker, A.: Intuitionistic description logic and legal reasoning. In: 2011 22nd International Workshop on Database and Expert Systems Applications, pp. 345–349 (2011), https://doi.org/10.1109/DEXA.2011.46

14. Haeusler, E.H., Rademaker, A.: On How Kelsenian jurisprudence and intuitionistic logic help to avoid contrary-to-duty paradoxes in Legal Ontologies. In: Lógica no Avião Seminars, vol. 1, pp. 44–59. Univeristy of Brasília (2019), http://doi.org/c768

15. Hayes-Roth, F., Waterman, D.A., Lenat, D.B.: Building Expert Systems. Addison-Wesley Longman Publishing Co., Inc, Boston (1983). https://dl.acm.org/doi/book/10.5555/6123

16. Heyting, A.: Intuitionism: An Introduction, 3rd edn. North-Holland Publishing, Amsterdam (1956), https://doi.org/10.2307/3609219

17. Jarke, M., Neumann, B., Vassiliou, Y., Wahlster, W.: KBMS Requirements of Knowledge-Based Systems, pp. 381–394. Springer-Verlag, Berlin, Heidelberg (1989). https://doi.org/10.1007/978-3-642-83397-7_17

18. Kelsen, H.: Pure Theory of Law. Lawbook Exchange (2005). ISBN: 9781584775782

19. Nickel, M., Murphy, K., Tresp, V., Gabrilovich, E.: A review of relational machine learning for knowledge graphs. In: Proceedings of the IEEE 104, December 2015. https://doi.org/10.1109/JPROC.2015.2483592

20. Prawitz, D.: Natural deduction: a proof-theoretical study. Ph.D. thesis, Almqvist & Wiksell (1965). https://doi.org/10.2307/2271676

21. Reiter, R.: A logic for default reasoning. Artif. Intell. **13**(1–2), 81–132 (1980). https://doi.org/10.1016/0004-3702(80)90014-4

22. Schneider, E.W., Human Resources Research Organization, Alexandria, V.: Course modularization applied [microform] :The interface system and its implications for sequence control and data analysis / E.W. Schneider. Distributed by ERIC Clearinghouse [Washington, D.C.] (1973). https://eric.ed.gov/?id=ED088424

23. Schwabe, D.: Trust and privacy in knowledge graphs. In: Companion Proceedings of The 2019 World Wide Web Conference (WWW 2019), pp. 722–728. Association for Computing Machinery, New York, NY, USA (2019). https://doi.org/10.1145/3308560.3317705

24. Seufert, S., Ernst, P., Bedathur, S.J., Kondreddi, S.K., Berberich, K., Weikum, G.: Instant espresso: interactive analysis of relationships in knowledge graphs. In: Bourdeau, J., Hendler, J., Nkambou, R., Horrocks, I., Zhao, B.Y. (eds.) Proceedings of the 25th International Conference on World Wide Web (WWW 2016), Montreal, Canada, April 11–15, 2016, Companion vol. pp. 251–254. ACM (2016). https://doi.org/10.1145/2872518.2890528

25. Valente, A., Breuker, J.: A model-based approach to legal knowledge engineering. In: Legal Knowledge Based Systems JURIX 92: Information Technology and Law, The Foundation for Legal Knowledge Systems. pp. 123–134. JURIX 1992 (The Hague, Netherlands), Koninklijke Vermande (1994). ISBN: 905-458-0313

Logical Comparison of Cases

Heng Zheng[1(✉)], Davide Grossi[1,2], and Bart Verheij[1]

[1] Artificial Intelligence, Bernoulli Institute, University of Groningen,
Groningen, The Netherlands
h.zheng@rug.nl
[2] ILLC/ACLE, University of Amsterdam, Amsterdam, The Netherlands

Abstract. Comparison between cases is a core issue in case-based reasoning. In this paper, we discuss a logical comparison approach in terms of the case model formalism. By logically generalizing the formulas involved in case comparison, our approach identifies analogies, distinctions and relevances. An analogy is a property shared between cases. A distinction is a property of one case ruled out by the other case, and a relevance is a property of one case, and not the other, that is not ruled out by the other case. The comparison approach is applied to HYPO-style comparison (where distinctions and relevances are not separately characterized) and to the temporal dynamics of case-based reasoning using a model of real world cases.

Keywords: Case-based reasoning · Cases · Case comparison

1 Introduction

Case-based reasoning, one of the main legal reasoning types, has been discussed in the Artificial Intelligence and Law community for years. It allows for a form of analogical reasoning [4,5], and a core issue is how to make decisions for a current case by comparing cases, namely the doctrine of *stare decisis*.

Case-based reasoning has been formalized using many different approaches. For instance, abductive logic programming [15], formal dialogue games [11], context-related frameworks [6,8], dialectical arguments [14], ontologies in OWL [18], the ASPIC+ framework [12], reason models [9], abstract argumentation [7], abstract dialectical frameworks [1] and case-based argumentation frameworks [10]. These works often discuss case comparison in terms of factors, following ideas developed in HYPO [3,13].

In [19–21], a formal approach to the modeling of case-based reasoning has been discussed using a formal logical language. It can be used for evaluating the

This paper extends the research abstract [20] and the informal workshop proceedings paper [21]. This research was partially funded by the Hybrid Intelligence Center, a 10-year programme funded by the Dutch Ministry of Education, Culture and Science through the Netherlands Organization for Scientific Research, https://hybrid-intelligence-centre.nl.

© Springer Nature Switzerland AG 2021
V. Rodríguez-Doncel et al. (Eds.): AICOL-XI 2018/AICOL-XII 2020/XAILA 2020, LNAI 13048, pp. 125–140, 2021.
https://doi.org/10.1007/978-3-030-89811-3_9

validity of arguments in legal reasoning [19] and the formal comparison of legal precedents [20,21]. The approach is based on the case model formalism [16,17].

The present paper is an extended version of [20,21], where we discuss the comparison between precedents that are represented as conjunctions of factors and outcomes. Here we define case comparison in the general setting of the case model formalism developed by Verheij [17]. This generalization is needed for the here newly presented application to the dynamics of case-based reasoning following the research developed by Berman and Hafner [6,8] as modeled in terms of case models [16].

In Sect. 2, we show the technical part of comparing cases using our formalism. Section 3 applies our comparison approach in a discussion of case comparison in HYPO-style case-based reasoning. Section 4 applies our approach to the development of precedential values in a series of legal cases. With these applications, we show that our approach can generalize case-based reasoning by comparing cases with general formulas and refine case-based reasoning by introducing the new notion of relevances. In this way, we show that comparing cases with respect to general properties, represented by general propositional formulas, offers a novel angle on case-based reasoning.

2   Theory: Case Comparisons

In this section, we present the case model formalism [16,17] and apply it to case comparison in case-based reasoning (also shown in [20,21]). The notions about case comparison are based on the analogies and distinctions defined in [17].

The formalism introduced in this paper uses a propositional logic language L generated from a finite set of propositional constants. We fix language L. We write $\neg$ for negation, $\wedge$ for conjunction, $\vee$ for disjunction, $\leftrightarrow$ for equivalence, $\top$ for a tautology, and $\bot$ for a contradiction. The associated classical, deductive, monotonic consequence relation is denoted $\models$.

Cases can be compared through the preference relation between cases in case models. A *case model* is a set of logically consistent, incompatible cases forming a total preorder (i.e., a transitive, total binary relation) representing a preference relation among the cases.

Definition 1. (Case models [16,17]). *A case model is a pair* $\mathcal{C} = (C, \geq)$ *with finite* $C \subseteq L$, *such that, for all* π, π' *and* $\pi'' \in C$:

1. $\not\models \neg\pi$;
2. *If* $\not\models \pi \leftrightarrow \pi'$, *then* $\models \neg(\pi \wedge \pi')$;
3. *If* $\models \pi \leftrightarrow \pi'$, *then* $\pi = \pi'$;
4. $\pi \geq \pi'$ *or* $\pi' \geq \pi$;
5. *If* $\pi \geq \pi'$ *and* $\pi' \geq \pi''$, *then* $\pi \geq \pi''$.

As customary, the asymmetric part of $\geq$ is denoted $>$. The symmetric part of $\geq$ is denoted $\sim$. Intuitively, $\geq$ means 'at least as preferred as'.

Example 1. Figure 1 shows a case model with case $\pi_0 = P \wedge Q \wedge R$ and $\pi_1 = P \wedge \neg Q$. π_0 is more preferred than π_1 as suggested by the size of boxes.

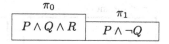

Fig. 1. A case model

We define now the notions of analogy, distinction and relevance. *Analogies* between two cases are the formulas that follow logically from both two cases. *Distinctions* are the formulas that only follow logically from one of the cases while their negation is logically implied by the other case. *Relevances* are, intuitively, formulas that are relevant to the analogies and distinctions between two cases. Relevances only follow from one of the cases but, unlike in distinctions, neither themselves nor their negation are logically implied by the other case. Intuitively, an analogy describes a shared property between two cases. Distinctions and relevances both describe unshared properties between cases, where distinctions are contradicting properties in the cases, and relevances are the properties that are not shared and not contradicted. As such distinctions and relevances clarify the two ways in which two cases may differ. Even though we present these notions in terms of cases, they can be defined in general for any given pair of propositional formulas, which is what we do now.

Definition 2 (Analogies, distinctions, relevances). *For any $\pi, \pi' \in L$, we define:*

1. *a sentence $\alpha \in L$ is an* analogy *between π and π' if and only if $\pi \models \alpha$ and $\pi' \models \alpha$. A most specific analogy between π and π' is an analogy that logically implies all analogies between π and π'.*
2. *a sentence $\delta \in L$ is a* distinction *in π with respect to π' (π-π' distinction) if and only if $\pi \models \delta$ and $\pi' \models \neg\delta$. A most specific π-π' distinction is a distinction that logically implies all π-π' distinctions.*
3. *a sentence $\rho \in L$ is a* relevance *in π with respect to π' (π-π' relevance) if and only if $\pi \models \rho$, $\pi' \not\models \rho$ and $\pi' \not\models \neg\rho$. ρ is a proper π-π' relevance if and only if ρ is a π-π' relevance that logically implies the most specific analogy between π and π'. A most specific π-π' relevance is a relevance that logically implies all π-π' relevances.*

Both π-π' distinctions and π'-π distinctions are called *distinctions between π and π'*. Both π-π' relevances and π'-π relevances are called *relevances between π and π'*. When a most specific analogy/distinction/relevance exists we consider it unique modulo logical equivalence, and we thus refer to it as *the* most specific analogy/distinction/relevance. Notice that when introducing relevances, we also define a special kind of this notion, namely proper relevances. These formally describe those relevances that logically imply the most specific analogy and are implied by the most specific distinction (if it exists).

Example 2. Comparing π_0 and π_1 in Fig. 1, we have:

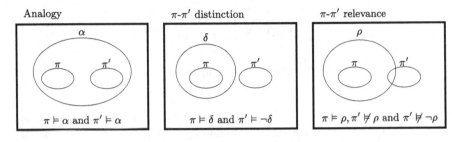

Fig. 2. Case comparison illustrated in terms of sets of worlds

- Analogies between π_0 and π_1: *e.g.*, P, $P \vee R$;
- The most specific analogy between π_0 and π_1: $(P \wedge Q \wedge R) \vee (P \wedge \neg Q)$;
- π_0-π_1 distinctions: *e.g.*, Q, $P \wedge Q$;
- The most specific π_0-π_1 distinction: $P \wedge Q \wedge R$;
- π_0-π_1 relevances: *e.g.*, R, $P \wedge R$;
- Proper π_0-π_1 relevances: *e.g.*, $P \wedge R$.

Now we further discuss the notions in Definition 2. Figure 2 illustrates analogies, distinctions and relevances using Venn diagrams representing the sets of worlds (or valuations) in which sentences are true (the so-called truth sets). As shown in Fig. 2, for any analogy α between cases π and π', the sets of π and π' worlds are subsets of the set of α worlds; for any π-π' distinction δ, the π worlds are a subset of the δ worlds, while the π' worlds and the δ worlds are disjoint; for any π-π' relevance ρ, the π worlds are a subset of the ρ worlds, while the π' worlds and the ρ worlds are not subsets of each other and the intersection of the π' worlds and the ρ worlds are always not empty. Notice that for any proper π-π' relevance ρ, not only the π worlds are a subset of the ρ worlds, but also the ρ worlds are a subset of the union of the π worlds and the π' worlds.

The following proposition shows the properties of analogies, distinctions and relevances between cases.

Proposition 1. *For any $\pi, \pi' \in L$:*

1. *The most specific analogy between π and π' always exists and is logically equivalent to $\pi \vee \pi'$.*
2. *There exists a π-π' distinction if and only if $\pi \wedge \pi' \models \bot$. If a π-π' distinction exists, then the most specific π-π' distinction exists and is logically equivalent to π.*
3. *A most specific π-π' relevance exists if and only if $\pi \wedge \pi' \not\models \bot$ and $\pi' \not\models \pi$. When it exists, the most specific π-π' relevance is logically equivalent to π.*
4. *If a π-π' distinction exists, then the most specific π-π' distinction logically implies each proper π-π' relevance. Each proper π-π' relevance logically implies the most specific analogy between π and π'.*

Proof. For any $\pi, \pi' \in L$:

$\boxed{\text{Property 1}}$ By Definition 2, for any analogy α, $\pi \models \alpha$ and $\pi' \models \alpha$. By propositional logic it follows that any analogy α is logically implied by $\pi \vee \pi'$. By Definition 2, $\pi \vee \pi'$ is therefore a most specific analogy.

$\boxed{\text{Property 2}}$ We prove the first claim first. Left to right. Assume a π-π' distinction δ exists. By Definition 2, $\pi \models \delta$ and $\pi' \models \neg\delta$. It follows by propositional logic that $\pi \wedge \pi' \models \bot$. Right to left. If $\pi \wedge \pi' \models \bot$, then by propositional logic $\pi' \models \neg\pi$. By Definition 2 and propositional logic, π is therefore a most most specific π-π' distinction. The second claim follows directly from the proof of the right to left direction of the previous claim.

$\boxed{\text{Property 3}}$ We prove the first claim first. Right to left. Assume $\pi \wedge \pi' \not\models \bot$ and $\pi' \not\models \pi$. Then we have that $\pi \models \pi$ (trivially), $\pi' \not\models \pi$ (by assumption) and $\pi' \not\models \neg\pi$ (by assumption). By Definition 2 π is therefore a relevance, and it is trivially most specific. Left to right. We proceed by contraposition and assume that either $\pi \wedge \pi' \models \bot$ or $\pi' \models \pi$. Clearly, if $\pi' \models \pi$ no relevance exists by Definition 2. So assume that $\pi \wedge \pi' \models \bot$. We show by a counterexample that a most specific relevance does not exist. Let $L = \{f_1, f_2\}$ and $\pi = f_1$, $\pi' = \neg f_1$. Consider then the two π-π' relevances $f_1 \vee f_2$, $f_1 \vee \neg f_2$. By propositional logic and Definition 2 there exists no π-π' relevance which entails both. So no most specific relevance exists in this example.

The second claim follows directly from the proof of the right to left direction of the previous claim.

$\boxed{\text{Property 4}}$ We prove the first claim first. By Property 2 if the most specific π-π' distinction exists, then it is logically equivalent to π. As to the second claim, by Definition 2, π logically implies all π-π' relevances, including proper ones, and proper π-π' relevances always logically imply the most specific analogy between π and π'. $\qquad\qquad\qquad\qquad\qquad\qquad\qquad\qquad\qquad\qquad\qquad\qquad\quad\square$

As shown in Proposition 1, $\pi \vee \pi'$ is the most specific analogy between π and π'. In legal case-based reasoning, this may seem counterintuitive. However, by the definition of case comparison in terms of propositional logic, we can see that the sentence $\pi \vee \pi'$ characterizes the properties shared exactly by the two cases (i.e., those implied by both).

Based on Property 2 and 3 in Proposition 1, we see there always exists a distinction between any pair of cases in a case model, since they are mutually incompatible. Recall that in any two cases π and π' in a case model are either identical or logically incompatible. By Property 3 then there cannot exist a most specific relevance between two cases in a case model: when π and π' are the same formula, no relevance exists between the two by the definition of case model; when they are not they need to be incompatible by the definition of case model, and hence by Property 3 no most specific relevance can exist between them.

Property 4 in Proposition 1 shows why we have singled out proper relevances: in the formally precise sense of the proposition, they are logically 'in between' the most specific distinction (if it exists) and the most specific analogy.

Two cases can be compared with a third case using the analogy relation defined below, which is similar to what is called on-pointness in HYPO [3]. The analogy relation is based on the shared formulas between cases. When comparing cases π and π' in terms of case π'', if the most specific analogy between π and π'' logically implies the most specific analogy between π' and π'', then we say that π is at least as analogous as π' with respect to π''. We define the analogy relation as follows:

Definition 3 (Analogy relation between cases). *For any π, π' and $\pi'' \in L$, we define:*
$$\pi \succeq_{\pi''} \pi' \text{ if and only if } \pi \vee \pi'' \models \pi' \vee \pi''.$$
Then we say π is at least as analogous as π' *with respect to π''.*

As customary, the asymmetric part of the relation is denoted as $\pi \succ_{\pi''} \pi'$, which means π is *more analogous* than π' with respect to π''. The symmetric part of the relation is denoted as $\pi \sim_{\pi''} \pi'$, which means π is *as analogous as π'* with respect to π''. If it is not the case that $\pi \succeq_{\pi''} \pi'$ and $\pi' \succeq_{\pi''} \pi$, then we say π and π' are *analogously incomparable with respect to π''*.

Example 3. Comparing π_0 and π_1 in Fig. 1 in terms of case $\pi_2 = P \wedge Q$, we have $\pi_0 \succ_{\pi_2} \pi_1$; If $\pi_2 = P$, then we have $\pi_0 \sim_{\pi_2} \pi_1$; If $\pi_2 = \neg R$, then π_0 and π_1 are analogously incomparable with respect to π_2.

In the following proposition, we show some interesting properties of the analogy relation.

Proposition 2. *For any π, π' and $\pi'' \in L$:*

1. *The analogy relation is reflexive and transitive, hence a preorder;*
2. *$\pi \succeq_{\pi''} \pi'$ if and only if $\pi \models \pi' \vee \pi''$;*
3. *For any $\alpha \in L$, if $\pi \succeq_{\pi''} \pi'$, and α is an analogy between π' and π'', then α is also an analogy between π and π''.*

Proof. | Property 1 | The relation is reflexive, since $\pi \vee \pi'' \models \pi \vee \pi''$. The relation is also transitive because of the transitivity of entailment in propositional logic. Assume $\pi = f_1 \wedge f_2$, $\pi' = f_1 \wedge f_3$ and $\pi'' = f_1 \wedge f_2 \wedge f_3$, π and π' are analogously incomparable with respect to π'', hence the relation is not in general total.

| Property 2 | From left to right, by Definition 3 we obtain $\pi \vee \pi'' \models \pi' \vee \pi''$, and by propositional logic $\pi \models \pi' \vee \pi''$. From right to left, from $\pi \models \pi' \vee \pi''$ and propositional logic, we obtain $\pi \vee \pi'' \models \pi \vee \pi''$, and by Definition 3 $\pi \succeq_{\pi''} \pi'$.

| Property 3 | Follows directly from Definition 2 and 3.   □

Notice that if $\pi \succeq_{\pi''} \pi'$, then it is still possible that $\pi \not\models \pi'$ and $\pi \not\models \pi''$. For instance, if $\pi = f_1$, $\pi' = f_1 \wedge f_2$, $\pi'' = f_1 \wedge \neg f_2$, then we have $\pi \succeq_{\pi''} \pi'$, but both π' and π'' are not logically implied by π. Also notice that if $\pi \succeq_{\pi''} \pi'$, it cannot be concluded that $\pi \models \pi'$. For instance, $\pi = f_1 \wedge f_2$, $\pi' = f_3$ and $\pi'' = f_1$. In this example, $\pi \succeq_{\pi''} \pi'$ but $f_1 \wedge f_2 \not\models f_3$.

HYPO

American Precision (Pla)

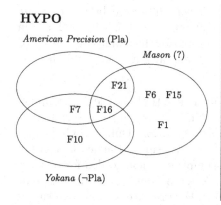

Case model formalism

Case model

$\pi_{American\_Precision}$	π_{Yokana}
F7 ∧ F16 ∧ F21 ∧ Pla	F7 ∧ F10 ∧ F16 ∧ ¬Pla

Current situation

π_{Mason}

F1 ∧ F6 ∧ F15 ∧ F16 ∧ F21

Fig. 3. The *Mason* problem in HYPO (the Venn diagram [4]) and in a case model [19, 21]

3   Application: HYPO-style Comparison

In this section, we apply our formalism to case comparison in HYPO-style case-based reasoning with an example from a real legal domain.

As shown in [3, 4], in HYPO, the set of shared factors between two cases are called *relevant similarity*, while the set of unshared factors are called *relevant difference*. Unshared factors can be used for pointing out the two cases should be decided differently. When comparing two cases in terms of a current situation, HYPO always makes sure that the cases are *on point* to the situation, namely the set of shared factors between any of the cases and the situation is not empty. If one of the cases shares more factors with the situation than the other one, then former case is *more on point* than the latter one with respect to the situation.

We take a set of legal cases from the United States trade secret law domain as an example. The cases has been discussed in [4, 20]. As shown in Fig. 3, the *Yokana* case[1] and the *American Precision* case[2] are considered as two decided cases. *Yokana* favors for defendants (represented by ¬Pla) and *American Precision* favors for plaintiffs (represented by Pla). The *Mason* case[3] is considered as a current undecided case in this example. HYPO considers F6, F7, F15, and F21 as pro-plaintiff factors and F1, F10, and F16 as con-plaintiff factors, which suggests that they favor for different sides in the court.

HYPO represents cases as sets of factors. When comparing *Mason* with *Yokana* in HYPO, we consider:

1. {F16} as the set of the relevant similarity between *Mason* and *Yokana*, since both *Mason* and *Yokana* contain F16;

---

[1] Midland-Ross Corp. v. Yokana, 293 F.2d 411 (3rd Cir.1961).

[2] American Precision Vibrator Co. v. National Air Vibrator Co., 764 S.W.2d 274 (Tex.App.-Houston [1st Dist.] 1988).

[3] Mason v. Jack Daniel Distillery, 518 So.2d 130 (Ala.Civ.App.1987).

2. {F6, F15, F21, F10} as the set of the relevant difference between *Mason* and *Yokana* that against the defendant's claim, since F6, F15, F21 are in *Mason*, but not in *Yokana*, and they are favorable for the plaintiff as suggested by HYPO. F10 is in *Yokana*, but not in *Mason*, and it is favorable for the defendant as suggested by HYPO.

In the case model shown in Fig. 3, we represent cases in HYPO by logical conjunctions of factors and outcomes. We consider both factors and outcomes as literals. A literal is either a propositional constant or its negation. Unlike in CATO [2], our use of factors does not assume that factors favor a side of the decision, either pro-plaintiff or pro-defendant, as such an assumption is not needed for our logical definitions of case comparison. Unlike HYPO [3], our factors do not come with a dimension that can express a magnitude. For instance, the *Yokana* case is represented as $\pi_{Yokana} = F7 \wedge F10 \wedge F16 \wedge \neg Pla$. Similarly for *American Precision* (as $\pi_{American\_Precision}$) and *Mason* (as π_{Mason}). The case model is with equal preference, as in HYPO's case base, all the cases are as preferred as each other. When comparing π_{Mason} with π_{Yokana} in the case model formalism, we consider:

1. F16 as an analogy between π_{Mason} and π_{Yokana} in the case model, since:
   (a) $\pi_{Mason} \models F16$; and
   (b) $\pi_{Yokana} \models F16$.
2. $F6 \wedge F15 \wedge F21$ as a π_{Mason}-π_{Yokana} relevance, since:
   (a) $\pi_{Mason} \models F6 \wedge F15 \wedge F21$;
   (b) $\pi_{Yokana} \not\models F6 \wedge F15 \wedge F21$ and $\pi_{Yokana} \not\models \neg(F6 \wedge F15 \wedge F21)$.
3. F10 as a π_{Yokana}-π_{Mason} relevance, since:
   (a) $\pi_{Yokana} \models F10$;
   (b) $\pi_{Mason} \not\models F10$ and $\pi_{Mason} \not\models \neg F10$.

Notice that there is no distinction between π_{Mason} and π_{Yokana}, as these two cases are not incompatible.

Compared to the notions in Definition 2, the relevant similarity between cases (in terms of sets of factors) corresponds to an analogy between the cases (in terms of logical sentences), in the sense that the conjunction of the factors in the relevant similarity are logically implied by each of the two conjunctions of factors that represent the cases. However, the relevant difference between cases (in terms of sets of factors) can not be simply considered as distinctions or relevances between cases (in terms of logical sentences), since HYPO does not consider the negation of factors, hence it cannot separate distinctions and relevances. For those unshared factors between cases as the relevance difference, they are implied by one of the cases, but not the other one. If the negation of the factor can be applied by the other one, then it is considered as a distinction, if not, it is considered as a relevance.

The relevant similarity is not the only analogy between π_{Mason} and π_{Yokana}. We also have sentences like $F16 \vee F21$, $(F7 \wedge F10 \wedge F16 \wedge \neg Pla) \vee (F1 \wedge F6 \wedge F15 \wedge F16 \wedge F21)$ as analogies between them. Notice that the latter one is the most specific analogy between π_{Mason} and π_{Yokana}. We also have $F1 \wedge F21$ as a *Mason*-*Yokana* relevance and $F16 \wedge \neg Pla$ as a *Yokana*-*Mason* relevance.

In the onpointness relation of HYPO, only sets of the shared factors are compared, and there is no outcome involved in the comparison. However, in the logical analogy relation, both factors and outcomes can be taken into account. Namely, when comparing cases in terms of onpointness, we compare the sets of factors in the conjunctions that represent cases, when comparing them in terms of the logical analogy relation, we use the most specific analogy between cases which can include outcomes. When comparing *American Precision* and *Yokana* in terms of *Mason*, *American Precision* is more on point than *Yokana* with respect to *Mason*, as the relevant similarity between *Yokana* and *Mason* ({F16}) is a subset of the relevant similarity between *American Precision* and *Mason* ({F16, F21}). However, according to the analogy relation, *American Precision* and *Yokana* are analogously incomparable with respect to *Mason*, which is determined by the most specific analogy between *Yokana* and *Mason* ($\pi_{Yokana} \vee \pi_{Mason}$) and between *American Precision* and *Mason* ($\pi_{American\_Precision} \vee \pi_{Mason}$):

1. $\pi_{Yokana} \vee \pi_{Mason} \not\models \pi_{American\_Precision} \vee \pi_{Mason}$; and
2. $\pi_{American\_Precision} \vee \pi_{Mason} \not\models \pi_{Yokana} \vee \pi_{Mason}$.

The above shows that, based on different comparison relations (analogy relation/onpointness), the selection of better case can be different. We observe that if two cases are onpointness comparable with respect to a third case, namely one of the two cases is either more on point or as on point than the other one with respect to the third case (otherwise, they are onpointness incomparable), the two cases are not always analogously comparable. For instance, when comparing *American Precision* and *Yokana* in terms of *Mason*.

For convenience, we now give an abstract example. We assume π_0, π_1 and π_2 are cases in HYPO, namely they are conjunctions of factors and outcomes (both are literals). When comparing π_0 and π_1 with respect to π_2, based on Proposition 2, we can further observe that:

1. If π_0 and π_1 are onpointness comparable with respect to π_2, then π_0 and π_1 are not always analogously comparable;
2. If π_0 and π_1 are onpointness incomparable with respect to π_2, then π_0 and π_1 are not always analogously incomparable;
3. If π_0 and π_1 are analogously comparable with respect to π_2, then π_0 and π_1 are always onpointness comparable:
   (a) If π_0 is more analogous than π_1 with respect to π_2, then π_0 is also more on point than π_1 with respect to π_2;
   (b) If π_0 is as analogous as π_1 with respect to π_2, then π_0 is also as on point as π_1 with respect to π_2.

The first observation has already been discussed above. For the second observation, assume that $\pi_0 = P \wedge Q$, $\pi_1 = \neg P$, and $\pi_2 = Q$, then π_0 and π_1 are onpointness incomparable with respect to π_2, but $\pi_0 \succeq_{\pi_2} \pi_1$, namely π_0 is more analogous than π_1 with respect to π_2. The last observation follows from Property 3 in Proposition 2.

4   Application: The Dynamics of Case-Based Reasoning

In our approach, we can formally distinguish the unshared part between cases in distinctions and relevances. We now apply this comparison approach to the development of precedential values in case-based reasoning by following a series of research [6,8,16]. The case model we analyzed has been studied in [16], and represents the series of New York car accident cases used in [6,8]. The focus is on the selection of the jurisdiction choice rules that applied in the cases:

1. **Smith v. Clute** 277 N.Y. 407, 14 N.E.2d 455 (1938): The claim was in tort law (driver negligence). The territorial rule applies.
2. **Kerfoot v. Kelley** 294 N.Y. 288, 62 N.E.2d 74 (1945): The claim was in tort law (driver negligence). The territorial rule applies.
3. **Auten v. Auten** 308 N.Y. 155, 124 N.E.2d 99 (1954): The claim was in contract law (enforce a child support agreement). The center-of-gravity rule applies.
4. **Kaufman v. American Youth Hostels** 5 N.Y.2d 1016 (1959): The claim was in tort law (travel guide negligence). The territorial rule applies.
5. **Haag v. Barnes** 9 N.Y.2d 554, 175 N.E.2d 441, 216 N.Y.S. 2d 65 (1961): The claim was in contract law (reopen a child support agreement). The center-of-gravity rule applies.
6. **Kilberg v. Northeast Airlines** 9 N.Y.2d 34, 172 N.E.2d 526, 211 N.Y.S.2d 133 (1961): The claim was in tort law (common carrier negligence). The territorial rule is overridden for reasons of public policy.
7. **Babcock v. Jackson** 12 N.Y.2d 473, 191 N.E.2d 279, 473 N.Y.S.2d 279 (1963): The claim was in tort law (driver negligence).

From the list of cases above, there are two kinds of cases, namely tort cases (represented as TORT) and contract cases (represented as CONTRACT). The jurisdiction choice rules can be: the territorial rule (represented as TERRITORY), the center-of-gravity rule (represented as GRAVITY), and exceptions (represented as EXCEPTION).

A case model (also discussed in [16]) can be generated from above cases. The model assumes that the kinds of cases exclude each other pairwise ($\neg$(TORT $\land$ CONTRACT)), and similarly for the choice rules ($\neg$(TERRITORY $\land$ EXCEPTION), etc.). We restrict the case model to the cases up and until a particular year. For instance, we write $\mathcal{C}(1945)$ for the case model with the set of cases that contains *Smith* and *Kerfoot* dating from 1945 or before. The cases in the model are represented as follows:

$$\pi_{Smith} = \text{TORT} \land \text{TERRITORY} \qquad \pi_{Haag} = \text{CONTRACT} \land \text{GRAVITY}$$
$$\pi_{Kerfoot} = \text{TORT} \land \text{TERRITORY}$$
$$\pi_{Auten} = \text{CONTRACT} \land \text{GRAVITY} \qquad \pi_{Kilberg} = \text{TORT} \land \text{EXCEPTION}$$
$$\pi_{Kaufman} = \text{TORT} \land \text{TERRITORY} \qquad \pi_{Babcock} = \text{TORT} \land \text{GRAVITY}$$

Now we compare a new, undecided tort case $\pi = \text{TORT}$ with the decided cases.

Suppose π arises before 1954 when only the *Smith* case and the *Kerfoot* case have been decided (i.e., $\mathcal{C}(1945)$), we can see TORT $\wedge$ TERRITORY is a relevance between in π_{Smith} (or $\pi_{Kerfoot}$) with respect to π in $\mathcal{C}(1945)$, as:

1. $\pi_{Smith} \models$ TORT $\wedge$ TERRITORY; but
2. $\pi \not\models$ TORT $\wedge$ TERRITORY and $\pi \not\models \neg$(TORT $\wedge$ TERRITORY).

Similarly, TORT $\wedge \neg$GRAVITY and TORT $\wedge \neg$EXCEPTION are also relevances between them in $\mathcal{C}(1945)$.

In 1954, *Auten* has been added into the model $\mathcal{C}(1954)$, which generally introduce the GRAVITY rule. If the undecided case π arises after 1954, $\top \wedge$GRAVITY can be a relevance in π_{Auten} with respect to π in $\mathcal{C}(1954)$. Formerly, in $\mathcal{C}(1945)$, $\top \wedge$ GRAVITY cannot be considered as a relevance between the cases, as neither π nor the decided cases (*Kerfoot* and *Smith*) imply the sentence.

Not only new jurisdiction choice rules can be considered as relevances between cases, but also new exceptions of these rules. In $\mathcal{C}(1961)$, *Kilberg* gives the territorial rule an exception, and sentence TORT $\wedge$ EXCEPTION for this new exception can be considered as a relevance in $\pi_{Kilberg}$ with respect to an undecided case π that arises after 1961. Before the exception occurred, TORT $\wedge$ EXCEPTION cannot be considered as a relevance or a distinction between them.

The general introduction of the GRAVITY rule cannot make the rule be a relevance when considering new cases in the tort law domain (represented by TORT$\wedge$GRAVITY). For instance, in $\mathcal{C}(1961)$, TORT$\wedge \neg$GRAVITY is a relevance between an undecided case π and any of the decided cases in the model, but not TORT $\wedge$ GRAVITY. This is changed when *Babcock* is added, which introduces the GRAVITY rule into the tort law domain, as in $\mathcal{C}(1963)$, TORT $\wedge$ GRAVITY can be considered as a relevance in $\pi_{Babcock}$ with respect to an undecided case π.

As shown above, the definition of relevances can support the analysis of the development of rules in the series of cases. When a new rule or exception is introduced in the case model, there is a new relevance for the future undecided case. For instance, the general introduction of the GRAVITY rule in 1954 by the *Auten* case makes $\top \wedge$ GRAVITY a relevance between *Auten* and undecided cases. For the cases that are decided after *Auten*, they need to consider whether the GRAVITY rule should be applied or not. Formally, they need to consider whether an undecided current case should imply sentence $\top \wedge$ GRAVITY or its negation $\neg (\top \wedge$ GRAVITY$)$.

Intuitively, for the unshared parts implied by decided cases but not by the undecided ones, their implied status in the undecided cases are unknown as yet, namely, these unshared sentences need to be considered by decision makers.

In contrast, the unshared parts between decided cases are not considered as relevances, but as distinctions in our formalism, since the status of these sentences in the cases are implied. For instance, TORT $\wedge$ GRAVITY and TORT $\wedge$ $\neg$GRAVITY are two distinctions between *Babcock* and *Smith* in $\mathcal{C}(1963)$, since:

1. $\pi_{Smith} \models$ TORT $\wedge \neg$GRAVITY; and
2. $\pi_{Babcock} \models$ TORT $\wedge$ GRAVITY.

Similarly, TORT∧TERRITORY and TORT∧¬TERRITORY are also distinctions between them. Recall that TORT ∧ GRAVITY and TORT ∧ TERRITORY are considered as relevances between $\pi_{Babcock}$ and an undecided case π in $\mathcal{C}(1963)$.

Therefore, the comparison approach in our formalism is able to formally identify the difference between the unshared sentences between two decided cases and between an undecided case and a decided case. For the unshared part between two decided cases, they are considered as distinctions in our approach, as they are known differences between the cases. For the unshared part in a decided case with respect to an undecided one, the status of this part in the undecided cases is unknown, and can be turned into analogies or distinctions in the further development of the cases. In this sense, they are different from the unshared part between two decided cases. The same sentence can play different roles in different comparisons of cases. As shown above, in $\mathcal{C}(1963)$ we can see sentences like TORT ∧ GRAVITY as a distinction between two decided cases ($\pi_{Babcock}$ and π_{Smith}), as a relevance between an undecided case π and a decided case $\pi_{Babcock}$. This cannot be achieved without first distinguishing distinctions and relevances. In this way, our approach refines the analysis of case-based reasoning.

5 Discussion

In this paper, we discuss case comparisons in case-based reasoning with the case model formalism, which is described in a formal propositional logic language. Unlike other case-based reasoning models, in which cases are represented as dimensions [3], sets of rules [11], sets of factors [12], combinations of rules, facts and outcomes [9] and hierarchies [1,2]. The formalism we present here represents cases using propositional logic sentences. Building on [17], we give a concrete account of the approach of comparison.

As an extension of [21], this paper defines the notions for case comparison in case models rather than in precedent models, a subclass of case models. We now discuss the comparison of cases in the general setting, not just in the precedents represented by conjunctions of factors and outcomes. In particular, the application shown in Sect. 4 is not in terms of the formal notion of precedents represented by factors and outcomes [21], and instead focuses only on the jurisdiction choice rules that applied in the legal cases modeled.

Case-based reasoning models following HYPO often discuss comparison between cases in terms of factors. In the formalism we present here, we generalize the comparison approach in case-based reasoning, namely comparing cases not only with factors, but also with more general propositional formulas.

Section 3 shows a key difference between our comparison approach and the research following HYPO [2,3,9]. Factors in HYPO-style comparison typically favor a side in the court case, showing which factors can strengthen or weaken the arguments given by the parties involved, hence constraining possible argument moves. However, the more general formulas used in our comparison approach may not favor a specific side in the court. For instance, F16 ∨ F21 is an analogy between π_{Mason} and π_{Yokana}, it does not favor a side in the court. The formulas we discuss are more general logical expressions than factors. It would be interesting to discuss the role of sides favored by factors in our formalism, for instance by investigating the modeling of argument moves in CATO, such as downplaying or emphasizing distinctions.

The comparison approach introduced here allows us to discuss general formulas beyond factors in case-based reasoning, such as conjunctions or disjunctions of factors, which can bring new discussion on case-based reasoning with the case model formalism. For instance, for future research we can discuss hierarchical factors shown in CATO [2], as higher level factors can be represented with compound formulas based on base-level factors. Therefore, it seems possible to compare abstract factors between cases directly in the formalism.

As shown in Sect. 3, when comparing *American Precision* and *Yokana* with respect to *Mason* in terms of the analogy relation defined in Definition 3, the result is different from the comparisons based on other relations, such as the preference relation in case models and onpointness in HYPO. By the preference relation of the case model shown in Fig. 3, *American Precision* and *Yokana* are as preferred as each other; according to onpointness, *American Precision* is more on point with respect to *Mason*; and according to the analogy relation, these two cases are analogously incomparable with respect to *Mason*. This is because the analogy relation discusses comparison in terms of the most specific analogy between cases, while other comparison relations are in terms of other notions. For instance, the onpointness is about the shared factors between cases, the conjunction of these factors is not always the most specific analogy between the cases, hence the comparison based on the analogy relation and based on the onpointness relation can have different results. In the above example we show that *American Precision* is a better precedent than *Yokana* based on the onpointness relation, however, the two cases are analogously incomparable by using the analogy relation. Things can be various in other examples. For instance, as we discussed in the observations shown in Sect. 3, cases can be onpointness incomparable but analogously comparable. Onpointness is for factor-based comparison built on sets, while the analogy relation is for logic-based comparison built on logical sentences. These two methods are from different perspectives and based on different theories. In legal case-based reasoning, users may pay more attention to the onpointness rather than the analogy relation, since the shared or absence of some factors can make the difference for winning a case. The formalism we develop is a general theory, which contributes to the further systematization of our formal understanding of case comparison in case-based reasoning. As the analogy relation can lead to a different selection of better cases, in the future, it will also be interesting to have a look at the comparison relation based on distinctions and relevances, the notions in the new refinement shown by our approach.

The representation we use can treat case-based reasoning from a perspective that is closer to logic, thereby allowing an analysis of the properties that are shared and not shared between cases, in terms of our notions of analogies, distinctions and relevances. In [17,19], we do not separate the distinctions and relevances between cases, nor do HYPO [3] and other case-based reasoning models [2,12]. In the formalism we present here, relevances between cases are distinguished from analogies and from distinctions. This refinement points to potential modification of case comparisons in case-based reasoning, in which the situation can change accordingly when new facts are found. While analogies between two cases refer to formulas that hold in both cases, and distinctions to formulas that hold in one case and are negated in the other, relevances are formulas that are not determined in a case and hence have the potential to turn out as an analogy or distinction once determined. Although both distinctions and relevances are related to unshared factors, relevances cannot be considered as distinctions directly, since if such relevant formulas are determined to hold in a situation, they will turn out as analogies rather than as distinctions between the case and the situation. For instance, in the *Mason* problem discussed in Sect. 3, we can see F7 as a relevance in *Yokana* with respect to *Mason*, since the conjunction of *Yokana* logically implies F7, but neither F7 nor ¬F7 is implied by the conjunction of *Mason*, in the sense that the status of F7 is unknown in *Mason*. Therefore, if F7 can be found in *Mason* later, then it will be considered as an analogy, if ¬F7 is found, then it will become a distinction. Therefore, our comparison approach has the potential to model heuristics for hypothetically modifying cases occurred in HYPO reasoning, such as heuristic H1 ("Make a near-miss Dimension apply" [3]).

As shown in Sect. 3, the relevant similarity and the relevant difference between cases in HYPO can be modeled in terms of analogies, distinctions and relevances defined in Definition 2. Although the relevant similarity is an analogy between cases, factors in the relevant difference are not always distinctions between cases, but can also be relevances. In this sense, our approach compares cases in a more specific way than HYPO. However, as we have not defined dimensions of factors in the formalism, it is unable to discuss the magnitude of factors in relevant differences, which means we cannot compare cases in terms of dimensions, such as finding a contrary case which has some factors with extreme magnitude. This needs further discussion in the future.

The refinement of case comparison can be further illustrated with the application in Sect. 4, which has not been discussed in [21]. The application shows that our comparison approach can support the discussion of the development of precedential values in case-based reasoning. By following [6,8,16], we have shown the difference of the unshared parts that exist between two decided cases and between an undecided case and a decided one, thereby we find the connection between the introduction of new rules and exceptions and the notion of relevances we define. When new rules and exceptions are introduced into the model, the number of relevances that need to be considered when comparing new cases with decided cases increases accordingly, which can make the decision making process harder than before.

6 Conclusion

In this paper, we have studied the logical comparison of cases. Continuing from [19–21], the comparison of cases is here not limited to precedents represented by conjunctions of factors and outcomes but is extended to the more general case model formalism. We also extend it with a discussion of the development of case-based reasoning in a temporal context.

With the formalism, we provide a way that refines comparisons in case-based reasoning. As shown in Sect. 3, we discuss not only the shared factors between cases, but also other logically compound formulas based on factors, which allows us to compare cases from a logical perspective and discuss other features among cases. We further distinguish the unshared formulas between cases into distinctions and relevances based on the implication of themselves and their negation in the cases. In this way, we show a refinement of comparisons in case-based reasoning. With the application about the dynamics of case-based reasoning shown in Sect. 4, we show how the refinement can support the analysis of the development of rules in cases.

The case model formalism has the potential to help analyze argument moves and applied status of legal rules in case-based reasoning and support the selection of good cases to cite in a court discussion. These topics could be investigated in future research.

References

1. Al-Abdulkarim, L., Atkinson, K., Bench-Capon, T.: A methodology for designing systems to reason with legal cases using Abstract Dialectical Frameworks. Artif. Intell. Law 24(1), 1–49 (2016). https://doi.org/10.1007/s10506-016-9178-1
2. Aleven, V.: Teaching case-based argumentation through a model and examples. Ph.D. thesis, University of Pittsburgh (1997)
3. Ashley, K.D.: Modeling Legal Arguments: Reasoning with Cases and Hypotheticals. MIT Press, Cambridge (1990)
4. Ashley, K.D.: Artificial Intelligence and Legal Analytics: New Tools for Law Practice in the Digital Age. Cambridge University Press, Cambridge (2017)
5. Bench-Capon, T.: HYPO's legacy: introduction to the virtual special issue. Artif. Intell. Law 25(2), 205–250 (2017)
6. Berman, D., Hafner, C.: Understanding precedents in a temporal context of evolving legal doctrine. In: Proceedings of the Fifth International Conference on Artificial Intelligence and Law, pp. 42–51. ACM, New York (1995)
7. Cyras, K., Satoh, K., Toni, F.: Abstract argumentation for case-based reasoning. In: Proceedings of the Fifteenth International Conference on Principles of Knowledge Representation and Reasoning (KR 2016), pp. 549–552. AAAI Press (2016)
8. Hafner, C., Berman, D.: The role of context in case-based legal reasoning: teleological, temporal, and procedural. Artif. Intell. Law 10(1), 19–64 (2002)
9. Horty, J., Bench-Capon, T.: A factor-based definition of precedential constraint. Artif. Intell. Law 20(2), 181–214 (2012)
10. Prakken, H.: Comparing alternative factor- and precedent-based accounts of precedential constraint. In: Araszkiewicz, M., Rodriguez-Doncel, V. (eds.) Legal Knowledge and Information Systems. JURIX 2019: The Thirty-Second Annual Conference, pp. 73–82. IOS Press, Amsterdam (2019)

11. Prakken, H., Sartor, G.: Modelling reasoning with precedents in a formal dialogue game. Artif. Intell. Law **6**, 231–287 (1998)
12. Prakken, H., Wyner, A., Bench-Capon, T., Atkinson, K.: A formalization of argumentation schemes for legal case-based reasoning in ASPIC+. J. Logic Comput. **25**(5), 1141–1166 (05 2013)
13. Rissland, E.L., Ashley, K.D.: A case-based system for trade secrets law. In: Proceedings of the 1st International Conference on Artificial Intelligence and Law, pp. 60–66. ICAIL 1987, ACM, New York (1987)
14. Roth, B., Verheij, B.: Dialectical arguments and case comparison. In: Gordon, T. (ed.) Legal Knowledge and Information Systems. JURIX 2004: The Seventeenth Annual Conference, pp. 99–108. IOS Press, Amsterdam (2004)
15. Satoh, K.: Translating case-based reasoning into abductive logic programming. In: Wahlster, W. (ed.) Proceedings of the 12th European Conference on Artificial Intelligence, ECAI 1996, pp. 142–146. Wiley, Chichester (1996)
16. Verheij, B.: Formalizing value-guided argumentation for ethical systems design. Artif. Intell. Law **24**(4), 387–407 (2016)
17. Verheij, B.: Formalizing arguments, rules and cases. In: Proceedings of the Sixteenth International Conference on Artificial Intelligence and Law, pp. 199–208. ICAIL 2017. ACM, New York (2017)
18. Wyner, A.: An ontology in OWL for legal case-based reasoning. Artif. Intell. Law **16**(4), 361 (2008)
19. Zheng, H., Grossi, D., Verheij, B.: Case-based reasoning with precedent models: preliminary report. In: Prakken, H., Bistarelli, S., Santini, F., Taticchi, C. (eds.) Computational Models of Argument. Proceedings of COMMA 2020, vol. 326, pp. 443–450. IOS Press, Amsterdam (2020)
20. Zheng, H., Grossi, D., Verheij, B.: Precedent comparison in the precedent model formalism: a technical note. In: Villata, S., Harašta, J., Kšemen, P. (eds.) Legal Knowledge and Information Systems. JURIX 2020: The Thirty-third Annual Conference, vol. 334, pp. 259–262. IOS Press, Amsterdam (2020)
21. Zheng, H., Grossi, D., Verheij, B.: Precedent comparison in the precedent model formalism: theory and application to legal cases. In: Proceedings of the EXplainable and Responsible AI in Law (XAILA) Workshop at JURIX 2020 (2020, to appear)

Explainable AI in Law and Ethics

A Conceptual View on the Design and Properties of Explainable AI Systems for Legal Settings

Martijn van Otterlo[1,2(✉)] and Martin Atzmueller[3]

[1] Computer Science Department, Open University, Heerlen, The Netherlands
[2] Computer Science Department, Radboud University, Nijmegen, The Netherlands
Martijn.vanOtterlo@ou.nl
[3] Osnabrück University, Semantic Information Systems Group, Osnabruck, Germany
martin.atzmueller@uni-osnabrueck.de

Abstract. The recent advances in AI have broad implications, e. g., onto explainability, accountability, and responsibility – in particular in legal settings. In this paper, we provide a conceptual view on design properties of such respective AI systems, where we focus on two specific approaches to explainable artificial intelligence (AI) for legal settings. The first approach aims at designing explainable AI using legal requirements – in a research and design methodology; the second one deals with a design strategy for ensuring requirements of computational ethical reasoning systems in and for which explainability is a core element.

Keywords: Artificial intelligence · Explainability · Law · System design · Values

1 Introduction

Ethical challenges of artificial intelligence (AI) are rising as technological advances are widely spread [50,77]. In addition to critiques from legal and sociology scholars who study the influence and regulation of algorithms, nowadays AI researchers themselves get actively involved by *creating* AI technology that is intrinsically *responsible, transparent* and especially *explainable* [30]. AI researchers contribute through, for example, *fair machine learning, value-based AI* and *ethical reasoning systems*, c. f., [75] and also the general collections in [23] and [41].

AI technology in legal settings has a long history [5], but recently, as in most other domains, *deep learning* is used for many tasks in which legal information is extracted by learning from many example texts [20]. Such approaches can, for example, be used to detect *unfair clauses* in legal contracts [43]. However, more in line with traditional approaches, AI can also be used for so-called *computational models of legal reasoning* (CLMR) [5], in which automated reasoning can be employed for statutory reasoning, or to deduce and predict legal outcomes. Better extraction procedures, for example induced by the success of modern deep learning, can then provide more and better opportunities for legal reasoning.

[†]This article is a substantially extended and adapted revision of [78].

ⓒ Springer Nature Switzerland AG 2021
V. Rodríguez-Doncel et al. (Eds.): AICOL-XI 2018/AICOL-XII 2020/XAILA 2020, LNAI 13048, pp. 143–153, 2021.
https://doi.org/10.1007/978-3-030-89811-3_10

In addition to deep learning, another prominent focus of the current AI field is *explainability* [4,39]: the capability of AI systems to give insight into their decisions, usually to humans, in order to support interpretation, decision making, and ultimately *computational sensemaking*, e. g., [10,16]. Concepts such as *transparency*, *interpretability* and especially *explainable* models have gained significant attention and momentum in the artificial intelligence, machine learning and data mining communities, e. g., [9,15,32]. Some methods focus on specific models, which are especially easy to understand and interpret, e. g., tree-based [69] or pattern-based approaches [24,44, 51] to better understand about a classifier or recommendation system – using local pattern mining techniques. Here, also, methods for associative classification, e. g., class association rules [8,40] and interpretable decision sets [25,38] can be applied for obtaining explicative, i. e., transparent, interpretable, and explainable models [6]. Then, individual steps of a classification/decision of the model can be traced-back, similar to a *reconstructive explanation*, c. f., [81]; the process can also be supported on several explanation dimensions [9]. For model agnostic explanation, e. g., [56], general directions are given by methods considering counterfactual explanation, e. g., [46,79], data perturbation and randomization techniques as well as interaction analysis methods, e. g., [12,34].

In the general AI community explainability has become a research focus in most sub-domains. For legal domains, explainability is very desirable given that the domain is all about humans, automated legal decisions by AI systems should always require full transparency, and explanations would give the means to be transparant about complex legal cases and decisions, possibly even through automated dialogue with the deciding AI system in court. Currently, several practical implementations exist, for example [17], but many challenges remain to fully bring explainability to the legal domain [14,84]

Given the rapid developments in AI and the emergence of explainability as one of the main desired directions to work on AI systems that are *trustworthy*, *human-centric* or even *ethically approved*, in particular in the legal domain, it is a natural question how to obtain such systems. Even though many guidelines, *codes of ethics* and rules of thumb exist, it is not easy to obtain legal AI systems that are ready to explain themselves. Therefore, suitable design and development strategies are required, for effectively designing and building such systems, while ensuring the respective principles of responsibility, transparency and explainability [4,18,36,37,62,84].

With that in mind, in this conceptual paper, we want to briefly highlight two possible routes to construct such systems, focusing on explainability. We discuss two methodological approaches – as tools for implementing the specific requirements.

1. First, we describe a design methodology, *KORA* that allows for ensuring particular properties at design time – in particular, for legal contexts, aligning non-functional and functional requirements in explainable AI system design.
2. Second, we outline a coherent set of dimensions, coined *IntERMeDIUM*, that will ensure that resulting systems will maximize the likelihood that they will be capable of explaining their legal decisions in human-centric terms.

The rest of the paper is structured as follows. Section 2 outlines and discusses the KORA methodology. Next, Sect. 3 presents IntERMeDIUM as a research strategy for ethical AI. Finally, Sect. 4 concludes this paper with a brief summary, and in particular outlines next (research) steps to take.

2 Designing Explainable AI Using Legal Requirements via KORA

One prominent method for integrating and matching legal (normative) and technical requirements and specific implementation choices is the KORA method [33,59,64]. It has been integrated and applied in several contexts, e. g., in a targeted approach for socio-technical design and development of ubiquitous computing, into system security evaluation [63], as well as the design of explicative applications and systems [7,11,28].

The basic KORA method aims at acquiring technical implementations based on legal requirements. It is built on a four step process model:

1. KORA starts with legal requirements (e. g., relating to transparency, privacy, etc.) that are mapped to legal criteria which are then matched with functional requirements. The legal requirements are typically derived from application specific legal provisions, e. g., given by the GDPR.
2. The obtained legal provisions are then made more concrete, also including technical functions as well as legal and social aspects. All these are ultimately formalized in specific so-called criteria.
3. These criteria are further mapped to functional requirements, e. g., supported by domain experts or based on specific design patterns, which relate to e. g., the interplay of application, user and/or system requirements as templates for the instantiation.
4. In the final step, the functional requirements map to specific implementation choices.

Relating to common process models in software engineering, the first three steps basically aim at requirements analysis, whereas the last step involves specific methods, patterns, and techniques relating to the concrete instantiations.

For enabling explainable AI systems, KORA provides an effective approach, by matching the (abstract) legal provisions with concrete implementation choices, e. g., by implementing/instantiating specific explainable models described above. For that, a categorization of these methods according to legal criteria (c. f., KORA/second step) is needed, connected to functional requirements. Then, a semi-automatic approach can be provided for mapping these criteria and its functional implementation.

Examples of (adapted) applications of KORA include, for example, the design of ubiquitous systems [21,29,58] with according explanation features [7] for specific functionality. Essentially, explainability capabilities are motivated by legal requirements (such as, e. g., transparency or fairness criteria), however, of course have to be transparently linked to the technical implementation choices. This is one specific advantage of applying the KORA methodology, following the steps as outlined above.

Furthermore, design patterns [27,57] for explainable systems can also be potentially captured, developed and incorporated, c. f., [19,52,71] i. e., by abstracting from specific design choices to more general classes of explanation patterns that are described in terms of their "explanation criteria" as well by the included "legal criteria". Then, this can be applied in all steps, of the process (described above), in principle. However, in particular techniques like this can specifically help in the step from criteria to functional requirements (steps 2–3), exemplifying the mapping and implementation, and thus making these choices and decisions more transparent and explainable in itself.

3   A General Research Strategy for Ethical AI – IntERMeDIUM

As a synthesis of a mix of ideas on learning, ethical codes and intentional agents [76,77] IntERMeDIUM is a research strategy to develop ethical AI systems. IntERMeDIUM refers to Joseph Licklider's connection between humans and *"the body of all knowledge"*, which is increasingly governed by AI in our society c. f., [77]. To unite human and machine ethics, a *code of ethics* can be seen as a *moral contract* between human and machine.

IntERMeDIUM is not a specific architecture, but a set of (technical) dimensions that can be instantiated in many ways, by choosing appropriate algorithms and representations. However, together the dimensions ensure the resulting AI system will be most likely ready to display properties such as transparency, explainability, and with that, trustworthiness. The acronym covers the main directions on which to focus implementation and research efforts on:

Intentional: The bridge between humans and machines consists of the right ontology of the (physical) world and the right level of description: beliefs, desires, intentions and goals. This concept was coined by Daniel Dennett [1] who provided arguments for what is a good way to *understand* the inner workings of intelligent systems, thereby laying the foundation for explainability in human-centric AI systems. For example, it has been shown that, in visual domains such as computer games, explanations in terms of *objects* can be more useful than in terms of pixels or general regions in images [35]. The intentional stance also entails the general principle that AI systems should be viewed as *rational agents*, maximizing particular goals that are measurable in objective terms. For legal systems, explanation *in the right, domain-specific legal concepts and relations* is vital too [14,17,65].

Executable: The beliefs and desires of the AI need to be embedded as code that can be *executed*. Instead of asking codes of ethics to be enforceable by punishing bad behavior after the fact, executable codes of ethics are biased by the code to ensure the right ethical behavior. [3](p16): *"Ethics must be made computable in order to make it clear exactly how agents ought to behave in ethical dilemmas"*. Recently we see an increased involvement of so-called *verification techniques* in general (adaptive) decision making AI systems, where the goal is to ensure and constrain particular properties (e. g., ethical aspects) of the behavior of an AI-system [2,18,36,37] and to use formal tools to *prove* particular properties are satisfied when executing the AI system. In legal systems this would coincide with *obeying the law, and legal procedures* in automated decisions, and also here we would like to, *ex ante*, prove *compliance*. Executability also provides the means for *computational* law, including *computational contracts* [83].

---

[1] https://en.wikipedia.org/wiki/Intentional_stance.

Reward-Based: AI systems' ethical reasoning is based on the human values in a particular domain. The core values come from the code of ethics used to bias the agent. In addition, AIs *finetune* their ethical behavior over time by adjusting relative values using data, feedback and experience. Experience from human actors is vital here, since they typically solved ethical dilemmas that arose thus far cf. [77]. This also makes *reinforcement learning* (RL) [74,82] the ideal learning setting (also for ethical behaviors [1]), whereas typically it needs to be extended towards *multi-objective* settings [72]. RL focuses on reward-based *behavior* learning through *evaluate feedback*, i.e. positive and negative *rewards*. This is contrasted with example-based learning settings such as in classification learning, or unsupervised learning tasks such as clustering.

Moral: IntERMeDIUMs focus of AI implementations here is on the moral (or, ethical) dimension [23,41,42,48,70,75]. Other skills will be developed elsewhere, including perception, mobile manipulation, reasoning with uncertainty, language interpretation and more, where technically similar methodologies can be used (e. g., deep learning).

Declarative: All ethical bias in the AI system is declarative knowledge and can be inspected at all times. Ethical inferences in specific circumstances can be *explained* in human-understandable terms. [3] (p17): *"What is critical in the explicit ethical agent distinction in our view, lies not only in who is making the ethical judgments (the machine versus the human programmer) but also in the ability to justify ethical judgments that only an explicit representation of ethical principles allows."* Ethical bias and learned ethical knowledge can be shared with other AIs and laws and regulations, such as the GDPR, can be implemented in the declarative bias to further bias the behavior of the AI towards legal compliance. For such ethical reasoning to work, declarative knowledge (for example about preferences [45]) is vital [48,68], and could even support *computational models of legal reasoning* [5] all the way to full *computational law* [83]. Deep learning is still useful as a component in any AI system (to learn various concepts from data) but it needs to be endowed with *common sense* [47] and *inspectable, declarative* knowledge for provide insightful explanations for humans [4]. Declarative knowledge can also help to (ethically) *constrain* [18,36] the learning process of *behavior* in AI systems, rendering behaviors *safe* [2]. In addition, they allow for explanations of behaviors in terms of the supplied constraints and knowledge [37], and could be carried over directly to legal settings [14,84].

Inductive: The AI is a *learning* agent. All knowledge that can not be injected as a declarative bias needs to be learned from experience or obtained from other AIs or humans. The AI's knowledge will typically not be complete, and learning should be continuing and life-long. Advanced machine learning needs to be implemented that allows the AI to ask human specialists for advice. Learning is needed for *perceptual* tasks such as reading and recognizing legal components [20,43]. However, for general decision making systems the so-called *reinforcement learning framework* (RL) [82] is needed, which is an excellent general framework for learning behaviors, and learning *ethical* behaviors [1], and for which explanation frameworks are being developed [54]. For many domains, including law, the need to insert commonsense knowledge [47] and structured *knowledge* [74] into RL is absolutely vital. In addition to learning about a domain (e. g., law), learning to make decisions and to explain them, a major (ethical) challenge for AI systems will be to learn about *human preferences* [60], which brings many conceptual and technical challenges [53,55]. Ethical behavior of the AI system is explicitly and implicitly related to whether humans *prefer* the behavior or explanations for it. Recent developments very relevant to law include techniques for learning ethical values and norms from *legal texts* such as constitutions of various nations [61].

Utilitarian: AI are utilitarian (collective consequentialist) moral reasoning agents. Protection of the rights of individuals is ensured by demanding that values and decision logic are declarative, open for inspection and transparent. Utilitarian frameworks fit the reward-based and reinforcement learning focus [82] in the other dimensions well, and have strong connections with *value alignment* [1,67,72] which is about creating AI systems that behave according to human values and norms. Finding out such values is difficult because there is a huge variance in the human population [13,60] but also because it is typically *multi-objective* [72] and (therefore) difficult to place all outcomes on the same *scale* [66]. Typical practical cases of utilitarian reasoning occur in selfdriving cars [42], but the underlying aspects immediately carry over to domains such as law. Once values are brought into play, systems can be *designed* [26] to better align with human values or, in general, to optimize the utilitarian value of decision outcomes all together to make AI systems (ethically) better [22]

Machine: The AI is a(n) (ethical) *machine* [3], and sometimes even a *physical* machine as in the case of robots [42]. The slow migration from human specialists to AI implementations in any domain, including law [5], requires that we should shift focus from humans to machines for the main operational aspects domains ranging from autonomous cars, libraries, and surely legal practices.

IntERMeDIUM is a general strategy for ethical, explainable AI systems. We claim that especially in the legal domain declarativeness, explainability and the AI's capability of engaging in a dialogue with humans to discuss and learn ethical and legal norms and values, is vital. Inspiration on how to translate human laws and ethical codes into declarative bias can come from e. g., medical [3] and autonomous driving [31] domains.

A first *instantiation* of IntERMeDIUM are *Declarative decision-theoretic ethical programs* (DDTEPs) [76], aligning with several aspects of earlier work [1]. They form a novel way to formalize *value-based ethical decision making* to help building understandable AI systems that are *value aligned* [67] in stochastic domains. The idea is to formalize what is known explicitly in the model, use *learning* to fill in knowledge gaps, and to use *reasoning* to obtain (optimal) decisions. DDTEPs fit into logical approaches for ethical (or: value-driven) reasoning [3] but also *relational reinforcement learning* [74] and provide novel opportunities for *explanation-focused* computations [73]. DDTEPs prove successful for toy ethical domains but could generally be applied to any kind of ethical reasoning where (some) domain knowledge is available, especially legal domains. The implementation of ethical AI systems has a history [48] but recently much effort is underway to implement ethical behaviors in computational systems [68].

4 Next Steps

With the advent of AI in basically every domain of society, it is no surprise AI will have a huge impact on the legal domain too. The societal consequences of the widespread use of AI can be mitigated, anticipated and turned for the better by paying attention to the *ethical* aspects of AI technology [23], equally so in law [65]. Before, we have described many aspects that are important for defining and designing ethical AI systems, and one aspects that stands out is the capability of AI systems to *explain their decisions.*

In general, explainable AI is a technical problem: given any AI algorithm and decision, how to create an algorithm that explains the decision automatically [4,54]. However, explanations are mainly used in interaction *with humans,* hence much can be learned what is known about explanatory patterns derived from other sciences [49].

In the preceding, we have outlined along conceptual lines two strategies to obtain AI systems that are ethically aligned with humans and explainable. Both approaches are ways to approach the construction of modern AI systems that could be employed in legal domains in a transparent way. Such work is only yet starting and progress needs to be made by answering a couple of questions in the process of designing and building: (i) What exactly *is* an explanation in the legal context? (ii) What are specific legal requirements and criteria for explainable AI systems? (iii) How to map legal criteria and functional criteria to each other for explainable AI? (iv) How to abstract functional requirements for legal AI into (legal) design patterns? (v) How to formalize existing codes of ethics, legal procedures, legal design patterns and legal practices into AI programs? (vi) How can AIs interactively learn and explain their functioning in *human-understandable* terms?

A big focus, especially in legal domains, should be *user-centric* explanations and empirical studies into their effectiveness [80]. In legal domains progress is made (e. g., [17]), but many remaining challenges await [14,84] before fully explainable legal AI systems can take over various tasks. Above all, AI and legal specialists need to engage in a dialogue when tackling explainability questions in legal AI. AI systems need to be explainable in order to be *trustworthy* [30] but AI researchers need to know *which* (types of) explanations are most useful to legal scholars. Much work awaits in this exciting field.

References

1. Abel, D., MacGlashan, J., Littman, M.L.: Reinforcement learning as a framework for ethical decision making. In: AAAI Workshop: AI, Ethics, and Society, vol. 16, p. 02. Phoenix (2016)
2. Alshiekh, M., Bloem, R., Ehlers, R., Könighofer, B., Niekum, S., Topcu, U.: Safe reinforcement learning via shielding. In: Proceedings AAAI Conference on Artificial Intelligence, AAAI, Palo Alto (2018)
3. Anderson, M., Anderson, S.: Machine ethics: creating an ethical intelligent agent. AI Mag. **28**, 15–26 (2007)
4. Arrieta, A.B., et al.: Explainable artificial intelligence (XAI): concepts, taxonomies, opportunities and challenges toward responsible AI. Inf. Fus. **58**, 82–115 (2020)
5. Ashley, K.D.: Artificial intelligence and legal analytics: new tools for law practice in the digital age. Cambridge University Press (2017)

6. Atzmueller, M.: Onto explicative data mining: exploratory, interpretable and explainable analysis. In: Proceedings Dutch-Belgian Database Day, TU Eindhoven (2017)
7. Atzmueller, M., et al.: Connect-U: a system for enhancing social networking. In: David, K., et al. (eds.) Socio-technical Design of Ubiquitous Computing Systems. Springer, Heidelberg (2014)
8. Atzmueller, M., Hayat, N., Trojahn, M., Kroll, D.: Explicative human activity recognition using adaptive association rule-based classification. In: Proceedings IEEE International Conference on Future IoT Technologies, IEEE, Boston (2018)
9. Atzmueller, M., Roth-Berghofer, T.: The mining and analysis continuum of explaining uncovered. In: Bramer M., Petridis M., Hopgood A. (eds.) Research and Development in Intelligent Systems XXVII, pp. 273–278. Springer, Heidelberg (2011)
10. Atzmueller, M.: Declarative aspects in explicative data mining for computational sensemaking. In: Seipel, D., Hanus, M., Abreu, S. (eds.) Proceedings International Conference on Declarative Programming, pp. 97–114. Springer, Heidelberg (2018)
11. Atzmueller, M.: Towards socio-technical design of explicative systems: transparent, interpretable and explainable analytics and its perspectives in social interaction contexts. In: Proceedings Workshop on Affective Computing and Context Awareness in Ambient Intelligence (AfCAI), UPCT, Cartagena (2019)
12. Atzmueller, M., Bloemheuvel, S., Kolepper, B.: A framework for human-centered exploration of complex event log graphs. In: Proceedings International Conference on Discovery Science (DS 2019), Springer, Berlin (2019)
13. Awad, E., et al.: The moral machine experiment. Nature **563**(7729), 59–64 (2018)
14. Bibal, A., Lognoul, M., de Streel, A., Frénay, B.: Legal requirements on explainability in machine learning. Artif. Intell. Law **29**(2), 149–169 (2020). https://doi.org/10.1007/s10506-020-09270-4
15. Biran, O., Cotton, C.: Explanation and justification in machine learning: a survey. In: IJCAI-17 Workshop on Explainable AI (2017)
16. Bloemheuvel, S., Kloepper, B., Atzmueller, M.: Graph summarization for computational sensemaking on complex industrial event logs. In: Proceedings Workshop on Methods for Interpretation of Industrial Event Logs, International Conference on Business Process Management, Vienna (2019)
17. Branting, L.K., et al.: Scalable and explainable legal prediction. Artif. Intell. Law **29**(2), 213–238 (2020). https://doi.org/10.1007/s10506-020-09273-1
18. Camacho, A., Icarte, R.T., Klassen, T.Q., Valenzano, R.A., McIlraith, S.A.: Ltl and beyond: Formal languages for reward function specification in reinforcement learning. In: IJCAI, vol. 19, pp. 6065–6073 (2019)
19. Cassens, J., Kofod-Petersen, A.: Designing explanation aware systems: the quest for explanation patterns. In: ExaCt, pp. 20–27 (2007)
20. Chalkidis, I., Kampas, D.: Deep learning in law: early adaptation and legal word embeddings trained on large corpora. Artif. Intell. Law **27**(2), 171–198 (2019)
21. Comes, D.E., et al.: Designing socio-technical applications for ubiquitous computing. In: Göschka, K.M., Haridi, S. (eds.) DAIS 2012. LNCS, vol. 7272, pp. 194–201. Springer, Heidelberg (2012). https://doi.org/10.1007/978-3-642-30823-9_17
22. Dignum, V., et al.: Ethics by design: necessity or curse? In: Proceedings of the 2018 AAAI/ACM conference on AI, ethics, and society, pp. 60–66 (2018)
23. Dubber, M.D., Pasquale, F., Das, S.: The Oxford Handbook of Ethics of AI. Oxford University Press, New York (2020)
24. Duivesteijn, W., Thaele, J.: Understanding where your classifier does (not) work - the SCaPE model class for EMM. In: Proceedings ICDM, pp. 809–814. IEEE (2014)
25. Filip, J., Kliegr, T.: Pyids-python implementation of interpretable decision sets algorithm by lakkaraju et al, 2016. In: RuleML+ RR (Supplement) (2019)

26. Friedman, B., Hendry, D.G.: Value Sensitive Design: Shaping Technology with Moral Imagination. MIT Press, Cambridge (2019)

27. Gamma, E., Helm, R., Johnson, R., Vlissides, J., Patterns, D.: Elements of Reusable Object-oriented Software. Addison-Wesley, Boston (1995)

28. Geihs, K., Leimeister, J., Roßnagel, A., Schmidt, L.: On socio-technical enablers for ubiquitous computing applications. In: Proceedings Workshop on Enablers for Ubiquitous Computing and Smart Services, pp. 405–408. IEEE, Izmir (2012)

29. Geihs, K., Niemczyk, S., Roßnagel, A., Witsch, A.: On the socially aware development of self-adaptive ubiquitous computing applications. IT-Inf. Technol. **56**(1), 33–41 (2014)

30. Nalepa, G.J., van Otterlo, M., Bobek, S., Atzmueller, M.: From context mediation to declarative values and explainability. In: Proceedings IJCAI Workshop on Explainable Artificial Intelligence (XAI), Stockholm (2018)

31. Goodall, N.J.: Machine ethics and automated vehicles. In: Meyer, G., Beiker, S. (eds.) Road Vehicle Automation. LNM, pp. 93–102. Springer, Cham (2014). https://doi.org/10.1007/978-3-319-05990-7_9

32. Guidotti, R., Monreale, A., Turini, F., Pedreschi, D., Giannotti, F.: A survey of methods for explaining black box models. arXiv preprint arXiv:1802.01933 (2018)

33. Hammer, V., Pordesch, U., Roßnagel, A.: Betriebliche Telefon- und ISDN-Anlagen rechtsgemäß gestaltet. Edition SEL-Stiftung, Springer, Verlag (1993)

34. Henelius, A., Puolamäki, K., Ukkonen, A.: Interpreting classifiers through attribute interactions in datasets. In: Proceedings ICML Workshop on Human Interpretability in Machine Learning, Sydney (2017)

35. Iyer, R., Li, Y., Li, H., Lewis, M., Sundar, R., Sycara, K.: Transparency and explanation in deep reinforcement learning neural networks. In: Proceedings of the 2018 AAAI/ACM Conference on AI, Ethics, and Society. pp. 144–150 (2018)

36. Kasenberg, D., Arnold, T., Scheutz, M.: Norms, rewards, and the intentional stance: Comparing machine learning approaches to ethical training. In: Proceedings of the 2018 AAAI/ACM Conference on AI, Ethics, and Society. pp. 184–190 (2018)

37. Kasenberg, D., Thielstrom, R., Scheutz, M.: Generating explanations for temporal logic planner decisions. In: Proceedings International Conference on Automated Planning and Scheduling, vol. 30, pp. 449–458 (2020)

38. Lakkaraju, H., Bach, S.H., Leskovec, J.: Interpretable decision sets: a joint framework for description and prediction. In: Proceedings ACM SIGKDD International Conference on Knowledge Discovery and Data Mining, pp. 1675–1684 (2016)

39. Langer, M., et al.: What do we want from explainable artificial intelligence (xai)?-a stakeholder perspective on xai and a conceptual model guiding interdisciplinary xai research. Artif. Intell. **296**, 103473 (2021)

40. Li, W., Han, J., Pei, J.: Cmar: Accurate and efficient classification based on multiple class-association rules. In: Proceedings IEEE International Conference on Data Mining, pp. 369–376. IEEE (2001)

41. Liao, S.M.: Ethics of Artificial Intelligence. Oxford University Press (2020)

42. Lin, P., Abney, K., Jenkins, R.: Robot ethics 2.0: From autonomous cars to artificial intelligence. Oxford University Press (2017)

43. Lippi, M., et al.: Claudette: an automated detector of potentially unfair clauses in online terms of service. Artif. Intell. Law **27**(2), 117–139 (2019)

44. Lonjarret, C., Robardet, C., Plantevit, M., Auburtin, R., Atzmueller, M.: Why should i trust this item? explaining the recommendations of any model. In: Proceedings IEEE International Conference on Data Science and Advanced Analytics (DSAA), pp. 526–535. IEEE (2020)

45. Loreggia, A., Mattei, N., Rossi, F., Venable, K.B.: Modeling and reasoning with preferences and ethical priorities in AI systems. In: Ethics of Artificial Intelligence, p. 127 (2020)

46. Mandel, D.R.: Counterfactual and causal explanation: from early theoretical views to new frontiers. In: The Psychology of Counterfactual Thinking, pp. 23–39. Routledge (2007)

47. Marcus, G., Davis, E.: Rebooting AI: Building Artificial Intelligence We Can Trust. Vintage, New York (2019)

48. McLaren, B.: Computational Models of Ethical Reasoning: Challenges, Initial Steps, and Future Directions. pp. 297–315, Machine ethics, Cambridge (2011)

49. Miller, T.: Explanation in artificial intelligence: insights from the social sciences. Artif. Intell. **267**, 1–38 (2019)

50. Mittelstadt, B., Allo, P., Taddeo, M., Wachter, S., Floridi, L.: The ethics of algorithms: mapping the debate. Big Data Soc. **3**(2) (2016)

51. Mollenhauer, D., Atzmueller, M.: Sequential exceptional pattern discovery using pattern-growth: An extensible framework for interpretable machine learning on sequential data. In: Atzmüller, M., Kliegr, T., Schmid, U. (eds.) Proceedings of the First International Workshop on Explainable and Interpretable Machine Learning (XI-ML 2020) co-located with the 43rd German Conference on Artificial Intelligence (KI 2020), Bamberg, September 21, 2020 (Virtual Workshop). CEUR Workshop Proceedings, vol. 2796. CEUR-WS.org (2020)

52. Naiseh, M.: Explainability design patterns in clinical decision support systems. In: Dalpiaz, F., Zdravkovic, J., Loucopoulos, P. (eds.) RCIS 2020. LNBIP, vol. 385, pp. 613–620. Springer, Cham (2020). https://doi.org/10.1007/978-3-030-50316-1_45

53. Petersen, S.: Machines learning values. In: Ethics of Artificial Intelligence, p. 413 (2020)

54. Puiutta, E., Veith, E.M.: Explainable reinforcement learning: a survey. In: Holzinger, A., Kieseberg, P., Tjoa, A.M., Weippl, E. (eds.) CD-MAKE 2020. LNCS, vol. 12279, pp. 77–95. Springer, Cham (2020). https://doi.org/10.1007/978-3-030-57321-8_5

55. Railton, P.: Ethical learning, natural and artificial. In: Ethics of Artificial Intelligence, p. 45 (2020)

56. Ribeiro, M.T., Singh, S., Guestrin, C.: Anchors: high-precision model-agnostic explanations. In: Proceedings of the AAAI Conference on Artificial Intelligence, AAAI, Palo Alto (2018)

57. Rising, L.: Design patterns: elements of reusable architectures. In: The Patterns Handbook: Techniques, Strategies and Applications, pp. 9–13 (1998)

58. Roßnagel, A., Jandt, S., Geihs, K.: Socially compatible technology design. In: David, K., et al. (eds.) Socio-technical Design of Ubiquitous Computing Systems, pp. 175–190. Springer, Cham (2014). https://doi.org/10.1007/978-3-319-05044-7_10

59. Roßnagel, A., Hammer, V.: KORA. Eine Methode zur Konkretisierung rechtlicher Anforderungen zu technischen Gestaltungsvorschlägen für Informations- und Kommunikationssysteme. Infotech 1, 21 ff. (1993)

60. Russell, S.: Human compatible: Artificial intelligence and the problem of control. Penguin (2019)

61. Schramowski, P., Turan, C., Jentzsch, S., Rothkopf, C., Kersting, K.: The moral choice machine. Front. Artif. Intell. **3**, 36 (2020)

62. Shin, D.: The effects of explainability and causability on perception, trust, and acceptance: implications for explainable AI. Int. J. Hum. Comput. Stud. **146**, 102551 (2021)

63. Simić-Draws, D., et al.: Holistic and law compatible it security evaluation: Integration of common criteria, ISO 27001/it-grundschutz and kora. In: Transportation Systems and Engineering: Concepts, Methodologies, Tools, and Applications, pp. 927–946. IGI Global (2015)

64. Spindler, M., Booz, S., Gieseler, H., Runschke, S., Wydra, S., Zinsmaier, J.: How to achieve integration? In: Das geteilte Ganze, pp. 213–239. Springer (2020)

65. Surden, H.: The ethics of artificial intelligence in law: basic questions. Forthcoming chapter in Oxford Handbook of Ethics of AI, pp. 19–29 (2020)

66. Sütfeld, L.R., Gast, R., König, P., Pipa, G.: Using virtual reality to assess ethical decisions in road traffic scenarios: applicability of value-of-life-based models and influences of time pressure. Front. Behav. Neurosci. **11**, 122 (2017)

67. Taylor, J., Yudkowsky, E., LaVictoire, P., Critch, A.: Alignment for advanced machine learning systems mIRI (unpublished) (2017). https://intelligence.org/2016/07/27/alignment-machine-learning/
68. Tolmeijer, S., Kneer, M., Sarasua, C., Christen, M., Bernstein, A.: Implementations in machine ethics: a survey. ACM Comput. Surv. (CSUR) 53(6), 1–38 (2020)
69. Tolomei, G., Silvestri, F., Haines, A., Lalmas, M.: Interpretable predictions of tree-based ensembles via actionable feature tweaking. In: Proceedings KDD, ACM (2017)
70. Turner, J.: Robot Rules: Regulating Artificial Intelligence. Springer, Heidelberg (2018)
71. Vallati, M., McCluskey, L.: In defence of design patterns for AI planning knowledge models. In: CEUR Workshop Proceedings, vol. 2745 (2020)
72. Vamplew, P., Dazeley, R., Foale, C., Firmin, S., Mummery, J.: Human-aligned artificial intelligence is a multiobjective problem. Ethics Inf. Technol. 20(1), 27–40 (2018)
73. van Otterlo, M.: Intensional Dynamic Programming: A Rosetta Stone for Structured Dynamic Programming. Journal of Algorithms 64, 169–191 (2009)
74. van Otterlo, M.: Solving Relational and First-Order Markov Decision Processes: A Survey. In: Wiering, M., van Otterlo, M. (eds.) Reinforcement Learning: State-of-the-art, chap. 8, pp. 253–292. Springer, Cham (2012)
75. van Otterlo, M.: Ethics and the value(s) of artificial intelligence. Nieuw Archief voor Wiskunde, 5(19), 206–209 (2018)
76. van Otterlo, M.: From algorithmic black boxes to adaptive white boxes: declarative decision-theoretic ethical programs as codes of ethics. In: Proceedings AAAI/ACM Conference on Artificial Intelligence, Ethics, and Society. ACM, New York (2018)
77. van Otterlo, M.: Gatekeeping Algorithms with Human Ethical Bias: The Ethics of Algorithms in Archives, Libraries and Society (2018). https://arxiv.org/abs/1801.01705
78. van Otterlo, M., Atzmueller, M.: On Requirements and Design Criteria for Explainability in Legal AI. In: Proceedings Workshop on Explainable AI in Law (XAILA), co-located with 31st International Conference on Legal Knowledge and Information Systems (JURIX), CEUR-WS (2018)
79. Wachter, S., Mittelstadt, B., Russell, C.: Counterfactual explanations without opening the black box: automated decisions and the GDPR. Harv. JL & Tech. 31, 841 (2017)
80. Wang, D., Yang, Q., Abdul, A., Lim, B.Y.: Designing theory-driven user-centric explainable AI. In: Proceedings of the 2019 CHI conference on human factors in computing systems, pp. 1–15 (2019)
81. Wick, M.R., Thompson, W.B.: Reconstructive Expert System Explanation. Artificial Intelligence 54(1–2), 33–70 (1992)
82. Wiering, M., Van Otterlo, M.: Reinforcement Learning: State-of-the-Art, Adaptation, Learning, and Optimization, vol. 12. Springer, Cham (2012)
83. Wolfram, S.: Computational law, symbolic discourse, and the AI constitution. In: Ethics of Artificial Intelligence, p. 155 (2020)
84. Yu, R., Alì, G.S.: What's inside the black box? AI challenges for lawyers and researchers. Legal Inf. Manag. 19(1), 2–13 (2019)

Towards Grad-CAM Based Explainability in a Legal Text Processing Pipeline. Extended Version

Łukasz Górski[1](✉) [iD], Shashishekar Ramakrishna[2,3] [iD], and Jędrzej M. Nowosielski[1] [iD]

[1] Interdisciplinary Centre for Mathematical and Computational Modelling, University of Warsaw, Warsaw, Poland
lgorski@icm.edu.pl
[2] Freie Universität Berlin, Berlin, Germany
[3] EY - AI Labs, Bangalore, India

Abstract. Explainable AI (XAI) is a domain focused on providing interpretability and explainability of a decision-making process. In the domain of law, in addition to system and data transparency, it also requires the (legal-) decision-model transparency and the ability to understand the model's inner working when arriving at the decision. This paper provides the first approaches to using a popular image processing technique, Grad-CAM, to showcase the explainability concept for legal texts. With the help of adapted Grad-CAM metrics, we show the interplay between the choice of embeddings, its consideration of contextual information, and their effect on downstream processing.

Keywords: Legal knowledge representation · Language models · Grad-CAM · HeatMaps · CNN

1 Introduction

Advancements in the domain of AI and Law have brought additional considerations regarding models development, deployment, updating and their interpretability. This can be seen with the advent of machine-learning-based methods, which naturally exhibit a lower degree of explainability than traditional knowledge-based systems. Yet, knowledge representation frameworks that handle legal information, irrespective of their origin, should cover the pragmatics or context around a given concept and this functionality should be easily demonstrable.

Explainable AI (XAI), is a domain which has focused on providing interpretability and explainability to a decision making process. In the domain of law, interpretability and explainability are more than dealing with information/data transparency or system transparency [1] (henceforth referred to as *ontological view*). It additionally requires the (legal-) decision-model transparency, the ability to understand the model's inner working when arriving at the decision (*epistemic view*). In this paper, we aim to present the system's user and architect

© Springer Nature Switzerland AG 2021
V. Rodríguez-Doncel et al. (Eds.): AICOL-XI 2018/AICOL-XII 2020/XAILA 2020, LNAI 13048, pp. 154–168, 2021.
https://doi.org/10.1007/978-3-030-89811-3_11

with a set of tools that facilitate the discovery of inputs that contribute to convolutional neural network's (CNN's) output to the greatest degree, by adapting the Grad-CAM method, which originated from the field of computer vision. We adapt this method to the legal domain and show how it can be used to achieve a better understanding of a given system's state and explain how different embeddings contribute to end result as well as to optimize this system's inner workings. While this work is concerned with the ontological perspective, we aim this as a stepping stone for another related perspective, where the legally-based positions are connected with explanation thus providing the ability to explain the decisions to its addressee. This paper addresses mainly the technical aspects, showing how Grad-CAMs can be applied to the legal texts, describing the text processing pipeline - taking this as a departing point for deeper analyses in future work. We aim to present this technical implementation as well as the quantitative comparison metrics as the main contribution of the paper.

The paper is structured as follows. State-of-the-art is described in Sect. 2. Section 3 describes the methodology, which includes the metrics used for results quantification. The architecture used for experiments is described in Sect. 4. Section 5 talks about the different datasets used and the experimental setup. The outcomes are described in Sect. 6. Finally, Sect. 7 provides a conclusion and future work. This paper extends our previous one [2], presented at XAILA'20 workshops during Jurix'20, by including survey-based user study and offering a more pronounced presentation of results.

2   Related Work

The feasibility of using different - contextual (e.g. BERT) and non-contextual (e.g. word2vec) - embeddings was already studied outside the domain of law. In [3], it was found that the usage of more sophisticated, context-aware methods is unnecessary in the domains where labelled data and simple language are present. As far as the area of law is concerned, the feasibility of using the domain-specific vs. general embeddings (based on word2vec) for the representation of Japanese legal texts was investigated, with the conclusion that general embeddings have an upper hand [4]. The feasibility of using BERT in the domain of law was also already put under scrutiny as well. In [5] its generic pretrained version was used for embeddings generation and it was found that large computational requirements may be a limiting factor for domain-specific embedding creation. The same paper concluded that the performance of the generic version is lower when compared with law-based non-contextual embeddings. On the other hand, in [6], BERT versions trained on legal judgments corpus (of 18000 documents) were used and it was found that training on in-domain corpus does not necessarily offer better performance compared to generic embeddings. In [7] contradictory conclusions were reached: the system's performance significantly improves when using pre-trained BERT on a legal corpus. Those results suggest that introduction of XAI-based methods might be a *condition sine qua non* for a proper understanding of general language embeddings and their feasibility in the domain.

Grad-CAMs are explainability method originating from computer vision [8]. It is a well established post-hoc explainability technique when CNNs are concerned. Moreover, Grad-CAM method passed independent sanity checks [9]. Whilst it is mainly connected with the explanations of deep learning networks used with image data, it has already been adapted for other areas of application. In particular, CNN architecture for text classification was described in [10], and there exists at least one implementation which extends this work with Grad-CAM support for explainability [11]. Grad-CAMs were already used in the NLP domain, for (non-legal) document retrieval [12]. Herein we build upon this work and investigate the feasibility of using this method for the legal domain, in particular allowing for the visualisation of context-dependency of various word embeddings. Legal language is a special register of everyday language and deservers investigation on its own. The evolution of legal vocabulary can be precisely traced to particular statutes and precedential judgments, where it is refined and its boundaries are tested [13]. Many terms have thus a particular legal meaning and efficacy and tools that can safeguard final black-box models' adherence to the particularities of legal language are valuable.

The endeavours aimed at using XAI methods in the legal domain, similar to this paper, have already been undertaken recently. In [14] an Attention Network was used for legal decision prediction - coupling it with attention-weight-based text highlighting of salient case text (though this approach was found to be lacking). The possibility of explaining the BERT's inner workings was already investigated by other authors, and it was already subject to static as well as dynamic analyses. An interactive tool for the visualisation of its learning process was implemented in [15]. Machine-learning-based evaluation of context importance was performed in [16]; therein it was found that accounting for the content of a sentence's context greatly improves the performance of legal information retrieval system.

However, the results mentioned hereinbefore do not allow for direct and easily interpretable comparison of different types of embeddings and we aim to explore an easy plug-in solution facilitating this aim.

3   Methodology

We study the interplay between the choice of embeddings, its consideration of contextual information, and its effect on downstream processing. For this work, a pipeline for comparison was prepared, with the main module being the embedder, classification CNN and metric-based evaluator. All the parts are easily pluggable, allowing for extendibility and further testing of a different combinations of modules.

The CNN used in the pipeline was trained for classification. We use two different datasets for CNN training (as well as testing)[1]:

1. The Post-Traumatic Stress Disorder (PTSD) [17] dataset [18], where rhetorical roles of sentences are classified.

---

[1] Section 5.1, provides a detailed discussion on the considered datasets.

2. Statutory Interpretation - Identifying Particular (SIIP) dataset [16], where the sentences are classified into four categories according to their usefulness for a legal provision's interpretation.

Whilst many methods have already been used for the analysis of aforementioned datasets (including regular expressions, Naive Bayes, Logistic Regression, SVMs [18], or Bi-LSTMs [19]), we are unaware of papers that use (explainable) CNNs for this tasks. On the other hand, usage of said CNN should not be treated as the main contribution of this paper, as the classification network is treated only as an exemplary application, warranting conclusions regarding the paper's main contribution, i.e. the context-awareness of various embeddings when used in the legal domain.

Further down the line, the embeddings are used to transform CNN input sentences into vectors, with vector representation for each word in a sentence concatenated. Herein our implementation is based on the prior work [10,11].

3.1 Comparison Metrics

Grad-CAM heatmaps are inherently visual tools for data analysis. In computer vision, they are commonly used for qualitative determination of input image regions that contribute to the final prediction of the CNN. While they are an attractive tool for a qualitative analysis of a single entity, they should be supplemented with other tools for easy comparison of multiple embeddings [20] and to facilitate quantitative analysis. Herein the following metrics are introduced and adapted to the legal domain:

1. Fraction of elements above relative threshold t ($\mathcal{F}(v,t)$)
2. Intersection over union with relative thresholds t_1 and t_2 ($\mathcal{I}(v_1, v_2, t_1, t_2)$)

The first metric, $\mathcal{F}(t)$, is designed to measure the CNN network attention spread over words present in the given input, i.e., what portion of the input is taken into account by CNN in the case of a particular prediction. It is defined as a number of elements in a vector that are larger than the relative threshold t multiplied by the maximum vector value divided by the length of this vector.

The second metric, $\mathcal{I}(v_1, v_2, t_1, t_2)$, helps to compare two predictions of two different models given the same input sentence. It answers the question of whether two models, when given the same input sentence, 'pay attention' to the same or different chunk(s) of the input sentence. It takes as arguments two Grad-CAM heatmaps (v_1 and v_2), binarizes them using relative thresholds (t_1 and t_2) and finally calculates standard intersection over union. It quantifies the relative overlap of words considered important for the prediction by each of two models.

4 System Architecture

The architecture, as shown in Fig. 1, is designed to implement the methodology described in Sect. 3 and comprises four main modules, i.e.: preprocessing module, embedding module, classification module and visualization module. The pre-processing module uses some industry *de facto* standard text processing libraries for spelling correction, sentence detection, irregular character removal, etc., enhanced with our own implementations which make them better-suited for legal texts. The embedding module houses a plug-in system to handle different variants of embeddings, in particular BERT and word2vec. The classification module houses simple 1D CNN which facilitates explainability method common in computer vision i.e. Grad-CAM. The visualization module is used for heatmap generation and metric computation.

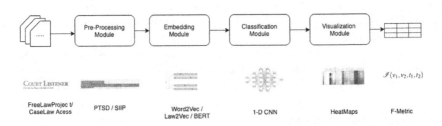

Fig. 1. System architecture

The output from the pre-processing module is fed into the embeddings module. The embeddings used are based on variants of BERT and word2vec. In addition to the pre-trained ones, raw data from CourtListener [21] dataset was used for training embeddings creation.

Within the frame of the classification module, the output from the embeddings module is fed into a 1D convolutional layer followed by an average pooling layer and fully-connected layers with dropout and softmax [10]. Although CNN architectures stem from computer vision where an image forms the input of the network, the use of CNN for the sequence of word vectors as an input is reasonable. In a sentence relative positions of words convey meaning. It is similar to an image where relative positions of pixels convey information, with the difference being about dimensionality. Standard image is 2D while a sentence is a 1D sequence of words, therefore we use the 1D CNN for the task of sentence classification.

With Grad-CAM technique it is possible to produce a class activation map (heatmap) for a given input sentence and predicted class. Each element of the class activation map corresponds to one token and indicates its importance in terms of the score of the particular (usually the predicted) class. The class activation map gives information on how strongly the particular tokens present in the input sentence influence the prediction of the CNN.

The software stack used for the development of this system was instrumented under Anaconda 4.8.3 (with Python 3.8.3). Tensorflow v. 2.2.0 was used for CNN instrumentation and Grad-CAMs calculations (with the code itself expanding prior implementation available at [11]). Spacy 2.1.8 and blackstone 0.1.15 were used for CourtListener text cleaning. Various BERT implementations and supporting codes were sourced from Huggingface libraries: transformers v. 3.1.0, tokenizers v. 0.8.1rc2, nlp v. 0.4.0. Two computing systems available at ICM University of Warsaw were exploited for the experiments. Text cleaning was performed using the okeanos system (Cray XC40) and main calculations were run on rysy GPU cluster (4x Nvidia Tesla V100 32 GB GPUs).

5 Experiments

5.1 Datasets

As stated in Sect. 3, we use two different datasets for experiments. The PTSD dataset is from the U.S. Board of Veterans' Appeals (BVA) from 2013 through 2017. The dataset deals with the decisions from adjudicated disability claims by veterans for service-related post-traumatic stress disorder (PTSD) [17]. The dataset itself is well-known and has already been studied by other authors. It annotates a set of sentences originating from 50 decisions issued by the Board according to their function in the decision [18,22,23]. The classification consists of six elements: *Finding Sentence, Evidence Sentence, Reasoning Sentence, Legal-Rule Sentence, Citation Sentence, Other Sentence.*

The SIIP dataset pertains to the United States Code 5 § 552a(a)(4) provision and aims to annotate the judgments that are most useful for interpretation of said provision. The seed information for annotation is collected from the court decisions retrieved from the Caselaw access project data. The sentences are classified into four categories according to their usefulness for the interpretation: *High Value, Certain Value, Potential Value, No Value* [16].

5.2 Embeddings/Language Modeling

We use pre-trained models as well as we train domain-specific models for the purpose of vector representation of texts. Many flavours of word2vec and BERT embedders were tested. The paper does not go into any details on the comparison of these pre-trained models (or other similar models) based on performance. This has been addressed in several other papers [15,24,25].

For the word2vec a (slimmed down) GoogleNews model was used, with a vocabulary of 300000 words [26]. In addition, Law2vec embeddings were also employed, which were trained on a large freely-available legal corpus, with 200 dimensions [27]. For BERT, bert-base-uncased model was used, a transformer model consisting of 12 layers, 768 hidden units, 12 attention heads and 110M parameters. In addition to that, a slimmed-down version of BERT, DistilBERT was also tried, due to its accuracy being on the par with vanilla BERT, yet offering better performance and smaller memory footprint.

In addition to pretrained models, we have also tried training our own word2vec and BERT models. For this aim, a CourtListener [21] database was sourced. However, due to the large computational requirements of BERT training, a small subset of this dataset was chosen, consisting of 180MiB of judgments. Moreover, while several legal projects provide access to a vast database of US case-laws, it was found that the judgments available therein need to be further processed, as the available textual representations usually contain unnecessary elements, such as page numbers or underscores, that hinder their machine processing. Our hand-written parser joined hyphenated words, removed page numbers and artifacts that were probably introduced by OCR-ing; furthermore, the text was split into sentences using spacy-based blackstone-parser. In line with other authors [28], we have found it to be imperfect and failing in segmenting the sentences that contained period-delimited legal abbreviations (e.g. *Fed.* - Federal). Thus it was supplemented with our own manually-curated list of abbreviations. The training was performed using DistilBERT model (for ca. 36 h), as well as word2vec in two flavours, 200-dimensional (in line with the dimensionality of Law2Vec) and 768-dimensional (in line with BERT embeddings dimensionality).

As far as the BERT-based embeddings go, there is a number of ways in which they can be extracted from the model. One of the ways is taking embeddings for special *CLS* token, which prefixes any sentence fed into BERT; another technique that was studied in the literature amounted to concatenating the model's final layer's values. The optimal technique is dependent on the task and the domain. Herein we have found the latter to offer better accuracy for downstream CNN training. The features for CNN processing consisted of tokenized sentences, together with embeddings for special BERT tokens (their absence would cause a slight drop in accuracy as well).

6   Results

6.1   Metric-Based Heatmap Comparison

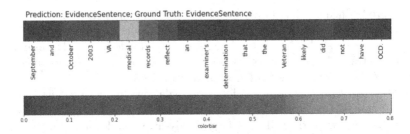

Fig. 2. A sample heatmap for correct prediction with word2vec (CourtListener, 768d) embedding

A sample heatmap can be referenced in Fig. 2 and Fig. 3, with a colorbar defining the mapping between the colors and values. Figure 2 clearly shows the area of CNN's attention, which can be quantified further down the line. This picture shows a properly classified sentence, a statement of evidence, defined by the PTSD dataset's authors as a description of a piece of evidence. CNN pays most attention to the phrase "medical records", which is in line with PTSD's authors'

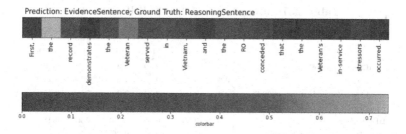

Fig. 3. A sample heatmap for failed prediction with word2vec (CourtListener, 768d) embedding

annotation protocols, where this kind of sentence describes a given piece of evidence (e.g. the records of testimony). We have found the sentence in Fig. 3 to be hard to classify for ourselves and it *prima facie* seemed for us to be an example of evidentiary sentence. In the case of CNN, no distinctive activations can be spotted (Table 4).

Table 1. Heatmap metrics for the PTSD dataset

	$\mathcal{F}(0.15)$		$\mathcal{F}(0.3)$		$\mathcal{F}(0.5)$	
	Mean	StdDev	Mean	StdDev	Mean	StdDev
Word2vec (pre-trained)	0.53	0.31	0.44	0.3	0.35	0.29
Law2vec	0.6	0.3	0.52	0.32	0.42	0.33
Word2vec (courtlistener, 200d)	0.49	0.28	0.39	0.27	0.29	0.26
Word2vec (courtlistener, 768d)	0.48	0.28	0.38	0.28	0.29	0.27
BERT (bert-base-uncased)	0.48	0.32	0.36	0.28	0.24	0.22
DistillBert (bert-base-uncased)	0.67	0.27	0.56	0.27	0.38	0.24
DistillBERT (courtlistener)	0.47	0.39	0.47	0.39	0.44	0.39

Yet, we did not perform any detailed analyses of such images. Instead, we focus on two types of comparison using metrics defined in Sect. 3.1. The comparisons are designed to capture differences between different embeddings, particularly in terms of context handling. First, for a given embedding we calculate CNN network attention spread over words quantified by metric $\mathcal{F}(t)$ averaged over all input sentences contained in the test set. Then we can compare the mean fraction of words (tokens) in the input sentences which contribute to prediction in the case of various embeddings. Criterion deciding if a particular word contributes to the prediction is, in fact, arbitrary and depends on class activation map (heatmap) binarization threshold. This is why we test a few thresholds, including 0.15 as suggested in [8] for weakly supervised localization. Essentially high value of the fraction $\mathcal{F}(t)$ indicates that most word vectors in input sentence are taken into account by CNN during inference. Conversely, the low value

Table 2. Heatmap metric $\mathcal{F}$ for the SIIP dataset

	$\mathcal{F}(0.15)$		$\mathcal{F}(0.3)$		$\mathcal{F}(0.5)$	
	Mean	StdDev	Mean	StdDev	Mean	StdDev
word2vec (GoogleNews)	0.51	0.23	0.39	0.2	0.26	0.16
Law2vec	0.5	0.25	0.41	0.24	0.27	0.2
word2vec (CourtListener, 200d)	0.47	0.24	0.34	0.2	0.2	0.15
word2vec (CourtListener, 768d)	0.45	0.2	0.31	0.17	0.19	0.11
BERT (bert-base-uncased)	0.5	0.26	0.39	0.23	0.27	0.18
DistilBert (distilbert-base-uncased)	0.61	0.26	0.51	0.26	0.37	0.23
DistilBERT (CourtListener)	0.31	0.35	0.31	0.34	0.3	0.35

Table 3. Heatmap metric $\mathcal{I}$ for the selected pairs of embeddings for the PTSD dataset

	$\mathcal{I}(0.15)$		$\mathcal{I}(0.3)$		$\mathcal{I}(0.5)$	
	Mean	StdDev	Mean	StdDev	Mean	StdDev
word2vec (GoogleNews) – BERT (bert-base-uncased)	0.49	0.25	0.41	0.24	0.3	0.21
Law2vec – BERT (bert-base-uncased)	0.51	0.26	0.43	0.25	0.34	0.25
word2vec (CourtListener, 200d) – Law2Vec	0.65	0.25	0.58	0.27	0.51	0.31
word2vec (CourtListener, 768d) – DistilBert (distilbert-base-uncased)	0.44	0.23	0. 35	0.22	0.26	0.21

Table 4. Test set accuracy.

	PTSD	SIIP
word2vec (GoogleNews)	0.7	0.9
Law2vec	0.69	0.85
word2vec (CourtListener, 200d)	0.78	0.93
word2vec (CourtListener, 768d)	0.79	0.94
BERT (bert-base-uncased)	0.84	0.94
DistilBERT (distilbert-base-uncased)	0.85	0.94
DistilBERT (CourtListener)	0.42	0.85

of the fraction $\mathcal{F}(t)$ indicates that most word vectors in the input sentence are ignored by CNN during inference. The comparison results for the PTSD dataset are shown in Table 1 and Table 3. Table 2 shows the results in terms of $\mathcal{F}$ metric for SIIP dataset. The outstanding similarity between word2vec and Law2Vec can be spotted in Table 3, due to both of those models belonging to the same class, as exhibited by the high value of $\mathcal{I}$ metric. The accuracy results are presented in Table 4.

Table 5. User evaluation study results the Reasoning sentences. Heatmaps generated from respondents' answers are presented.

	$\mathcal{I}(0.15)$	$\mathcal{I}(0.3)$	$\mathcal{I}(0.5)$
Further, as discussed below, none of the medical evidence indicates that a psychiatric disorder had its onset during service, and psychiatric disorders are complex matters requiring medical evidence for diagnosis: they are not the kind of disorders that subject to lay observation.			
word2vec (GoogleNews)	0.62	0.27	0.21
Law2vec	0.37	0.24	0.15
word2vec (CourtListener, 200d)	0.21	0.13	0
word2vec (CourtListener, 768d)	0.46	0.29	0.13
BERT (bert-base-uncased)	0.3	0.29	0.3
DistilBert (distilbert-base-uncased)	0.43	0.34	0.5
DistilBERT (CourtListener)	0.052	0.08	0
Given the inconsistencies between the Veteran's reports and the objective evidence of record, the Veteran's credibility is diminished.			
word2vec (GoogleNews)	0.1	0.13	0.1
Law2vec	0.78	0.64	0.35
word2vec (CourtListener, 200d)	0.67	0.47	0.46
word2vec (CourtListener, 768d)	0.56	0.41	0.25
BERT (bert-base-uncased)	1	0.81	0.29
DistilBert (distilbert-base-uncased)	1	0.84	0.69
DistilBERT (CourtListener)	1	0.84	0.65

6.2   User-Based Study

In addition to the results presented hereinbefore, a user study was performed. This involved asking the legal professionals to identify the most significant words in a number of sentences originating from the PTSD dataset. The following categories of PTSD sentences were chosen: legal rule sentence (3 sentences), reasoning sentence (2 sentences), citation sentence (1 sentence). We have abstained from including other categories, as their classification is not always obvious. Only 1 citation sentence was chosen due to the similarity of all sentences in this class. In all the cases, the accuracy of classification was out of the scope of our study.

Six respondents participated in this part of this study. Each underlined the most important words in presented sentences. Subsequently, a binary vector was prepared. Each identified important word was mapped to one and those that were left out by respondent were treated as 0. After collecting individual answers from each of the participants, the results were summarized by preparing

Table 6. User evaluation study results for the Legal rule sentences. Heatmaps generated from respondents' answers are presented.

	$\mathcal{I}(0.15)$	$\mathcal{I}(0.3)$	$\mathcal{I}(0.5)$
Service connection for PTSD requires medical evidence diagnosing the condition in accordance with 38 C.F.R. 4.125(a); a link, established by medical evidence, between current symptoms and an in-service stressor, and credible supporting evidence that the in-service stressor occurred.			
word2vec (GoogleNews)	0.67	0.51	0.26
Law2vec	0.76	0.64	0.38
word2vec (CourtListener, 200d)	0.78	0.33	0.11
word2vec (CourtListener, 768d)	0.63	0.33	0.13
BERT (bert-base-uncased)	0.75	0.52	0.32
DistilBert (distilbert-base-uncased)	0.73	0.46	0.19
DistilBERT (CourtListener)	0.03	0.02	0
There must be 1) medical evidence diagnosing PTSD; 2) a link, established by medical evidence, between current symptoms of PTSD and an in-service stressor; and 3) credible supporting evidence that the claimed in-service stressor occurred.			
word2vec (GoogleNews)	0.67	0.24	0.14
Law2vec	0.57	0.3	0.15
word2vec (CourtListener, 200d)	0.07	0.05	0.06
word2vec (CourtListener, 768d)	0.17	0.05	0.06
BERT (bert-base-uncased)	0.81	0.49	0.31
DistilBert (distilbert-base-uncased)	0.58	0.38	0.3
DistilBERT (CourtListener)	0.05	0.06	0.03
The Federal Circuit has held that 38 U.S.C.A. 105 and 1110 preclude compensation for primary alcohol abuse disabilities and secondary disabilities that result from primary alcohol abuse.			
word2vec (GoogleNews)	0.4	0.22	0.09
Law2vec	0.57	0.35	0.22
word2vec (CourtListener, 200d)	0.07	0.05	0.05
word2vec (CourtListener, 768d)	0.14	0.05	0.05
BERT (bert-base-uncased)	0.85	0.58	0.34
DistillBert (distilbert-base-uncased)	0.59	0.38	0.34
DistilBERT (CourtListener)	0.06	0.07	0.04

ground truth vectors, one for each of the six sentences. Those ground truth vectors were obtained by averaging out respondents' vectors for each of the sentences separately. Subsequently, they were used to generate user assessment-based heatmaps. Those in-turn were compared with the results of our classifier using the metrics presented hereinbefore.

Tables 5, 6, 7 can be consulted for comparison results. In general BERT and Law2vec were found to conform to expert's opinion to the largest extent.

Table 7. User evaluation study results for the Citation sentences. Heatmaps generated from respondents' answers are presented.

	$\mathcal{I}(0.15)$	$\mathcal{I}(0.3)$	$\mathcal{I}(0.5)$
(heatmap: See also Mittleider v. West, 11 Vet. App. 181, 182 (1998) (in the absence of medical evidence that does so, VA is precluded from differentiating between symptomatology attributed to a nonservice-connected disability and symptomatology attributed to a service-connected disability).)			
word2vec (GoogleNews)	0.1	0.17	0.18
Law2vec	0.67	0.3	0
word2vec (CourtListener, 200d)	0.25	0.39	0.4
word2vec (CourtListener, 768d)	0.25	0.43	0.67
BERT (bert-base-uncased)	0.78	0.28	0.03
DistilBert (distilbert-base-uncased)	0.32	0.43	0.4
DistilBERT (CourtListener)	0.03	0.05	0

6.3 Grad-CAM Guided Context Extraction

The analysis of heatmaps and metrics presented hereinbefore proves that only a part of a given sentence contributes to a greater extent to final results. We have hypothesized that it is possible to decrease the amount of CNN's input data to those important parts without compromising the final prediction. In this respect, Grad-CAM was treated as a helpful heuristic that allows to identify the most important words for a given CNN in its training phase. For this experiment, the value of $\mathcal{F}$, for the threshold of 0.15 was used to select a percentage of the most important words from a given training example. This in turn was used to compose a vocabulary (or white-list) of the most important words that were encountered during the training. Further down the line, this white-list was used during the inference and only the words present on the list were passed as input to the CNN. Nevertheless, the number of white-listed words allowed coherent sentences to be still passed into CNN (for example, the PTSD sentence

However, this evidence does not make it clear and, before white-listing amounted to *However, this evidence does not make it clear and unmistakable.*).

We have managed to keep accuracy up to the bar of an unmodified dataset using this procedure (e.g. 0.7 for PTSD-word2vec (GoogleNews) and 0.85 for PTSD-DistilBERT (distilbert-base-uncased)).

7   Conclusion and Future Work

We presented the first approach to using a popular image processing technique, Grad-CAMs to showcase the explainability concept for legal texts. Few conclusions which we can be drawn from the presented methodology are:

– The mean value of $\mathcal{F}(t)$ is higher in the case of DistilBERT embedding than in the cases of word2vec and Law2vec embeddings. It suggests that CNN trained and utilised with this embedding tends to take into account a relatively larger chunk of input sentence while making prediction.
– Described metrics and visualizations provide a peek into the complexity of context handling aspects embedded in a language model.
– It enables an user to identify and catalog attention words in a sentence type for data optimization in downstream processing tasks.

Some issues which need further investigation are:

– Training of these domain-specific models requires time and resources. Apart from algorithmic optimization, data optimization also plays an important role. Extension of methodology presented herein can be used to remove tokens that do not contribute to the final outcome of any downstream processing tasks. A systematic analysis of the method presented in Sect. 6.3 is warranted.
– Mapping of metrics from our methodology to standard machine learning metrics could allow us to infer the quality of language models in a given domain (i.e. legal domain). This could allow to measure the quality of a model when there is not sufficient gold data which can be used for effective training of models (inline to the concept of semi-supervised learning).
– An extension of our approach could become a part of some argumentation systems. Lets consider the argumentation schemes proposed by Walton *et al.* [29] and Douglas [30], which deal with base premise identification, similarity premise and conclusion. Regarding the base premise identification, our method could be used for analysis of context-drift (a phenomenon in which a change in concept used induces more or less radical changes to context and thus changes the target concept inferred) [31]. Thus, minimizing such drift would help to build better arguments to start with. Regarding the similarity premise, which is based on concept semantic similarity technique, it would help in identifying instances where the source case argument is similar to the target case argument thereby helping in drawing a similar conclusion.

Acknowledgment. This research was carried out with the support of the Interdisciplinary Centre for Mathematical and Computational Modelling (ICM), University of Warsaw, under grant no GR81-14.

Disclaimer. The views reflected in this article are the views of the authors and do not necessarily reflect the views of the global EY organization or its member firms.

References

1. Bibal, A., Lognoul, M., de Streel, A., Frénay, B.: Legal requirements on explainability in machine learning. Artif. Intell. Law **29**(2), 149–169 (2020). https://doi.org/10.1007/s10506-020-09270-4
2. Gorski, L., Ramakrishna, S., Nowosielski, J.M.: Towards grad-cam based explainability in a legal text processing pipeline. arXiv preprint arXiv:2012.09603 (2020)
3. Arora, S., May, A., Zhang, J., Ré, C.: Contextual embeddings: when are they worth it? (2020)
4. Tang, L., Kageura, K.: An examination of the validity of general word embedding models for processing Japanese legal texts. In: Proceedings of the Third Workshop on Automated Semantic Analysis of Information in Legal Texts, Volume 2385 of CEUR Workshop Proceedings, Montreal, QC, Canada, 21 June 2019 (2019)
5. Condevaux, C., Harispe, S., Mussard, S., Zambrano, G.: Weakly supervised one-shot classification using recurrent neural networks with attention: application to claim acceptance detection. In: JURIX, pp. 23–32 (2019)
6. Rossi, J., Kanoulas, E.: Legal search in case law and statute law. In: JURIX, pp. 83–92 (2019)
7. Elwany, E., Moore, D., Oberoi, G.: BERT goes to law school: quantifying the competitive advantage of access to large legal corpora in contract understanding. arXiv preprint arXiv:1911.00473 (2019)
8. Selvaraju, R.R., Cogswell, M., Das, A., Vedantam, R., Parikh, D., Batra, D.: Grad-CAM: visual explanations from deep networks via gradient-based localization. Int. J. Comput. Vis. **128**(2), 336–359 (2019)
9. Adebayo, J., Gilmer, J., Muelly, M., Goodfellow, I., Hardt, M., Kim, B.: Sanity checks for saliency maps. In: Bengio, S., Wallach, H., Larochelle, H., Grauman, K., Cesa-Bianchi, N., Garnett, R. (eds.) Advances in Neural Information Processing Systems, vol. 31, pp. 9505–9515. Curran Associates Inc. (2018)
10. Kim, Y.: Convolutional neural networks for sentence classification. In: Proceedings of the 2014 Conference on Empirical Methods in Natural Language Processing (EMNLP), Doha, Qatar, October 2014, pp. 1746–1751. Association for Computational Linguistics (2014)
11. Grad-cam for text. https://github.com/HaebinShin/grad-cam-text. Accessed 05 Aug 2020
12. Choi, J., Choi, J., Rhee, W.: Interpreting neural ranking models using grad-CAM. arXiv preprint arXiv:2005.05768 (2020)
13. Rissland, E.L., Ashley, K.D., Loui, R.P.: AI and Law: a fruitful synergy. Artif. Intell. **150**(1–2), 1–15 (2003)
14. Branting, L.K., et al.: Scalable and explainable legal prediction. Artif. Intell. Law **29**(2), 213–238 (2020). https://doi.org/10.1007/s10506-020-09273-1
15. Hoover, B., Strobelt, H., Gehrmann, S.: exBERT: a visual analysis tool to explore learned representations in transformers models (2019)

16. Savelka, J., Xu, H., Ashley, K.D.: Improving sentence retrieval from case law for statutory interpretation. In: Proceedings of the Seventeenth International Conference on Artificial Intelligence and Law, pp. 113–122 (2019)
17. Moshiashwili, V.H.: The downfall of Auer deference: veterans law at the federal circuit in 2014 (2015)
18. Walker, V.R., Pillaipakkamnatt, K., Davidson, A.M., Linares, M., Pesce, D.J.: Automatic classification of rhetorical roles for sentences: comparing rule-based scripts with machine learning. In: Proceedings of the Third Workshop on Automated Semantic Analysis of Information in Legal Texts, Volume 2385 of CEUR Workshop Proceedings, Montreal, QC, Canada, 21 June 2019 (2019)
19. Ahmad, S.R., Harris, D., Sahibzada, I.: Understanding legal documents: classification of rhetorical role of sentences using deep learning and natural language processing. In: 2020 IEEE 14th International Conference on Semantic Computing (ICSC), pp. 464–467 (2020)
20. Krakov, D., Feitelson, D.G.: Comparing performance heatmaps. In: Desai, N., Cirne, W. (eds.) JSSPP 2013. LNCS, vol. 8429, pp. 42–61. Springer, Heidelberg (2014). https://doi.org/10.1007/978-3-662-43779-7_3
21. Free Law Project. Courtlistener (2020)
22. Walker, V.R., Han, J.H., Ni, X., Yoseda, K.: Semantic types for computational legal reasoning: propositional connectives and sentence roles in the veterans' claims dataset. In: ICAIL 2017, pp. 217–226. Association for Computing Machinery, New York (2017)
23. Savelka, J., Walker, V.R., Grabmair, M., Ashley, K.D.: Sentence boundary detection in adjudicatory decisions in the United States (2017)
24. Martin, L., et al.: CamemBERT: a tasty French language model. In: Proceedings of the 58th Annual Meeting of the Association for Computational Linguistics, pp. 7203–7219. Association for Computational Linguistics, July 2020
25. Sanh, V., Debut, L., Chaumond, J., Wolf, T.: DistilBERT, a distilled version of BERT: smaller, faster, cheaper and lighter (2019)
26. Word2vec-slim. https://github.com/eyaler/word2vec-slim. Accessed 21 Sept 2020
27. Law2vec: Legal word embeddings. https://archive.org/details/Law2Vec. Accessed 21 Sept 2020
28. Choi, E., Brassil, G., Keller, K., Ouyang, J., Wang, K.: Bankruptcy map: a system for searching and analyzing US bankruptcy cases at scale (2020)
29. Walton, D., Reed, C., Macagno, F.: Argumentation Schemes. Cambridge University Press, Cambridge (2008)
30. Walton, D.: Legal reasoning and argumentation, pp. 47–75, July 2018
31. Widmer, G., Kubat, M.: Learning in the presence of concept drift and hidden contexts. Mach. Learn. **23**(1), 69–101 (1996). https://doi.org/10.1023/A: 1018046501280

Making Things Explainable vs Explaining: Requirements and Challenges Under the GDPR

Francesco Sovrano[1]([✉])[iD], Fabio Vitali[1][iD], and Monica Palmirani[2][iD]

[1] DISI, University of Bologna, Bologna, Italy
francesco.sovrano2@unibo.it
[2] CIRSFID-AI, University of Bologna, Bologna, Italy

Abstract. The European Union (EU) through the High-Level Expert Group on Artificial Intelligence (AI-HLEG) and the General Data Protection Regulation (GDPR) has recently posed an interesting challenge to the eXplainable AI (XAI) community, by demanding a more user-centred approach to explain Automated Decision-Making systems (ADMs). Looking at the relevant literature, XAI is currently focused on producing explainable software and explanations that generally follow an approach we could term *One-Size-Fits-All*, that is unable to meet a requirement of centring on user needs. One of the causes of this limit is the belief that *making things explainable* alone is enough to have *pragmatic explanations*. Thus, insisting on a clear separation between *explainabilty* (something that can be explained) and *explanations*, we point to explanatorY AI (YAI) as an alternative and more powerful approach to win the AI-HLEG challenge. YAI builds over XAI with the goal to collect and organize explainable information, articulating it into something we called user-centred explanatory discourses. Through the use of explanatory discourses/narratives we represent the problem of generating explanations for Automated Decision-Making systems (ADMs) into the identification of an appropriate path over an explanatory space, allowing explainees to interactively explore it and produce the explanation best suited to their needs.

Keywords: Trustworthy AI · explanatorY AI (YAI) · XAI · HCI

1 Introduction

The academic interest in Artificial Intelligence (AI) [11] has grown together with the attention of Countries and people towards the possibly disruptive effects of ADM [38] in industry and the public administration (e.g., COMPAS [13], or in Italy the case-law "Buona Scuola"[1]), effects that may affect the lives of billions

---

[1] Cons. stato, sez. VI, sent. 8 aprile 2019, n. 2270, Cons. Stato, sez. VI, sent. del 13 dicembre 2019, n. 8472, Cons. Stato, sez. VI, sent. del 4 febbraio 2020, n. 881.

ⓒ Springer Nature Switzerland AG 2021
V. Rodríguez-Doncel et al. (Eds.): AICOL-XI 2018/AICOL-XII 2020/XAILA 2020, LNAI 13048, pp. 169–182, 2021.
https://doi.org/10.1007/978-3-030-89811-3_12

of persons [20]. Therefore governments are starting to act towards the establishment of ground rules of behaviour from complex systems, for instance through the enactment of the European GDPR[2], which identifies *fairness, lawfulness,* and in particular *transparency* as basic principles for every data processing tools handling personal data; even identifying a new *right to explanation* for individuals whose legal status is affected by a solely-automated decision. As a result, several expert groups, including those acting for the European Commission, have started asking the AI industry to adopt ethics code of conducts as quickly as possible [8,14], drawing a set of expectations to meet in order to guarantee a *right to explanation*. These expectations define the goal of explanations under the GDPR and thus describe the requirements for explanatory content. Many interpretations have been given of what qualifies an explanation in this context, but among them we mention the one by the AI-HLEG, for its relevance and prominence. The AI-HLEG was established in 2018, by the European Commission, with the explicit purpose of applying the principles of the GDPR specifically to AI software, and produced a list of fundamental ethical principles for *Trustworthy AI* tools that include *fairness* and *explicability*. The *explicability* principle, in particular, means to provide alternative measures in case of "black box" algorithms like "traceability, auditability and transparent communication on system capabilities", in order to respect the fundamental rights. So it is important to provide information about *how* the ADM works, *what* is the final decision, *why* the ADM provides such conclusion, *which* data are used for training the AI and for the concrete real case processing. *Explicability* concerns the *ex-post* processing but also the *ex-ante* informative communication. Most importantly, according to the AI-HLEG, explanations should be "adapted to the expertise of the stakeholder concerned (e.g. layperson, regulator or researcher)" and more over it "highly dependent on the context" [17], putting individual's needs at the centre, in a challenging way.

Notwithstanding these quite recent efforts, understanding what constitutes an explanation is a long-standing open problem. In literature there are various efforts in this direction and a long history of debates and philosophical traditions, often rooted in Aristotle's works and those of other philosophers. Among the many models proposed over the last few centuries some are now considered fallacious, albeit historically useful (e.g. Hempel et al.'s one [16]), in favour of more pragmatic (user-centred) ones (e.g. Achinstein's [2]). Despite this, Hempel et al.'s theory and Salmon's *Causal Realism* are probably the most (implicitly) mentioned and adopted models for explanations in AI, raising the question of whether technology is really aligned to the understandings of regulators and society or it is just acting conveniently. In fact, most of the literature on AI and explanations (e.g. eXplainable AI [3]) is currently focused on one-size-fits-all approaches usually able to produce only one type of explanations, defined through causal lens. Additional literature is focused on argumentation theory [9] or on sub-symbolic methodologies [7] for providing a deductive or inductive explanation.

---

[2] Regulation (EU) 2016/679.

It appears that this focus on pursuing one-size-fits-all explanations in XAI is justified by convenient definitions framing an explanation as the product of an act of making things explainable rather than a pragmatic (user-centred) act of explaining based on explainability. In other terms, there is no clear distinction between *making things explainable* and actually *explaining*. The exceptions to this pattern seem to be still too rare to be representative of disciplines like XAI. In this paper we take a strong stand against the idea that static, one-size-fits-all approaches to explanation have a chance of being pragmatic, thus meeting the AI-HLEG guidelines, and we propose to adopt a strong logical separation between *explainability* and *explaining*. In fact, we argue that explaining to humans is *computationally irreducible* and one-size-fits-all approaches (in the most generic scenario) may suffer the curse of dimensionality as soon as the complexity of the explanandum surpasses a fairly trivial threshold. For example, a complex big-enough *explainable* software can be super hard to *explain*, even to an expert, and the optimal (or even sufficient) explanation might change from expert to expert. In this specific example, an explainable software is necessary but not sufficient for explaining. This is why we first draw a clear separation between XAI and explanatorY AI (YAI), which refers to systems that (given a "traditional" XAI system) are actually able to produce a satisfactory explanation ready to be delivered to a human user interested in examining the complex working and output of the system. Subsequently, we propose a model for YAI shaped on *discursive explanations*. Discursive explanations give a strong background of principles and means to create an interactive explanatory system that is able to produce user-centred explanations, by providing an explanatory space that is amenable to exploration by the users in order to create the explanation that best suits each one's background, needs and objectives.

This paper is structured as follows. In Sect. 2 we provide an introduction to the GDPR and the *Right to Explanation*, and we also provide a brief summary of the AI-HLEG Guidelines for Trustworthy AI. In Sect. 3, taking off from the GDPR and the AI-HLEG guidelines, we give a motivation of why user-centred explanatory tools are a key ingredient for Trustworthy AI. In this section we discuss the most prominent XAI issues to this end and the problem of *computational irreducibility* in explanations. In Sect. 4 we give an high-level overview of a possible model of User-Centred Explanatory Tool, defining YAI as a Explanatory Discursive Process responsible to collect and structure explainable information articulating it into user-centred explanations. Finally, in Sect. 5 we conclude with a brief recap, pointing to a proof of concept.

2   Background: The Right to Explanation

The General Data Protection Regulation (GDPR) is an important 2016 EU regulation on personal data protection and the connected freedoms and rights. Since the GDPR is technology-neutral, it does not directly refer to AI, but several provisions are highly relevant to the use of AI for Automated Decision-Making system (ADM). For instance [19]:

- Principle 1. (a) requires personal data processing to be fair, lawful, transparent, necessary and proportional (Articles 5).
- Article 12 defines the obligations to fulfil a transparent information, communication and the modalities for the exercise of the data subject's rights.
- Articles 13-14-15 give individuals the right to be informed of the existence of solely automated decision-making, meaningful information about the logic involved, and the significance and envisaged consequences for the individual.
- Article 22 gives individuals the right not to be subject to a solely automated decision producing legal or similarly significant effects.
- Article 22(3) obliges organizations to adopt suitable measures to safeguard individuals when using solely automated decisions, including the right to obtain human intervention, to express his or her view, and to contest the decision.

Art. 22 defines the right to claim of a human intervention when a completely Automated Decision-Making systems (ADMs) may affect the legal status of a citizen. Art. 22 includes also several exceptions that derogate "to be subject to a decision based solely on automated processing" when the legal basis are supported by contract, consent or law. These conditions significantly limit the potential applicability of the right to explanation. For this reason in case of contract or consent the art. 22, paragraph 3 introduces the "right to obtain human intervention on the part of the controller, to express his or her point of view and to contest the decision". Here explanations seem to be provided only after decisions have been made (*ex-post* explanations), and are not a required precondition to protest decisions. This is not completely true: in arts. 13-14-15 there is the obligation to inform about the "the existence of automated decision-making, including profiling, referred to in Article 22(1) and (4) and, at least in those cases, meaningful information about the logic involved (Recital 63), as well as the significance and the envisaged consequences of such processing for the data subject." (*ex-ante* explanations). This combination of articles make the right of explanation very articulated and composed of different stages. Additionally, the recent White Paper on Artificial Intelligence [10] emitted by the European Commission stressed the need to monitor and audit not only the Automated Decision-Making system (ADM) algorithms but also the data records used for training, developing, running, the AI systems in order to fight the opacity and to improve transparency. From a technical point of view, there are technology-specific information to consider in order to fully meet the explanation requirements of the GDPR, for a more detailed overview refer to [35]. The qualities of explanations are listed in different works [25], but the EU Parliament [31] lists the following as a good summary of the current state of the art: intelligibility, understandability, fidelity, accuracy, precision, level of detail, completeness, consistency.

Article 22 is open to several interpretations [28,29,36] about whether providing individualised explanations is mandatory or just a good practice. To this end, Recital 71 provides interpretative guidance of Article 22. Two items are missing in Article 22 relative to Recital 71: the provision of "specific information" and the "right to obtain an explanation of the decision reached after such assessment". The second omission in particular raises the issue of whether controllers are really required by law to provide an individualised explanation. This issue is partially tackled by the AI-HLEG guidelines (endorsed by the EU Commission), giving further reason to believe that there is the intention to prefer user-centred explanations as soon as the technology is mature enough to guarantee them. At contrary Recital 63 requires *ex-ante* that the data subject should have the right to know and obtain communication in particular with regard to "the logic involved in any automatic personal data processing". The AI-HLEG tries to extend the GDPR expectations, targeting AI and giving further guidelines: accessibility and universal design should be a requirement for Trustworthy AI, with user-centrality at the core. This idea of a user-centred explanatory process find its roots in philosophy, for example in:

- Ordinary Language Philosophy [1,22]: the act of explanation as the illocutionary attempt to produce understanding in another by answering questions in a pragmatic way.
- Cognitive Science [18,22]: explaining as a process of belief revision, etc.

3   Problem Statement

Some of the limits in the current generation of XAI approaches have already been identified and spelt out by existing literature:

- "XAI has produced algorithms to generate explanations as short rules, attribution or influence scores, prototype examples, partial dependence plots, etc. However, little justification is provided for choosing different explanation types or representations" [37].
- "Research on explanation is typically focused on the person (or system) producing the explanation. [...] Does the explainee understand the system, concepts, or knowledge?" [25].
- "Much of XAI research tended to use the researchers' intuition of what constitutes a good explanation. There exist vast and valuable bodies of research in philosophy, psychology, and cognitive science of how people define, generate, select, evaluate, and present explanations, which argues that people employ certain cognitive biases and social expectations to the explanation process." [23]
- "XAI systems are built for developers, not users." [24,25]
- etc.

To summarize, despite several efforts (e.g. [12,23]) to tackle these issues, we can notice a majority of XAI tools lacking:

1. A broader vision: XAI should not involve only computer science, but also philosophy, psychology, cognitive science, etc.
2. Focus on user-centrality.
3. A consistent approach to evaluate the quality of explanations.

We claim that the cause of these limits are in the misunderstanding that explainability is enough for explaining. Indeed, by insisting on a clear logical separation between explainable systems and actual explanations, we argue that XAI is necessary but not sufficient for Trustworthy AI. In fact, XAI seems to be currently focused on producing explainable software and explanations that generally follow only a One-Size-Fits-All approach, failing to meet the user-centrality requirements. In the most generic scenario, explanations following a One-Size-Fits-All approach (*OSFA explanations*) should be considered not user-centred, by construction. For example, static representations where all aspects of a fairly long and complex computation are described and explained are one-size-fits-all explanations.

OSFA explanations have intuitively at least two problems:

1. if they are small enough to be simple, then in a complex enough domain they would not be able to generate an explanation containing enough information to satisfy the explanation appetite of every user, as the quantity of details required for satisfying every user would be necessarily larger than any small explanation in a few words.
2. if they contain all the necessary information, in a complex-enough domain they would contain an enormous amount of content and users interested in a specific aspect of the explanation would need to look for it within the whole explanation in hundreds or thousands of explanatory items mostly irrelevant to their purposes.

OSFA explanations could be useful for simple domains, but the complexity of a domain is exactly what motivates the need for explanations. In other terms, usefulness of explanations is obviously greater in complex domains.

An interesting parallel, to show the second problem, is that of surveillance cameras in front of a bank door. Surveillance cameras continuously record and make available to the investigators hundreds and hundreds of hours of excellent quality videos that allow the precise identification of thousands of people passing under the cameras. But our investigator is not interested in hundreds of hours of video, but only in those three seconds in which a suspect person in need to be identified was under the cameras. The relevance of these few seconds (out of hundreds of hours) is entirely based on the specific investigative task, which depends on the function that the investigator gives to the identification of the person, and this function depends on the purpose of identification (i.e. Is he the robber? A possible accomplice? A witness?). The purpose of the investigation is known to the investigator but not to the surveillance system, and in many cases it cannot be decided in advance but it becomes clear only during the evolution of the investigation. Similarly, the interest of a user in the output of an explanation system often may lie on a few short statements out of the hundreds of thousands that the explanation system may be able to generate, and these few ones depend on the function that the user gives to the explanation. This is why we must assume that in general the purpose of the explanation is known to the user but not to the explanation system, and it cannot be decided in advance but it becomes clear only during the evolution of the task in which the explanation is required. This phenomenon is known also as *computational irreducibility* [39] and it is typical of emerging phenomena, such as physical, biological and social ones [5].

A user-centred explanatory tool requires to provide goal-oriented explanations. Goal-oriented explanations implies explaining facts that are relevant to the user, according to her/his background knowledge, interests and other peculiarities that make her/him a unique entity with unique needs that may change over time. The computational irreducibility issue raises the following questions:

1. How to model and create a *user-centred* explanatory process, without rewriting the tool for every different user?
2. How to evaluate the quality of an explanatory process?

4   Proposed Solution

In order to answer the first question we propose to:

- Disentangle *explainability* from *explaining*: that is separate the presentation logic (*explaining*) from the application logic (*explainability*). In fact, only *explaining* has to be user-centred.
- Design a presentation logic that would allow personalised explanations given the same explainable information.

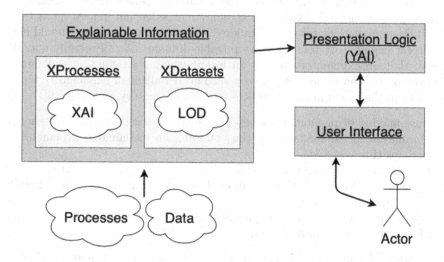

Fig. 1. XAI vs YAI: an abstract model of Explanatory Tool for Trustworthy AI. This model shows how to decompose the flow of explanatory information that moves from raw representations of processes/data to the explainee (or actor). Raw data are refined into explainable datasets - e.g. Linked Open Data (LOD), etc. Raw processes are refined into explainable processes. Explainable information can be used by YAI to generate pragmatic explanations.

In Fig. 1 we show a simple model of an Explanatory Tool for Trustworthy AI, obtained by our own need to clearly separate between explainabilty and explanations. More in detail, to increase the overall cohesion of the system, in this model we require an explicit logical separation between the functionalities related to *producing explainable information*, and those related to *producing pragmatic explanations*. In addition, we envision another logical separation in the production of actual explanations between *building explanations* (i.e. the presentation logic) and *interfacing with users*. Independently, producing explainable information should be separated in *generating explainable processes* and *producing explainable data-sets*. Thus, the main modules involved in the model are:

- The Explainable Information (EI) module, made of the eXplainable Processes (XP) and the eXplainable Datasets (XD) sub-modules.
- The YAI or Presentation Logic module.
- The User Interface (UI) module.

In other terms, we propose to distinguish between eXplainable AI (XAI) and explanatorY AI (YAI), considering them as different components of Trustworthy AI. We like to say that Trustworthy AI needs both the Xs and the Ys of AI[3].

The YAI module is the module responsible to collect and structure explainable information articulating it into user-centred explanations. In other terms, defining the YAI module is the same of defining a *user-centred explanatory process*. We are interested in defining a user-centred explanatory process aligned to the GDPR and the AI-HLEG guidelines. Speaking of user-centrality, we may assume that different types/groups of users exist: lay person, expert, legal operators, etc. each one with its own background knowledge and unique characteristics. If the explanations have to be tailored, does this imply that we should have a different explanatory tool for every possible different user? Probably not. We believe that an explanatory tool is an instrument for articulating explainable information into an *explanatory discourse*. This definition of explanatory tool is drawn from the essential best-practices of scientific inquiry, involving [6]:

- Sense-making of phenomena: classical question answering to collect enough information for understanding, thus building an explainable explanandum (perhaps through XAI).
- Articulating understandings into discourses: re-ordering and aggregation of explainable information to form an explanatory narrative or more generally a discourse to answer research questions.
- Evaluating: pose and answer questions about the quality of the presented information; e.g., argument them in a public debate.

Therefore we define a user-centred explanatory discourse as: "A sequence of information (explanans) to increase understanding over explainable data and processes (explanandum), for the satisfaction of a specified explainee that efficiently and effectively interacts with the explanandum (interaction) having specific goals in a specified context of use". Our definition takes inspiration from [21, 26, 32],

---

[3] XX and XY are human chromosomes responsible for gender.

integrating concepts of usability defined in ISO 9241 (Ergonomics of Human System Interaction [15]), such as the insistence on the term "specific", the triad "explainee", "goal" and "context of use", as much as the identification of specific quality metrics, which in our case are *effectiveness*, *efficiency* and *satisfaction*.

Similarly to how *satisfaction* has increased in importance in user experience studies in recent years, we believe that satisfaction should be considered one of the most important metrics for the assessment of the quality of explanations, too. The qualities of the explanation that provides the explainee with the necessary *satisfaction*, using the categories provided by [26], can be summarized in a good choice of narrative appetite, structure and purpose. To understand "narrative appetite" we have to consider that "in order for a narrative discourse to flourish, both parties (the narrator and the reader) have to find engagement in this social transaction interesting enough to prevail over competing activities. Thus, stories must not only be accounts of events, but accounts of events that someone cares to know more about; we must want to know what happened if we are to continue reading or listening." This appetite can be quenched by the proper structuring of the narration: "Narrative, we have shown, is a narrator's recounting of *events structured in time*. The elements of both time and structure are associated in many descriptions of narrative". In addition, "The element of *connectability* [...] structures different texts. Connectability [...] must be strictly observed in expository texts where an argument is to be developed or information is to be conveyed. In such texts, the writer aims for a precise interpretation where a multiplicity of possible meanings must be constantly narrowed down". Finally, the identification of purpose in narratives is central: "stories are constructed to help us understand the world we live in: to help comprehend the life that is in me and around me. [...] it is through narrative that we are able to accommodate the new within that which is familiar to us. In these descriptions of purpose, narrative can be interpreted as helping us better understand the natural as well as the human world".

The problems of a user-centred approach to explanations is that fully-automated explanatory processes are unlikely to target quality parameters that guarantee the satisfaction of all specified explainees, as described above, due to the *computational irreducibility* of the process of explaining. Even if an AI could be used to generate such user-centred explanations, in the context of explanations under the GDPR this would only shift the problem of explaining from the original ADM to another ADM (the explanatory AI that explains the original ADM). As such we believe that (at least for the explanations under the GDPR), the most straightforward solution is to encourage readers (explainees) and narrators (explainers) to become one, *users* generating the narration for themselves by selecting and organizing narratives of individual event-tokens according to the structure that best caters their appetite and purpose. In this sense, a tool for creating explanatory discourses would allow users to build intelligible sequences of information, containing arguments that support or attack the claims underlying the goal of an *explanatory narrative process*. This idea of data controllers and data subjects "becoming one" can be understood in a twofold way. First, at its best possible light, such tool should convince and dissuade data subjects to ask for human intervention, e.g. Art 22(3) of the GDPR. Second, the tool should help data controllers to abide by the law, by illustrating the decision that can be contested by data subjects.

An *explanatory narrative* is always only one of the many possible narratives that can be built to shed light on an explanandum. All the possible narratives for an explanandum form a complex network of information that we call *Explanatory Space*. In this sense, an explanatory discourse is a path within an Explanatory Space. As analogy, we might see the Explanatory Space as a sort of manifold space where every point within it is interconnected information about one or more aspects of the explanandum. So that every point of the Explanatory Space is not user-centred locally, but globally as an element of a sequence of information that can be chosen by a user according to its interest drift while exploring the space.

As mentioned in Sect. 3, the amount of information forming such Explanatory Spaces can be overwhelming, given any complex-enough explanandum. Thus, in order to answer our research question, what we need is to design a process to effectively allow users to extract explanatory narratives from an Explanatory Space. In [35] we present our model of Explanatory Narrative Process making specific references to the GDPR and the AI-HLEG guidelines, modelling a generic explanatory process, giving a formal definition of explanandum, explanans and Explanatory Space. Hereafter we show a plausible example of YAI in action.

4.1 Example

Let's consider the following example where a user-centred explanatory tool is used to explain the decision taken by an ADM on a case concerning the GDPR, art. 8. The aforementioned case is about the conditions applicable to child's consent in relation to information society services. The art. 8 of GDPR fixes at 16 years old the maximum age for giving the consent without the parent-holder authorization. This limit could be derogated by the domestic law. In Italy the legislative decree 101/2018 defines this limit at 14 years. In this situation we could model legal rules in LegalRuleML [4, 30] using defeasible logic, in order to be able to represent that the GDPR art. 8 rule (16 yearsOld) is overridden by the Italian's (14 yearsOld). The SPINDle legal reasoner processes the correct rule according to the jurisdiction (e.g., Italy) and the age. Suppose that Marco (a 14 years old Italian teenager living in Italy) uses Whatsapp, and his father, Giulio, wants to remove Marco's subscription to Whatsapp because he is worried about the privacy of Marco when online. In this simple scenario, the Automated Decision-Making system (ADM) system would reject Giulio's request to remove Marco's profile, because of the Italian legislative decree 101/2018. What if Giulio wants to know the reasons why his request was rejected? Figure 2 shows a possible view of a user-centred explanatory tool based on our model. Thanks to the user-centred explanatory tool Giulio can actually choose what information to expand and consider, building its own personalised explanatory discourse out of a predefined Explanatory Space.

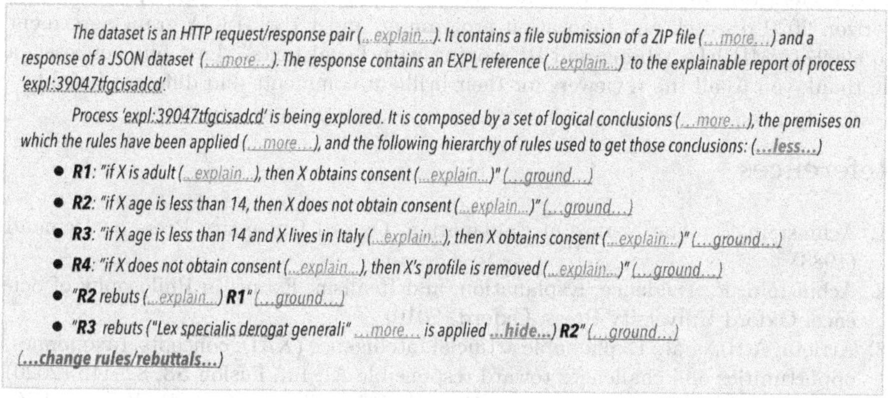

The dataset is an HTTP request/response pair (...explain...). It contains a file submission of a ZIP file (...more...) and a response of a JSON dataset (...more...). The response contains an EXPL reference (...explain...) to the explainable report of process 'expl:39047tfgcisadcd'.

Process 'expl:39047tfgcisadcd' is being explored. It is composed by a set of logical conclusions (...more...), the premises on which the rules have been applied (...more...), and the following hierarchy of rules used to get those conclusions: (...less...)

- **R1**: "if X is adult (...explain...), then X obtains consent (...explain...)" (...ground...)
- **R2**: "if X age is less than 14, then X does not obtain consent (...explain...)" (...ground...)
- **R3**: "if X age is less than 14 and X lives in Italy (...explain...), then X obtains consent (...explain...)" (...ground...)
- **R4**: "if X does not obtain consent (...explain...), then X's profile is removed (...explain...)" (...ground...)
- "**R2** rebuts (...explain...) **R1**" (...ground...)
- "**R3** rebuts ("Lex specialis derogat generali" ...more... is applied ...hide...) **R2**" (...ground...)

(**...change rules/rebuttals...**)

Fig. 2. Example of explainer: underlined coloured words represent different possible actions a user can operate to explore the Explanatory Space, extracting its own narrative. For example, clicking on a "...more..." button the user can expand the explanans.

5   Conclusions

In this paper we analysed some of the limits in the current generation of XAI approaches, with respect to the goals of Trustworthy AI set by the GDPR and the AI-HLEG guidelines, identifying the cause of these limits in the misunderstanding that *making things explainable* is enough for *pragmatically explaining*. Indeed, by insisting on a clear logical separation between explainable systems and actual explanations, we argued that XAI is necessary but not sufficient for Trustworthy AI, therefore presenting an abstract model of explanatorY AI (YAI). In our model, YAI builds over XAI and it is intended to be a set of tools for organising the presentation logic of a user-centred explanatory software in a way that would allow personalised explanations about complex-enough explananda by generating *discursive explanations* out of an Explanatory Space. In this paper we take a strong stand against the idea that static, one-size-fits-all approaches to explanation have a chance of being pragmatic, thus meeting the AI-HLEG guidelines. For a concrete proof of concept of YAI (including software and experiment analysis) we point the reader to our most recent works, e.g. [34].

Finally, it is clear that the solution we proposed avoids the problem which relates to balancing between what is possible in terms of formal explainability and what is required as to the level of detail of information regarding the "logic of processing". In other words, we assumed that systems in question can be both formally explainable and pragmatically able to be explained. So, we leave as future work an analysis of what are the minimum requirements for information to be considered explainable enough for pragmatic explanations with a proper degree of exactness, detail and fruitfulness[4]. This might help also to perform a reasonable impact assessment of the ADM, as defined by art. 35 of the GDPR.

Acknowledgements. This work was partially supported by the European Union's Horizon 2020 research and innovation programme under the MSCA grant agreement No 690974 "MIREL: MIning and REasoning with Legal texts". Last but not least, a big thank you to all the reviewers for their brilliant comments and different insights.

References

1. Achinstein, P.: The Nature of Explanation. Oxford University Press on Demand (1983)
2. Achinstein, P.: Evidence, Explanation, and Realism: Essays in Philosophy of Science. Oxford University Press, Oxford (2010)
3. Arrieta, A.B., et al.: Explainable artificial intelligence (XAI): concepts, taxonomies, opportunities and challenges toward responsible AI. Inf. Fusion **58**, 82–115 (2020)
4. Athan, T., Boley, H., Governatori, G., Palmirani, M., Paschke, A., Wyner, A.Z.: Oasis legalruleml. In: ICAIL, vol. 13, pp. 3–12 (2013)
5. Beckage, B., Kauffman, S., Gross, L.J., Zia, A., Koliba, C.: More complex complexity: exploring the nature of computational irreducibility across physical, biological, and human social systems. In: Zenil, H. (ed.) Irreducibility and Computational Equivalence. ECC, vol. 2, pp. 79–88. Springer, Heidelberg (2013). https://doi.org/10.1007/978-3-642-35482-3_7
6. Berland, L.K., Reiser, B.J.: Making sense of argumentation and explanation. Sci. Educ. **93**(1), 26–55 (2009)
7. Omicini, A., Calegari, R., Ciatto, G.: On the integration of symbolic and subsymbolic techniques for XAI: a survey. Intelligenza Artificiale **14**(1), 7–32 (2020)
8. Cath, C., Wachter, S., Mittelstadt, B., Taddeo, M., Floridi, L.: Artificial intelligence and the 'good society': the US, EU, and UK approach. Sci. Eng. Ethics **24**(2), 505–528 (2018). https://doi.org/10.1007/s11948-017-9901-7
9. Cocarascu, O., Rago, A., Toni, F.: Explanation via machine arguing. In: Manna, M., Pieris, A. (eds.) Reasoning Web 2020. LNCS, vol. 12258, pp. 53–84. Springer, Cham (2020). https://doi.org/10.1007/978-3-030-60067-9_3
10. EU Commission: White paper - on artificial intelligence - a European approach to excellence and trust, COM(2020) 65 final (2020)
11. European Commission: COM(2018) 237 final Brussels, Artificial Intelligence for Europe. European Commission (2018)
12. DARPA: Broad agency announcement explainable artificial intelligence (XAI). DARPA-BAA-16-53, pp. 7–8 (2016)
13. Flores, A.W., Bechtel, K., Lowenkamp, C.T.: False positives, false negatives, and false analyses: a rejoinder to machine bias: there's software used across the country to predict future criminals. And it's biased against blacks. Fed. Probation **80**, 38 (2016)
14. Floridi, L., et al.: AI4People–an ethical framework for a good AI society: opportunities, risks, principles, and recommendations. Minds Mach. **28**(4), 689–707 (2018). https://doi.org/10.1007/s11023-018-9482-5

---

[4] Perhaps drawing from Carnap's theory [27].

15. International Organization for Standardization: Ergonomics of human-system interaction: part 210: human-centred design for interactive systems. ISO (2010)
16. Hempel, C.G., et al.: Aspects of Scientific Explanation (1965)
17. AI HLEG: Ethics guidelines for trustworthy AI (2019)
18. Holland, J.H., Holyoak, K.J., Nisbett, R.E., Thagard, P.R.: Induction: Processes of Inference, Learning, and Discovery. MIT Press, Cambridge (1989)
19. ICO: Project explain interim report (2019). https://ico.org.uk/about-the-ico/research-and-reports/project-explain-interim-report/. Accessed 05 Jan 2020
20. Kuhn, R., Kacker, R.: An application of combinatorial methods for explainability in artificial intelligence and machine learning (draft). Technical report, National Institute of Standards and Technology (2019)
21. Lipton, P.: What good is an explanation? In: Hon, G., Rakover, S.S. (eds.) Explanation. SYLI, vol. 302, pp. 43–59. Springer, Dordrecht (2001). https://doi.org/10.1007/978-94-015-9731-9_2
22. Mayes, G.R.: Theories of explanation. In: The Internet Encyclopedia of Philosophy (2005)
23. Miller, T.: Explanation in artificial intelligence: insights from the social sciences. Artif. Intell. **267**, 1–38 (2018)
24. Miller, T., Howe, P., Sonenberg, L.: Explainable AI: beware of inmates running the asylum or: how I learnt to stop worrying and love the social and behavioural sciences. arXiv preprint arXiv:1712.00547 (2017)
25. Mueller, S.T., Hoffman, R.R., Clancey, W., Emrey, A., Klein, G.: Explanation in human-AI systems: a literature meta-review, synopsis of key ideas and publications, and bibliography for explainable AI. arXiv preprint arXiv:1902.01876 (2019)
26. Norris, S.P., Guilbert, S.M., Smith, M.L., Hakimelahi, S., Phillips, L.M.: A theoretical framework for narrative explanation in science. Sci. Educ. **89**(4), 535–563 (2005)
27. Dutilh Novaes, C., Reck, E.: Carnapian explication, formalisms as cognitive tools, and the paradox of adequate formalization. Synthese **194**(1), 195–215 (2015). https://doi.org/10.1007/s11229-015-0816-z
28. Pagallo, U.: Algoritmi e conoscibilità. Rivista di filosofia del diritto **9**(1), 93–106 (2020)
29. Palmirani, M.: Big data e conoscenza. Rivista di filosofia del diritto **9**(1), 73–92 (2020)
30. Palmirani, M., Governatori, G.: Modelling legal knowledge for GDPR compliance checking. In: JURIX, pp. 101–110 (2018)
31. EU Parliament: Understanding algorithmic decision-making: opportunities and challenges (2019)
32. Passmore, J.: Explanation in everyday life, in science, and in history. Hist. Theory **2**(2), 105–123 (1962)
33. Salmon, W.C.: Scientific Explanation and the Causal Structure of the World. Princeton University Press, Princeton (1984)
34. Sovrano, F., Vitali, F.: From philosophy to interfaces: an explanatory method and a tool based on Achinstein's theory of explanation. In: Proceedings of the 26th International Conference on Intelligent User Interfaces (2021)
35. Sovrano, F., Vitali, F., Palmirani, M.: Modelling GDPR-compliant explanations for trustworthy AI. In: Kő, A., Francesconi, E., Kotsis, G., Tjoa, A.M., Khalil, I. (eds.) EGOVIS 2020. LNCS, vol. 12394, pp. 219–233. Springer, Cham (2020). https://doi.org/10.1007/978-3-030-58957-8_16

36. Wachter, S., Mittelstadt, B., Floridi, L.: Why a right to explanation of automated decision-making does not exist in the general data protection regulation. Int. Data Priv. Law **7**(2), 76–99 (2017)
37. Wang, D., Yang, Q., Abdul, A., Lim, B.Y.: Designing theory-driven user-centric explainable AI. In: Proceedings of the 2019 CHI Conference on Human Factors in Computing Systems, p. 601. ACM (2019)
38. WP29: Guidelines on automated individual decision-making and profiling for the purposes of regulation 2016/679 (wp251rev.01). European Commission (2016)
39. Zwirn, H., Delahaye, J.P.: Unpredictability and computational irreducibility. In: Zenil, H. (ed.) Irreducibility and Computational Equivalence. ECC, vol. 2, pp. 273–295. Springer, Heidelberg (2013). https://doi.org/10.1007/978-3-642-35482-3_19

Explaining Arguments at the Dutch National Police

AnneMarie Borg[1]([✉])[ID] and Floris Bex[1,2][ID]

[1] Department of Information and Computing Sciences, Utrecht University,
Utrecht, The Netherlands
{a.borg,f.j.bex}@uu.nl
[2] Department of Law, Technology, Markets and Society, Tilburg University,
Tilburg, The Netherlands

Abstract. As AI systems are increasingly applied in real-life situations, it is essential that such systems can give explanations that provide insight into the underlying decision models and techniques. Thus, users can understand, trust and validate the system, and experts can verify that the system works as intended. At the Dutch National Police several applications based on computational argumentation are in use, with police analysts and Dutch citizens as possible users. In this paper we show how a basic framework of explanations aimed at explaining argumentation-based conclusions can be applied to these applications at the police.

Keywords: Explainable AI · Computational argumentation

1 Introduction

Recently *explainable AI* (XAI) has received much attention, mostly directed at new techniques for explaining decisions of machine learning algorithms [21]. However, explanations also play an important role in (symbolic) knowledge-based systems [12]. One area in symbolic AI which has seen a number of real-world applications lately is formal or computational argumentation [1]. Two central concepts in formal argumentation are *abstract argumentation frameworks* [7] – sets of arguments and the attack relations between them – and *structured* or *logical argumentation frameworks* [2] – where arguments are constructed from a knowledge base and a set of rules and the attack relation is based on the individual elements in the arguments. Common for argumentation frameworks, abstract and structured, is that we can determine their extensions, sets of arguments that can collectively be considered as acceptable, under different semantics [7].

The Dutch National Police employs several applications based on structured argumentation frameworks (a variant of ASPIC$^+$ [20]). One such application concerns complaints by citizens about online trade fraud (e.g., a product bought

This research has been partly funded by the Dutch Ministry of Justice and the Dutch National Police. We thank Daphne Odekerken for her help with designing the examples and the anonymous reviewers for their valuable comments.

V. Rodríguez-Doncel et al. (Eds.): AICOL-XI 2018/AICOL-XII 2020/XAILA 2020, LNAI 13048, pp. 183–197, 2021.
https://doi.org/10.1007/978-3-030-89811-3_13

through a web-shop or on eBay turns out to be fake). The system queries the citizen for various observations, and then determines whether the complaint is a case of fraud [3,19]. Another related example is a classifier for checking fraudulent web-shops, which gathers information about online shops and thus tries to determine whether they are real (bone fide) or fake (mala fide) shops [18]. These applications are aimed at assisting the police at working through high volume tasks, leaving more time for tasks that require human attention.

Argumentation is often considered to be inherently transparent and explainable. A complete argumentation framework and its extensions is a *global* explanation [8]: what can we conclude from the model as a whole? Such global explanations can be used by argumentation experts to check whether the model works as intended. However, as we have noticed when deploying argumentation systems to be used by lay-users (e.g., citizens, police analysts) at the police, more natural and compact explanations are needed. Firstly, we need ways to explain the (non-)acceptability of *individual arguments*, that is, *local* explanations [8] for particular decisions or conclusions. Secondly, explanations should be *compact*, and contain only the *relevant arguments* which are needed in order to draw a conclusion. Finally, explanation should be *tailored to the receiver*. For example, in the case of online trade fraud, for a citizen the system should return only the observations provided in the report ("this is presumably a case of fraud because you provided the following facts in your report: ..."), but for a police analyst the system should also show which (legal) rules were applied and why there were no exceptions in this case ("this is presumably (not) a case of fraud because the following legal rules are not applicable: ...").

In this paper, we show how a variety of different local explanations can be derived from an argumentation framework and we provide motivations for the design options. We start with the basic explanations from [4], which are based on concepts from formal argumentation (Sect. 3.1). We then discuss how explanations can be selected based on sufficiency and necessity (Sect. 3.2) and how our explanations can be used to create contrastive explanations (i.e., "why P rather than Q") (Sect. 3.3). Each of the discussed explanations is based on underlying formal definitions that we cannot introduce here in full detail. We refer the interested reader to [4], [6] and [5] respectively.

Our informal exploration has clear ties to recent more formal work on methods to derive explanations for specific conclusions [9–11,13,22]. We apply and extend the framework from [4] here for several reasons. Often, explanations are only defined for a specific semantics [9,10] and can usually only be applied to abstract argumentation [10,13,22],[1] while our framework can be applied on top of any argumentation setting (structured or abstract) that results in a Dung-style argumentation framework. Furthermore, when this setting is a structured one based on a knowledge base and set of rules (like ASPIC$^+$ or logic-based argumentation [2]), the explanations can be further adjusted (something which is not considered at all in the literature). Moreover, explanations from the literature

---

[1] These explanations do not account for the sub-argument relation in structured argumentation. For example, in structured argumentation one cannot remove specific arguments or attacks without influencing other arguments/attacks.

are usually only for acceptance [9,13] or non-acceptance [10,22], while with this framework both acceptance and non-acceptance explanations can be derived in a similar way.[2] Finally, to the best of our knowledge, this is the first approach to local explanations for formal argumentation in which necessary, sufficient and contrastive explanations are considered.

The paper is structured as follows: in the next section we recall some of the most basic and important concepts from formal argumentation. Then, in Sect. 3, the internet trade fraud scenario and the different possible explanations for the derived conclusions are discussed. We conclude in Sect. 4.

2   Argumentation Preliminaries

We focus in this paper on the intuition behind the explanations introduced in [4,6] and the motivation for some of the choices that can be made in the derivation of these explanations. We therefore keep the formal definitions and results limited, leaving more space for an informal discussion.

An *abstract argumentation framework* (AF) [7] is a pair $\mathcal{AF} = \langle Args, \mathcal{A} \rangle$, where Args is a set of *arguments* and $\mathcal{A} \subseteq Args \times Args$ is an *attack relation* on these arguments. An AF can be viewed as a directed graph, in which the nodes represent arguments and the arrows represent attacks between arguments (see, e.g., Fig. 1 on page 6). Dung-style semantics can be applied to an AF, to determine what combinations of arguments can collectively be accepted.

Definition 1. *For $\mathcal{AF} = \langle Args, \mathcal{A} \rangle$, $A \in Args$ attacks $B \in Args$ if $(A, B) \in \mathcal{A}$ and $S \subseteq Args$ attacks B if there is some $C \in S$ such that $(C, B) \in \mathcal{A}$; A defends B if A attacks an attacker of B and S defends B if it attacks every attacker of B;[3] S is conflict-free if there are no $A_1, A_2 \in S$ such that $(A_1, A_2) \in \mathcal{A}$; and S is admissible if it is conflict-free and it defends all of its arguments.*

A $\subseteq$-maximal admissible set is a preferred extension (Prf) of $\mathcal{AF}$. The set of all preferred extensions of $\mathcal{AF}$ will be denoted by $\mathsf{Prf}(\mathcal{AF})$.

There are different ways in which the conclusions can be drawn from the extensions of a framework. At the police, when drawing a definite conclusion (e.g., someone is guilty) it is important to be completely certain. This means that the application uses a very skeptical approach towards drawing conclusions: only arguments that are part of every complete set are considered conclusions (i.e., the grounded semantics from [7] is used). When considering whether there is the possibility of the conclusion (e.g., it could be a case of fraud), a more credulous approach can be taken. We follow the latter approach here: an argument that is part of some preferred extension can be considered a conclusion or *accepted*.

---

[2] An exception to this might be [11]. However, we consider our framework more easily applicable, since it returns sets of arguments rather than sets of dialectical trees, which might contain many arguments.

[3] In [7], attack and defense are defined from a set of arguments to an argument. In this paper we will mainly rely on attack and defense between arguments, since we are interested in the *arguments that defend* a certain argument, rather than whether that argument is *defended by the set of arguments*.

For example, in the AF from Fig. 1 we have that all arguments are accepted (while under the grounded semantics only C_1 would be accepted). In particular, we have the following preferred extensions: $\{A_1, A_2, A_3\}$, $\{A_1, A_2, A_5\}$, $\{A_1, A_3, A_4\}$, and $\{A_3, A_4, A_6\}$.

In abstract argumentation, as defined above, the arguments are abstract entities and the attack relation is pre-defined. In contrast, in structured argumentation, the arguments are derived from a knowledge base and a set of rules and the attack relation is based on the structure of the arguments. Each of the applications that is in use, is based on a variation of ASPIC$^+$, one of the best-known approaches to structured argumentation [20]. In particular, the notions of a language, axioms and defeasible rules are taken from ASPIC$^+$. See [19] for the formal details.[4] In this paper we only provide the preliminaries that are necessary for the explanations. As we will show in the next section, the AF from Fig. 1 is based on a structured setting.

Argumentation frameworks in ASPIC$^+$ are constructed from an *argumentation theory*: $AT = \langle AS, \mathcal{K} \rangle$, where $AS = \langle \mathcal{L}, \mathcal{R}, n \rangle$, an *argumentation system*, is a triple of a formal language, a set of defeasible rules and a naming function for these rules, and $\mathcal{K} = \mathcal{K}_n \cup \mathcal{K}_p$ is the *knowledge base* containing the disjoint sets of axioms ($\mathcal{K}_n$) and ordinary premises ($\mathcal{K}_p$). Arguments are constructued from an argumentation theory as follows:

Definition 2. *An argument A on the basis of an argumentation theory $AT = \langle AS, \mathcal{K} \rangle$, where $AS = \langle \mathcal{L}, \mathcal{R}, n \rangle$ is:*

- ϕ *if $\phi \in \mathcal{K}$, where* $\mathsf{Prem}(A) = \mathsf{Sub}(A) = \{\phi\}$, $\mathsf{Conc}(A) = \phi$ *and* $\mathsf{TopRule}(A) =$ *undefined;*
- $A_1, \ldots, A_n \Rightarrow \psi$, *if $A_1, \ldots, A_n$ are arguments such that there is a rule* $\mathsf{Conc}(A_1), \ldots, \mathsf{Conc}(A_n) \Rightarrow \psi \in \mathcal{R}$.
  $\mathsf{Prem}(A) = \mathsf{Prem}(A_1) \cup \ldots \cup \mathsf{Prem}(A_n)$, $\mathsf{Sub}(A) = \mathsf{Sub}(A_1) \cup \ldots \cup \mathsf{Sub}(A_n) \cup \{A\}$, $\mathsf{Conc}(A) = \psi$, $\mathsf{TopRule}(A) = \mathsf{Conc}(A_1), \ldots, \mathsf{Conc}(A_n) \Rightarrow \psi$ *additionally, we denote* $\mathsf{Ant}(\mathsf{TopRule}(A)) = \{\mathsf{Conc}(A_1), \ldots, \mathsf{Conc}(A_n)\}$. *Moreover, where* S *is a set of arguments* $\mathsf{Prem}(\mathsf{S}) = \bigcup \{\mathsf{Prem}(A) \mid A \in \mathsf{S}\}$.

Attacks between arguments are based on the premises and conclusions of these arguments.

Definition 3. *An argument A attacks an argument B iff, (where $\phi = -\psi$ iff $\phi = \neg\psi$ or $\psi = \neg\phi$)*

- $\mathsf{Conc}(A) = \neg n(d_i)$, *where there is some $B' \in \mathsf{Sub}(B)$ such that $\mathsf{TopRule}(B') = d_i$, it denies a rule; or*
- $\mathsf{Conc}(A) = -\phi$, *where there is some $B' \in \mathsf{Sub}(B)$ such that $\mathsf{Conc}(B') = \phi$, it denies a conclusion; or*
- $\mathsf{Conc}(A) = -\phi$, *for some $\phi \in \mathsf{Prem}(B) \setminus \mathcal{K}_n$, it denies a premise.*

---

[4] The corresponding demo of [19], demonstrating the argumentation-based part of the application, is available at https://nationaal-politielab.sites.uu.nl/estimating-stability-for-efficient-argument-based-inquiry/.

Dung-style semantics can be applied to argumentation frameworks based on argumentation theories as defined in Definition 1. We will say that a formula ϕ in an argumentation framework $\mathcal{AF}(\text{AT})$ is *accepted* if there is some $\mathcal{E} \in \text{Prf}(\mathcal{AF}(\text{AT}))$ with $A \in \mathcal{E}$ such that $\text{Conc}(A) = \phi$ and *non-accepted* if there is some $\mathcal{E} \in \text{Prf}(\mathcal{AF}(\text{AT}))$ such that there is no $A \in \mathcal{E}$ with $\text{Conc}(A) = \phi$.

These basic preliminaries on formal argumentation are enough to illustrate the different possibilities for explaining argumentation-based conclusions derived from the internet trade fraud application at the police.

3   Deriving Explanations

Suppose that the following knowledge base is provided: a citizen has *ordered a product* through an online shop, *paid* for it and *received* a package. However, it is the *wrong product*, it seems *suspicious* as if it might be a replica, rather than a real product. Yet an *investigation* cannot find a problem with the product. Still, the citizen wants to file a complaint of internet trade fraud.

While the citizen provides the information from the described scenario, the system constructs further arguments from this, based on the Dutch law.[5] In particular, the following rules are applied:

R_1 If the complainant *paid* then usually the *complainant delivered*;

R_2 If the *wrong product* was *received* then usually this is *not a case of fraud*;

R_3 If the *wrong product* was *received* then usually the *counter party has delivered*;

R_4 If the product seem *suspicious* then usually the product is *fake*;

R_5 If the product is *fake* then usually the *counter party did not deliver*;

R_6 If an *investigation* shows that there is no problem with the product then usually the product is *not fake*;

R_7 If the *complainant delivered* and the *counter party did not deliver* it is usually a *case of fraud*.

From this we obtain arguments for:[6]

C_1 :   the complainant *paid* + R_1 ⇒ the *complainant delivered*

A_1 :   the *wrong product* was *received* + R_2 ⇒ it is *not a case of fraud*

A_2 :   the *wrong product* was *received* + R_3 ⇒ the *counter party has delivered*

A_3 :   the product seems *suspicious* + R_4 ⇒ the product is *fake*

A_4 : A_3 + R_5 ⇒ the *counter party did not deliver*

A_5 :   an *investigation* shows no problems + R_6 ⇒ the product is *not fake*

A_6 : C_1 + A_4 + R_7 ⇒ it is a *case of fraud*.

---

[5] In order to make the argumentation framework and corresponding explanations more interesting the rules that are applied here are only inspired by the law. The real application is based on slightly different rules [19].

[6] We do not state the arguments based on the knowledge base explicitly, since these neither attack other arguments nor can be attacked themselves and do therefore not influence the acceptability of other arguments.

Note that the argument A_5 which has conclusion *not fake* will attack any argument with the conclusion *fake* (and vice versa), as well as any argument based on the conclusion *fake* (i.e., A_5 and A_3 attack each other and A_5 attacks A_4 and A_6 because they have *fake* as a sub-conclusion). The graphical representation of the AF, which we will refer as $\mathcal{AF}_1 = \langle \mathrm{Args}_1, \mathcal{A}_1 \rangle$ can be found in Fig. 1.

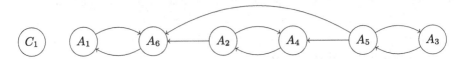

Fig. 1. Graphical representation of the argumentation framework $\mathcal{AF}_1$ constructed based on information provided in the complaint.

As the aim of the system is to determine whether a particular situation is a case of fraud, we will focus here on the arguments A_1 (*not fraud*) and A_6 (*fraud*). Note that, from an argumentative perspective, both arguments can be accepted, though not simultaneously. For A_1 this is the case since A_1 attacks any argument by which it is attacked (i.e., $(A_1, A_6) \in \mathcal{A}$). For A_6 additional conclusions have to be accepted as well. In particular, one can accept the argument for *fraud* when also accepting the arguments for the *counter party did not deliver* (A_4) and that the *product is fake* (A_3). This follows since $\{A_3, A_4, A_6\}$ is a preferred extension and A_3 and A_4 attack attackers of A_6 that A_6 would otherwise not be defended against. In what follows we will consider for both A_1 and A_6 explanations for why one could (not) accept them.

3.1 Basic Explanations

In [4] skeptical and credulous acceptance and non-acceptance explanations for abstract and structured argumentation were introduced. These explanations are defined in terms of two functions: $\mathbb{D}$, which determines the arguments that are in the explanation and $\mathbb{F}$, which determines what elements of these arguments the explanation presents. For the basic explanations in this paper, we instantiate $\mathbb{D}$ with the following functions, let $A \in \mathrm{Args}$ and $\mathcal{E} \in \mathsf{Prf}(\mathcal{AF})$ for some AF $\mathcal{AF} = \langle \mathrm{Args}, \mathcal{A} \rangle$:

- Defending$(A) = \{B \in \mathrm{Args} \mid B \text{ defends } A\}$ denotes the set of arguments that defend A and Defending$(A, \mathcal{E}) = \text{Defending}(A) \cap \mathcal{E}$ denotes the set of arguments that defend A in $\mathcal{E}$.
- NotDefAgainst$(A, \mathcal{E}) = \{B \in \mathrm{Args} \mid B \text{ attacks } A \text{ and } \mathcal{E} \text{ does not defend } A \text{ against this attack}\}$ denotes the set of all attackers of A that are not defended by $\mathcal{E}$.

The explanations are defined for arguments and formulas.

Definition 4. *Let* $\mathcal{AF} = \langle Args, \mathcal{A} \rangle$ *be an AF and suppose that* $A \in Args$ *[resp.* $\phi \in \mathcal{L}]$ *is accepted. Then:*

$\mathsf{Acc}(A) = \{\mathsf{Defending}(A, \mathcal{E}) \mid \mathcal{E} \in \mathsf{Prf}(\mathcal{AF}) \text{ and } A \in \mathcal{E}\}.$

$\mathsf{Acc}(\phi) = \{\mathbb{F}(\mathsf{Defending}(A, \mathcal{E})) \mid \mathcal{E} \in \mathsf{Prf}(\mathcal{AF}) \text{ such that } A \in \mathcal{E} \text{ and } \mathsf{Conc}(A) = \phi\}.$

An acceptance explanation, for an argument or formula, contains all the arguments that defend the argument (for that for that formula) in an extension. If it is an explanation for a formula explanation, the function $\mathbb{F}$ can be applied to it.

Definition 5. *Let* $\mathcal{AF} = \langle Args, \mathcal{A} \rangle$ *be an AF and suppose that* $A \in Args$ *[resp.* $\phi \in \mathcal{L}]$ *is non-accepted. Then:*

$$\mathsf{NotAcc}(A) = \bigcup_{\mathcal{E} \in \mathsf{Prf}(\mathcal{AF}) \text{ and } A \notin \mathcal{E}} \mathsf{NotDefAgainst}(A, \mathcal{E}).$$

$$\mathsf{NotAcc}(\phi) = \bigcup_{A \in Args \text{ and } \mathsf{Conc}(A)=\phi} \bigcup_{\mathcal{E} \in \mathsf{Prf}(\mathcal{AF}) \text{ and } A \notin \mathcal{E}} \mathbb{F}(\mathsf{NotDefAgainst}(A, \mathcal{E})).$$

A non-acceptance explanation contains all the arguments that attack the argument [resp. an argument for the formula] and to which no defense exists in some preferred extension. For a formula $\mathbb{F}$ can be applied again.

The function $\mathbb{F}$ can be instantiated in different ways. We recall here some of the variations introduced in [4]. These will be motivated in the discussions on the different explanations.

- $\mathbb{F} = \mathsf{id}$, where $\mathsf{id}(\mathsf{S}) = \mathsf{S}$. Then explanations are sets of arguments.
- $\mathbb{F} = \mathsf{Prem}$. Then explanations only contain the premises of arguments (i.e., knowledge base elements).
- $\mathbb{F} = \mathsf{AntTop}$, where $\mathsf{AntTop}(A) = \langle \mathsf{TopRule}(A), \mathsf{Ant}(\mathsf{TopRule}(A)) \rangle$. Then explanations contain the last applied rule and its antecedents.
- $\mathbb{F} = \mathsf{ConcSub}$, where $\mathsf{ConcSub}(A) = \{\mathsf{Conc}(B) \mid B \in \mathsf{Sub}(A), \mathsf{Conc}(B) \notin \mathcal{K} \cup \{\mathsf{Conc}(A)\}\}$. Then the explanation contains the sub-conclusions that were derived in the construction of the argument.

We can now turn to a discussion on explanations for the (non-)acceptance of (not) fraud.

It is a Case of Fraud (Acceptance of A_6/*Non-acceptance of* A_1). The basic explanation here is that A_6 *can be accepted, when* A_3 *and* A_4 *are accepted as well*. In terms of the conclusions of the arguments, we say that it is *a case of fraud* (A_6), because the product is *fake* (A_3) and the *counter party did not deliver* (A_4). When considering the variations of $\mathbb{F}$, further explanations can be considered. For example, it is a case of fraud, because:

- the *complainant delivered* (C_1) and the *counter party did not deliver* (A_4) and there is a rule (R_7) that states that from these conclusions it can be derived that it is a *case of fraud* (A_6), i.e., $\mathbb{F} = \mathsf{AntTop}$. Such an explanation can be used by an analyst at the police, who is familiar with the rules and wants to understand what parts of the law were applied.

- the *complainant paid* and the product seems *suspicious*, i.e., $\mathbb{F}$ = Prem. At the moment, the system returns this type of explanation, which can be used by the complainant, to understand what parts of the report made the system derive this conclusion.
- the *complainant delivered* (C_1), the *counter party did not deliver* (A_4) and the product is *fake* (A_3), i.e., $\mathbb{F}$ = ConcSub. Explanations like this provide insight into the reasoning process of the system: it shows the sub-steps that were taken. It might be useful for an analyst at the police, who wants more insight into the reasons than only the last step, but also for the complainant, who might not be convinced by an explanation that only contains information provided in the complaint itself.

Similar explanations can be given for not(it is *not a case of fraud*), i.e., that A_1 is not accepted. This follows since the main reason that A_1 cannot be accepted is the fact that A_6 is accepted.

It is Not a Case of Fraud (Acceptance of A_1/Non-acceptance of A_6). While A_1 can be explained by the acceptance of A_1 (since it can defend itself against the attack from A_6), additional arguments defend A_1 as well (i.e., A_2 and A_5 defend A_1 against the attack from A_6 as well). To give an overview of the possible explanations, we consider here the most extensive set of arguments: A_1, A_2 and A_5. In terms of the conclusions of the arguments, it follows that it is not a case of fraud, because the *counter party has delivered* and the product is *not fake*. Similarly as above, we can also consider other explanations based on elements of arguments: It is not a case of fraud, because:

- the *wrong product* was delivered and there is a rule (R_2) that states that usually, when the wrong product is delivered, it is not a case of fraud, i.e., $\mathbb{F}$ = AntTop. Note that this explanation is the same, whether we consider A_1 to be an explanation for its own acceptance, or the arguments A_2 and A_5 are considered as well.
- the *wrong product* was delivered and an *investigation* shows that there is no problem with the product, i.e., $\mathbb{F}$ = Prem. If A_5 is not a part of the explanation, then this explanation only contains the information that the *wrong product* was delivered.
- the *counter party has delivered* (A_2) and the product is *not fake* (A_5), i.e., $\mathbb{F}$ = ConcSub. Note that, in the case A_1 is its own acceptance explanation, no sub-conclusions are derived in the process.

Like in the case above, the explanations that it is not(*a case of fraud*) is similar to the explanations for *not a case of fraud*. This follows since the argument for *a case of fraud* (A_6) is attacked by each of the arguments considered here (i.e., A_6 is attacked by A_1, A_2 and A_5).

The suggested explanations above are not too extensive for the given example. However, a rule might have many antecedents, a conclusion might be based on many knowledge base elements or the derivation might be long, resulting in many sub-conclusions. It is therefore useful to consider how we can reduce the size of explanations. To this end, it has been argued that humans select their explanations in a biased manner. Selection happens based on e.g., simplicity, generality, robustness – see [17] for an overview on findings for the social sciences on how humans come to their explanations and how this could be applied in artificial intelligence. In the next section we will consider two ways of reducing the size of explanations. Given the space restrictions and since the basic explanations were similar for acceptance and non-acceptance, we only discuss acceptance explanations.

3.2   Necessity and Sufficiency

Necessity and sufficiency in the context of philosophy and cognitive science are discussed in, for example, [14,15,23]. Intuitively, an event Γ is *sufficient* for Δ if no other causes are required for Δ to happen, while Γ is *necessary* for Δ, if in order for Δ to happen, Γ has to happen as well.[7]

Sufficiency. In terms of arguments, one could say that a set of arguments is *sufficient* for the acceptance of some argument, if by accepting those arguments the argument can also be accepted (i.e., that the set of arguments defends the argument against all its attackers). For example, in the cases above:

- it was already mentioned that the acceptance of A_1 (that it is *not a case of fraud*) can be explained by the argument itself, but also by $\{A_1, A_2\}$, by $\{A_2, A_5\}$ and by $\{A_1, A_2, A_5\}$. Each of these sets is sufficient for the acceptance of A_1. If one were interested in *minimal sufficiency*, then the argument itself would be enough.
- for the argument A_6 (that it *is a case of fraud*) the arguments A_3 and A_4 have to be accepted. Thus there is only one sufficient set: $\{A_3, A_4, A_6\}$.

Formally, given $\mathcal{AF} = \langle \mathrm{Args}, \mathcal{A} \rangle$ and accepted argument $A \in \mathrm{Args}$:

- $\mathsf{S} \subseteq \mathrm{Args}$ is *sufficient for the acceptance* of A if for each $B \in \mathsf{S}$, there is an attack-path from B to A,[8] S is conflict-free and S defends A against all its attackers.

We denote by $\mathsf{Suff}(A) = \{\mathsf{S} \subseteq \mathrm{Args} \mid \mathsf{S}$ is sufficient for the acceptance of $A\}$ the set of all sufficient sets of arguments for the acceptance of A. With this sufficient explanations can be defined:

Definition 6. *Let* $\mathcal{AF} = \langle \mathrm{Args}, \mathcal{A} \rangle$ *be an AF and suppose that* $A \in \mathrm{Args}$ *is accepted. Then:* $\mathsf{Acc}(A) \in \mathsf{Suff}(A)$.

For minimally sufficient explanations $\mathsf{Acc}(A) \in \min \mathsf{Suff}(A)$, *where minimality can be taken w.r.t.* $\subseteq$ *or the number of arguments in a set.*

---

[7] See [6] for the technical details, in this paper we focus on the application of necessary and sufficient explanations.

[8] There is an attack path from B to A if there are $C_1, \ldots, C_k \in \mathrm{Args}$ such that $(B, C_1), (C_1, C_2), \ldots, (C_{k-1}, C_k), (C_k, A) \in \mathcal{A}$.

The resulting explanations for $\mathcal{AF}_1$ are as described before the formal definitions.

When the structure of the arguments is known we can again look at explanations in terms of the elements of the arguments. Note that when explanations should contain minimal sufficient sets of elements (e.g., minimal sufficient sets of premises or sub-conclusions) one should not simply take the elements of the minimal sufficient set of arguments, but rather compare the sets of elements obtained from each sufficient set and compare those sizes.

Definition 7. *Let $\mathcal{AF}(AT) = \langle Args, \mathcal{A} \rangle$ be an AF, based on an argumentation theory AT and suppose that $\phi \in \mathcal{L}$ is accepted. Then:*

$$\mathsf{Acc}(\phi) \in \bigcup \{\mathbb{F}(\mathsf{Suff}(A)) \mid A \in Args \text{ and } \mathsf{Conc}(A) = \phi\}.$$

$$\mathsf{Acc}(\phi) \in \min \bigcup \{\mathbb{F}(\mathsf{Suff}(A)) \mid A \in Args \text{ and } \mathsf{Conc}(A) = \phi\}.$$

In our example we have that:

- receiving the *wrong product* is sufficient for that it is *not a case of fraud*, if $\mathbb{F} = \mathsf{Prem}$ and, combined with the rule that usually when the wrong product is received it is not a case of fraud, when $\mathbb{F} = \mathsf{AntTop}$.
- the premises that the *complainant paid* and that the product seems *suspicious* are sufficient for that it is *a case of fraud*. When $\mathbb{F} = \mathsf{AntTop}$, the rules from A_3 (if the product seem suspicious then usually the product is fake), A_4 (if the product is fake then usually the counter party did not deliver) and A_6 (if the complainant delivered and the counter party did not deliver it is usually a case of fraud) form the explanation, together with their antecedents that the product seems *suspicious*, the product is *fake*, the *complainant delivered* and the *counter party did not deliver*.

Given the structure of $\mathcal{AF}_1$, there is not much difference between the basic explanations and sufficient explanations. Therefore, we introduce the following example, this time not based on a scenario from the police.

Example 1. Let $AT_2 = \langle AS_2, \mathcal{K}_2 \rangle$, where the rules in AS_2 are such that, with $\mathcal{K}_2 = \{r, s, t, v\}$, the following arguments can be derived:[9]

$$A: \; s, t \overset{d_1}{\Rightarrow} u \qquad\qquad B: \; p, \neg q \overset{d_2}{\Rightarrow} \neg n(d_1) \qquad\qquad C: \; r, s \overset{d_3}{\Rightarrow} q$$

$$D: \; v \overset{d_4}{\Rightarrow} \neg q \qquad\qquad E: \; r, t \overset{d_5}{\Rightarrow} \neg p \qquad\qquad F: \; v \overset{d_6}{\Rightarrow} p$$

See Fig. 2 for a graphical representation of the correspond AF $\mathcal{AF}_2$. Note that, like for $\mathcal{AF}_1$, all arguments can be accepted.

On an abstract level, in order to accept A either C or E should be accepted as well. To accept B, one has to accept both D and F. Sufficient explanations for the acceptance of A are $\{C\}$, $\{E\}$, $\{C, E\}$, but also $\{C, F\}$ and $\{D, E\}$ (since these still include C resp. E). Minimally sufficient explanations are $\{C\}$ and $\{E\}$ and $\{D, F\}$ is the only (minimally) sufficient explanation for the acceptance of B.

---

[9] We ignore again the arguments based on the elements of $\mathcal{K}_2$.

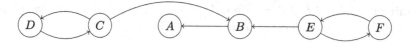

Fig. 2. Graphical representation of the abstract argumentation framework $\mathcal{AF}_2$.

When looking at the structure of the arguments, taking $\mathbb{F} = \mathsf{Prem}$, we have that $\{r, s\}$, $\{r, t\}$ and $\{r, s, t\}$ are some of the sufficient sets for the acceptance of u and $\{v\}$ is sufficient to accept an exception to the rule d_1.

Necessity. In terms of arguments, an argument can be understood as *necessary* if without that argument, the considered argument could not be accepted. For $\mathcal{AF}_1$, the (minimal) sufficient sets of arguments are also the necessary arguments: A_1 is the only necessary argument for the acceptance of A_1, while there are three arguments necessary for the acceptance of A_6: A_3, A_4 and A_6.

Formally, given an AF $\mathcal{AF} = \langle \text{Args}, \mathcal{A} \rangle$ and $A \in \text{Args}$ an accepted argument:

- $B \in \text{Args}$ is *necessary for the acceptance* of A if there is an attack-path from B to A and if $B \notin S$ for some admissible set $S \subseteq \text{Args}$, then $A \notin S$.

We denote by $\mathsf{Nec}(A) = \{B \in \text{Args} \mid B \text{ is necessary for the acceptance of } A\}$ the set of all arguments that are necessary for the acceptance of A. With this necessary explanations can be defined:

Definition 8. *Let* $\mathcal{AF} = \langle \text{Args}, \mathcal{A} \rangle$ *be an AF and suppose that* $A \in \text{Args}$ *is accepted. Then:* $\mathsf{Acc}(A) = \mathsf{Nec}(A)$.

For an illustration of the difference between sufficiency and necessity, consider the argument A_2. Then $\{A_2\}$ is sufficient for its own acceptance, but $\{A_5\}$ is also sufficient for its acceptance. Therefore, there is no argument that is necessary for the acceptance of A_2 (see also Proposition 2).

Similar reasoning as in the case of sufficiency applies to necessary explanations based on the elements of the arguments. One can collect premises, rules and sub-conclusions from the necessary arguments. However, in terms of elements we can be more detailed. For this we need the following results.

Proposition 1. *Let* $\mathcal{AF} = \langle \text{Args}, \mathcal{A} \rangle$ *and let* $A \in \text{Args}$ *be accepted. Then* $\mathsf{Acc}(A) = \emptyset$ *iff there is no* $B \in \text{Args}$ *such that* $(B, A) \in \mathcal{A}$, *where* Acc *can be defined as in Definition 4 or 6.*

Proposition 2. *Let* $\mathcal{AF} = \langle \text{Args}, \mathcal{A} \rangle$ *and let* $A \in \text{Args}$ *be accepted. Then* $\mathsf{Nec}(A) = \emptyset$ *if* $\bigcap \mathsf{Suff}(A) = \emptyset$.

While, in view of the above results, a necessary explanation for arguments might be empty, one could still collect necessary premises, rules and sub-conclusions. We therefore define:

Definition 9. *Let $\mathcal{AF}(AT) = \langle Args, \mathcal{A} \rangle$ be an AF, based on an argumentation theory AT and suppose that $\phi \in \mathcal{L}$ is accepted. Then:*

$$\mathsf{Acc}(\phi) = \bigcap \{\mathbb{F}(\mathsf{Suff}(A)) \mid A \in Args \text{ and } \mathsf{Conc}(A) = \phi\}.$$

To illustrate the difference between necessary and sufficient explanations and the application of the above definition, we return to the AF $\mathcal{AF}_2$ from Example 1.

Example 2. For the AF $\mathcal{AF}_2$ we have that for the acceptance of A no argument is necessary. But, when $\mathbb{F} = \mathsf{Prem}$ we have that r is necessary. For the acceptance of B both D and F are necessary and, when $\mathbb{F} = \mathsf{Prem}$, v is necessary.

3.3   Contrastive Explanations

Another relevant way in which humans structure and select their explanations is *contrastiveness* [14,16,17]: when people ask 'why P?', they often mean 'why P rather than Q?' – here P is called the fact and Q is called the foil [14]. The answer to the question is then to explain as many of the differences between fact and foil as possible.[10]

When humans provide a contrastive explanation, the foil is not always explicitly stated. While humans are capable of detecting the foil based on context and the way the question is asked, AI-based systems struggle with this.

When the foil is not explicitly stated, formal argumentation has an advantage over some other approaches to AI because it comes with an explicit notion of conflict (i.e., the attack relation). This allows us to derive a foil when none is provided. For example, given an argument one could take as the foil:

- all the arguments that directly attack or defend it;
- all the arguments that directly or indirectly attack or defend it.

In the context of structured arguments, one can also look at the claims of the arguments and take the foil to be arguments with conflicting conclusion.

Given an argument of which the acceptance status should be explained (the fact) and a foil, a contrastive explanation contains those arguments that explain:

- the acceptance of the fact and the non-acceptance of the foil;
- the non-acceptance of the fact and the acceptance of the foil.

Definition 10. *Let $\mathcal{AF} = \langle Args, \mathcal{A} \rangle$ be an AF, let $A \in Args$ be the fact and $S \subseteq Args$ be the foil (for example defined by the direct attacking arguments of A). Suppose that A is accepted [resp. non-accepted] and that each $B \in S$ is non-accepted [resp. accepted]. Then:*

$$\mathsf{Cont}(A, S) = \mathsf{Acc}(A) \cap \left(\bigcup_{B \in S} \mathsf{NotAcc}(B) \right)$$

$$\mathsf{ContN}(A, S) = \mathsf{NotAcc}(A) \cap \left(\bigcup_{B \in S} \mathsf{Acc}(B) \right).$$

---

[10] See [5] for the technical details, in this paper we focus on the application of contrastive explanations.

When $\mathsf{Cont}(A, \mathsf{S}) = \emptyset$ *(the case for* ContN *is similar) the explanation will return a pair:* $\mathsf{Cont}(A, \mathsf{S}) = \langle \mathsf{Acc}(A), \bigcup_{B \in \mathsf{S}} \mathsf{NotAcc}(B) \rangle$.

Thus, given explanations for the acceptance [resp. non-acceptance] of the fact and the non-acceptance [resp. acceptance] of the foil the contrastive explanation returns the intersection of these explanations when it is not empty (otherwise it would simply return those two explanations). An empty contrastive explanation rarely happens. In particular:

Proposition 3. *Let* $\mathcal{AF} = \langle \mathit{Args}, \mathcal{A} \rangle$ *be an AF. Let* $A \in \mathit{Args}$ *and let* $\mathsf{S} \subseteq \mathit{Args}$ *be such that for each* $B \in \mathsf{S}$, $(B, A) \in \mathcal{A}$. *Then* $\mathsf{Cont}(A, \mathsf{S}) = \emptyset$ *[resp.* $\mathsf{ContN}(A, \mathsf{S}) = \emptyset$] *implies that* $\mathsf{Acc}(A) = \emptyset$ *[resp.* $\bigcup_{B \in \mathsf{S}} \mathsf{Acc}(B) = \emptyset$].

Intuitively, this shows that a contrastive explanation is only empty if the fact is not attacked at all [resp. no argument in the set of foils is attacked]. To illustrate contrastive explanations we introduce another scenario, this time about a possible malafide webshop, based on the application in [18].

Example 3. Consider a language $\mathcal{L}_3$, containing the atoms *cf* (a complaint was filed), *m* (the webshop is malafide), *iw* (an investigation is done), *sa* (the url is suspicious), *rc* (the complaint is retracted), *kp* (the webshop owner is known by the police), *ka* (the address is registered at the chamber of commerce), *rr* (the registration was recently retracted) and their negations.

Let $\mathrm{AT}_3 = \langle \mathrm{AS}_3, \mathcal{K}_3 \rangle$, where the rules in AS_2 are such that, with the language $\mathcal{L}_3$ and $\mathcal{K}_3 = \{ \mathit{cf}, \mathit{rc}, \mathit{sa}, \mathit{ka}, \mathit{kp}, \mathit{rr} \}$, the following arguments can be derived:

$$A_1 : \mathit{cf} \quad A_2 : \mathit{rc} \quad A_3 : \mathit{sa} \quad A_4 : \mathit{ka} \quad A_5 : \mathit{kp} \quad A_6 : \mathit{rr}$$

$$B_1 : A_1 \overset{d_1}{\Rightarrow} \mathit{iw} \qquad B_2 : A_2 \overset{d_2}{\Rightarrow} \neg n(d_1) \qquad B_3 : A_5 \overset{d_5}{\Rightarrow} \neg \mathit{rc}$$

$$B_4 : B_1, A_3 \overset{d_3}{\Rightarrow} m \qquad B_5 : A_4 \overset{d_4}{\Rightarrow} \neg n(d_3) \qquad B_6 : A_6 \overset{d_6}{\Rightarrow} \neg \mathit{ka}.$$

See Fig. 3 for a graphical representation of the corresponding AF $\mathcal{AF}(\mathrm{AT}_3)$. As in our previous examples, each of the arguments can be accepted.

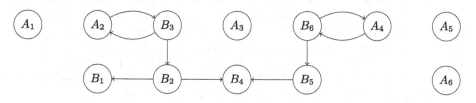

Fig. 3. Graphical representation of the AF $\mathcal{AF}(\mathrm{AT}_3)$.

To start with, we have the following basic explanation for the acceptance of *m* (i.e., the webshop is malafide): the owner of the webshop is known by the police (*kp*) and the registration at the chamber of commerce was recently retracted (*rr*), from which it follows that no exceptions could be derived.

Basic explanations are exhaustive: all the reasons why the webshop is malafide are provided. With our contrastive explanations, the explanation can focus on an explicit contrastive question. For example: the webshop is malafide rather than that there is an exception to rule d_1, since the owner is known by the police (kp); and the webshop is malafide rather than that there is an exception to rule d_3, since the registration was recently retracted (rr). Thus, the contrastive explanations are better tailored to one question and result in smaller explanations.

4   Conclusion

In this paper we have discussed how a general framework for explaining conclusions derived from an argumentation framework can be applied on top of the argumentation systems in use at the Dutch National Police. As an example we took the system in use to assist in the processing of complaints on online trade fraud. The ideas presented in this paper can also be applied to the other systems in use at the police as well as any other system based on argumentation frameworks as introduced in [7].

Recall from the introduction that, unlike other approaches to local explanations of argumentation-based conclusions [9–11,13,22], the framework that we applied can capture both acceptance and non-acceptance explanations, is not based on one specific semantics (although we only considered preferred semantics here) and allows to take the structure of arguments into account (i.e., explanations can be sets of premises or rules, rather than just sets of arguments). Moreover, we have shown how our framework can be used to study how findings from the social sciences (those collected in, e.g., [17]) can be implemented. The presented studies of sufficiency, necessity and contrastiveness are just the beginning. On the one hand, especially in the case of contrastive explanations, much more can be said about the individual concepts than we could present here. On the other hand, there are many other aspects of human explanation that have not been investigated yet.

In future work we will continue our study of integrating findings from the social sciences into our explanations. For example, we will study the notion of contrastiveness further, we will look into the robustness of explanations and we will consider further selection criteria. Additionally, for the applications at the Dutch National Police, we will implement the framework and conduct a user study on the best explanations for these specific applications and, possibly, the best explanations for other argumentation-based applications.

References

1. Atkinson, K., et al.: Towards artificial argumentation. AI Mag. **38**(3), 25–36 (2017)
2. Besnard, P., et al.: Introduction to structured argumentation. Arg. Comp. **5**(1), 1–4 (2014)

3. Bex, F., Testerink, B., Peters, J.: AI for online criminal complaints: from natural dialogues to structured scenarios. In: Workshop Proceedings of Artificial Intelligence for Justice at ECAI 2016, pp. 22–29 (2016)
4. Borg, A., Bex, F.: A basic framework for explanations in argumentation. IEEE Intell. Syst. **36**(2), 25–35 (2021)
5. Borg, A., Bex, F.: Contrastive explanations for argumentation-based conclusions. arXiv/CoRR abs/2107.03265 (2021). https://arxiv.org/abs/2107.03265
6. Borg, A., Bex, F.: Necessary and sufficient explanations for argumentation-based conclusions. In: Vejnarová, J., Wilson, N. (eds.) Symbolic and Quantitative Approaches to Reasoning with Uncertainty. Proceedings of ECSQARU 2021, pp. 45–58. Springer, Cham (2021). https://doi.org/10.1007/978-3-030-86772-0_4
7. Dung, P.M.: On the acceptability of arguments and its fundamental role in non-monotonic reasoning, logic programming and n-person games. Artif. Intell. **77**(2), 321–357 (1995)
8. Edwards, L., Veale, M.: Slave to the algorithm: why a 'right to an explanation' is probably not the remedy you are looking for. Duke Law Technol. Rev. **16**(1), 18–84 (2017)
9. Fan, X., Toni, F.: On computing explanations in argumentation. In: Bonet, B., Koenig, S. (eds.) Proceedings of AAAI 2015, pp. 1496–1502. AAAI Press (2015)
10. Fan, X., Toni, F.: On explanations for non-acceptable arguments. In: Black, E., Modgil, S., Oren, N. (eds.) TAFA 2015. LNCS (LNAI), vol. 9524, pp. 112–127. Springer, Cham (2015). https://doi.org/10.1007/978-3-319-28460-6_7
11. García, A., Chesñevar, C., Rotstein, N., Simari, G.: Formalizing dialectical explanation support for argument-based reasoning in knowledge-based systems. Expert Syst. Appl. **40**(8), 3233–3247 (2013)
12. Lacave, C., Diez, F.J.: A review of explanation methods for heuristic expert systems. Knowl. Eng. Rev. **19**(2), 133–146 (2004)
13. Liao, B., van der Torre, L.: Explanation semantics for abstract argumentation. In: Prakken, H., Bistarelli, S., Santini, F., Taticchi, C. (eds.) Proceedings of COMMA 2020, pp. 271–282. IOS Press (2020)
14. Lipton, P.: Contrastive explanation. Roy. Inst. Philos. Suppl. **27**, 247–266 (1990)
15. Lombrozo, T.: Causal-explanatory pluralism: how intentions, functions, and mechanisms influence causal ascriptions. Cogn. Psychol. **61**(4), 303–332 (2010)
16. Miller, T.: Contrastive explanation: a structural-model approach. CoRR abs/1811.03163 (2018). http://arxiv.org/abs/1811.03163
17. Miller, T.: Explanation in artificial intelligence: insights from the social sciences. Artif. Intell. **267**, 1–38 (2019)
18. Odekerken, D., Bex, F.: Towards transparent human-in-the-loop classification of fraudulent web shops. In: Villata, S., Harašta, J., Křemen, P. (eds.) Proceedings of JURIX 2020, pp. 239–242. IOS Press (2020)
19. Odekerken, D., Borg, A., Bex, F.: Estimating stability for efficient argument-based inquiry. In: Prakken, H., Bistarelli, S., Santini, F., Taticchi, C. (eds.) Proceedings of COMMA 2020, pp. 307–318. IOS Press (2020)
20. Prakken, H.: An abstract framework for argumentation with structured arguments. Arg. Comput. **1**(2), 93–124 (2010)
21. Samek, W., Wiegand, T., Müller, K.R.: Explainable artificial intelligence: understanding, visualizing and interpreting deep learning models. CoRR abs/1708.08296 (2017). http://arxiv.org/abs/1708.08296
22. Saribatur, Z., Wallner, J., Woltran, S.: Explaining non-acceptability in abstract argumentation. In: Proceedings of ECAI 2020, pp. 881–888. IOS Press (2020)
23. Woodward, J.: Sensitive and insensitive causation. Philos. Rev. **115**(1), 1–50 (2006)

Like Circles in the Water: Responsibility as a System-Level Function

Giovanni Sileno[1](✉), Alexander Boer[2], Geoff Gordon[1], and Bernhard Rieder[1]

[1] University of Amsterdam, Amsterdam, The Netherlands
g.sileno@uva.nl
[2] KPMG, Amsterdam, The Netherlands

Abstract. What eventually determines the semantics of algorithmic decision-making is not the program artefact, nor—if applicable—the data used to create it, but the preparatory (enabling) and consequent (enabled) practices holding in the environment (computational and human) in which such algorithmic procedure is embedded. The notion of responsibility captures a very similar construct: in all human societies actions are evaluated in terms of the consequences they could reasonably cause, and of the reasons that motivate them. But to what extent does this function exist in computational systems? The paper aims to sketch links between several of the approaches and concepts proposed for *responsible computing*, from AI to networking, identifying gaps and possible directions for operationalization.

Keywords: Responsibility · Responsible computing · Responsible AI · Responsible networking · Contextual integrity · Conditional contextual disparity

1 Introduction

The various research tracks denoted as *responsible, ethical, fair,* and *trustworthy AI* can be overall divided in two main families. On the one hand, works contributing to the discussion of what (ethical) principles should be applied, in all phases from conception to deployment, to algorithmic decision-making systems. On the other, works attempting to operationally define open concepts as e.g. "fairness" or "privacy" to be embedded during training or deployment of AI modules. The distance existing between these two approaches raises critical concerns on whether they can be bridged at all. This paper argues for a change of perspective. What eventually determines the semantics (meaning and performativity) of algorithmic decision-making is not the program artefact in itself, nor the data used to create it, but consists of preparatory (enabling) and consequent (enabled) practices holding in the environment in which the algorithmic procedure is embedded. For this reason, computational components need to be seen in

This research was partly supported by the UvA (RPA Human(e) AI seed grant) and by NWO (DL4LD project, no. 628.009.001).

V. Rodríguez-Doncel et al. (Eds.): AICOL-XI 2018/AICOL-XII 2020/XAILA 2020, LNAI 13048, pp. 198–211, 2021.
https://doi.org/10.1007/978-3-030-89811-3_14

an "ecological" perspective, i.e. whose working is entrenched with the operations of other computational modules, and whose deployment is driven-by and applies-on heterogeneous human social systems, characterized by competing interests, distinct socio-economic positions and possibly incompatible preferences.

In parallel work [20], we started exploring methods to investigate how "values" are generated, distributed, and translated between contextualized social processes and automatic/automated decision-making components; inspired by the idea of *encircling* introduced in security studies [4], we are studying how to approach *de facto* inaccessible or opaque entities by looking at what is occurring in their background (practices, ambient knowledge, etc.). The present paper, instead, is meant to take a position in the debate concerning the *system-design* part of the problem. Even acknowledging the primacy of (highly contextual and dynamic) human factors in setting the premises and the consequences of the system's activity, system designers and developers still need solutions to identify and reduce frictions deemed (or feared) to occur between computational and societal dimensions. With this objective in mind, the paper organizes insights coming from different domains, aiming to be "minimally complete" in highlighting the functions required to achieve a sound infrastructure for *responsible computing*.

The paper proceeds as follows. Section 1 contrasts a *data-flow* perspective against the most common data-centric ones. Section 2 reviews under a data-flow perspective two non-technical frameworks highlighting the role of context: *contextual integrity* [18], and *contextual demographic disparity* [28]. Section 3 elaborates on the function and functioning of *responsibility* as a cognitive mechanism, proposing the concept of "agentive responsibility". Section 4 considers a recent proposal on *responsible Internet* [12] revisiting the *accountability-responsibility-transparency* (ART) principles for AI [5] in the domain of networking, and elaborates on how to extend it to take into account what presented in the previous sections.

2   From Data to Data-Flow Problems

Most approaches emerging in responsible AI and related fields with respect to problems of *fairness* (non-discrimination) focus primarily on the problem of selecting or producing adequate data. Following the overview given in [8], one can for instance:

(1) purge the input data from sensitive elements at runtime,
(2) debias the sample data used during the training process,
(3) correct the network parameters used in the inferential model, or
(4) add an external module to produce unbiased output at aggregate level.

2.1   Computational Reflection

A relevant framework through which to look at these interventions is provided by the notion of *computational reflection*, i.e. the ability of a system to inspect and modify itself in order to improve its performance (see e.g. [2]), generally further distinguished in:

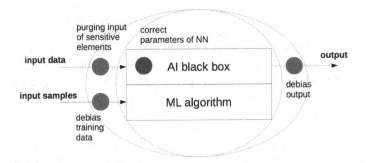

Fig. 1. Most common interventions proposed by algorithmic fairness solutions, illustrated in terms of computational reflection: behavioural (blue circles) or structural (red circle). (Color figure online)

a. *structural reflection*, concerned by non-contingent properties of the system (e.g. data structures, procedures);
b. *behavioural reflection*, concerned by the overall activity of the system, as described e.g. by requests/invocations.

In order to be effective, behavioural reflection requires the system to be aware of its "semantics" with respect to its environment, i.e. of the processes it may initiate, inhibit or may be involved into, whereas structural reflection only needs the system to be able to look at its own components (data structures, procedures) and modify them according to some criteria.

Through the lens of computational reflection, interventions of the types (1), (2), (4) become examples of behavioural reflection: they introduce additional modules to process the input before the output and/or after the core module, without modifying it structurally; (3) is instead an example of structural reflection, concerning in particular the neural network parameters (see Fig. 1). In all these cases the focus is clearly on *data*, either input data, output data or data relative to the model. The knowledge used to guide behavioural reflection does not go beyond the qualification of which types of data are sensitive/protected.

2.2   From Data to Data-Flow

Methods based on behavioural reflection (e.g. the blue circles in Fig. 1) suggests that, alternatively, one can see fairness as a problem of *data-flow*: i.e. of intervening or constraining adequately the connections between the data processing components.

On a fundamental level, any computational component can seen as an assemblage of lower-level computational components. Even a Turing machine can be mapped to a functionally equivalent distributed system, whose individual components activate other components, performing in turn activating actions, and so on. In fact, parallel models of computation have been proven to be more general than traditional sequential/procedural models (e.g. [13]). Other computational models, like those applied for computer networks or for neural networks, can be directly looked through actor-based lenses. See for instance the recent introduction of *agnostic networks* [9], based on the intuition that training could be based not on modifying the weights of the network components, but on tinkering with their connections. This transforms our perspective on the machine learning problem from being *metrical* in nature (e.g. targeting an adequate latent space given a fixed network topology), to being explicitly *topological*.

This change of perspective facilitates the convergence of various problems into one of *responsible processing of informational flows*.[1] For instance, privacy can be seen as a set of limited rights and abilities controlling disclosure-of (i.e. channels transmitting) self-information. *Differential privacy* methods [7], introduced to protect against the reconstruction of data of individuals by intersection of a sufficient number of queries, work by adding a certain amount of noise to the output provided by a curator (that is, a module responding to queries), or by means of a stochastic curator. Both differential privacy templates can be functionally interpreted as an addition of external noise channels, destroying part of the information by interference. Opposite to protective measures, there exist initiatives and solutions that support or facilitate the construction of informational connections, e.g. as those driven by the FAIR data principles (*findable-accessible-interoperable-reusable*) [29], and, with distinct bases and purposes, the various Open Data initiatives (for governmental data, research data, etc.).

2.3 Responsible Computing as Responsible Disclosure

On a functional level, a data-flow perspective highlights the pivotal role of the *control of information disclosure*, which can be—in terms of information—*negative* (i.e. restricting, limiting disclosure) or *positive* (i.e. enabling, granting it). However, this qualification does not say anything about the practical (i.e. non-informational) effects of processing, or more precisely, of the use of such processing.

Let us consider a machine learning application used in support for decision-making, schematized in Fig. 2. Let us separate the computational module running the machine learning method from the module producing inferences by means of the model parameters extracted by the first. The sample data used for training typically follows a different routing than the data used for prediction; for instance, they come from two distinct data-providers, which in turn collect data belonging to different data-subjects. The inferential module can be used by several data-processors, in turn controlled by distinct data-users, that possibly utilize the output produced by the processors for their specific purposes. This derived information may intervene in the users' decision-making, determining *counterfactually* a certain decision resulting in positive or negative action. Depending on the actual environmental disposition, each individual decision (suppose e.g. about recruitment) would produce effects affecting not only

---

[1] We include in the term "flow" interface aspects: how outputs are presented, and how user-side interventions are able to act on processing.

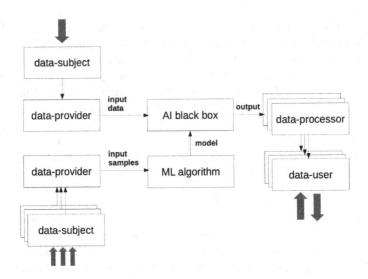

Fig. 2. Schematic data and value flows associated to training and use of a machine-learning application processing personal data in support to decision-making.

the data-user (e.g. the recruiter), but typically also the data-subject whose data was processed (e.g. the target candidate), and other parties (e.g. competitor candidates, the families of candidate and competitors, their communities, etc.). Potentially, it may indirectly affect also the data-subjects that provided the data for the training, depending on the ramification of the consequences.

This schematization serves two purposes here. First, it highlights that contemporary technological challenges are not only a matter of responsible machine learning, but rather of *responsible computing* (including processing, data-sharing, networking, etc.). Under this lens, privacy can be seen as a set of limited rights or abilities to control disclosure of self-information when this information has the potential to affect outcomes for the individual. Dually, excess of control on information disclosure opens up to abilities to obtain unfair outcomes, as (direct or indirect) *discrimination*, or even *fraudulent schemes*.[2] Second, removing the boundaries between components internal and external to the system, and looking at them as a network in which certain information flows (or does not flow) producing some (positive, negative) impact on the social participants, unveils that the main difference between structural and behavioural reflection concerns the amount of observability/controllability on the network required for the reflection to be applied. As soon as we consider humans or other artificial components producing and further processing that data, the depth required for

---

[2] Administrative, corporate and other types of frauds are typically conducted by exploiting an overlap of roles over the same identity (a physical person, an organization, etc.) [1,24]; such an overlap enables access to information that should be otherwise inaccessible.

a proper behavioural reflection naturally increases, creating figuratively "circles in the water" of components interfering/interacting with each other.

3 The Role of Context

At face value, technical solutions as those proposed for *algorithmic fairness* or *differential privacy* do not bring to the foreground that the legitimacy of a certain query or computation is not a problem of the processing in itself, but of the context in which such a processing is performed. For instance, the use of sensitive data such as ethnicity (or proxies of it) is deemed unfair in tasks that produce effects of social discrimination (e.g. deciding the premium for an insurance policy), but not necessarily in other tasks (e.g. deciding the colour/style of a dress in an e-shop). As a paradoxical situation, would we need differential privacy when we are querying our own personal data?

More in detail, interventions for algorithmic fairness are meant primarily for three purposes [3]:

- *anti-classification*: decisions are taken without considering explicitly sensitive or protected attributes (ethnicity, gender, etc. or any proxies of those);
- *classification parity*: performance of prediction as measured e.g. by false positive and false negative rates are equal across the groups selected by protected attributes;
- *calibration*: outcomes of prediction is independent of protected attributes.

These purposes are reflected in distinct definitions that are incompatible amongst each other, and, furthermore, they can produce effects which are still detrimental to the protected classes [3]. Even at a technical level, it is recognized that something is missing in the picture. Indeed, the requirement of adding context in issues about privacy and discrimination is the starting point in several works looking at the problem from a higher-level (typically socio-legal) perspective. We will consider here two examples in this respect.

3.1 Contextual Integrity

The well-known framework of *contextual integrity* by Nissenbaum [18] makes clear that privacy can not be defined in absolute terms, but depends on several parameters, including the actors involved (data subject, sender, recipient), the type of information, the basis for disclosure/transmission, and various contextual elements (e.g. interests of parties, societal norms and practices). For instance, consent acts as a basis for disclosure of personal data (e.g. biometrical information) for a specific purpose (e.g. healthcare research), and any other use (e.g. marketing) would be a breach of contextual integrity. However, in some cases (e.g. for medical necessity), the processing of the same personal data without consent will not count as a breach of contextual integrity, because there are legal or even moral norms making clear the presence of a situation (e.g. where survival is at stake) providing a distinct basis for disclosure. Context is defined not only by purpose, but also by *domain knowledge*, associated with that purpose in the current situation (e.g. norms and practices, and roles related to those), which is used by the subject and other parties to form their expectations. The "ecological" nature of all these contextual elements makes it difficult if not impossible to capture them *monistically* within the informational artefacts which are target of directives about disclosure. In other words, context maps to layers well beyond the system boundaries, but still, it provides crucial terms to define the system "semantics" required for an appropriate behavioural reflection.

3.2 Contextual Demographic Disparity

Recent work by Wachter et al. [28] analyzes the concept of *contextual demographic (dis)parity* (CDD) (based on the measure of *conditional (non-)discrimination* proposed by Kamiran et al. in [15]), evaluating it with respect to the decisions of the European Court of Justice on cases of discrimination. The authors highlight the complexity of automatizing decisions about discrimination, caused by the diverse "composition of the disadvantaged and advantaged group, the severity and type of harm suffered, and requirements for the relevance and admissibility of evidence". They suggest therefore to separate (a) the assessment of automated discrimination (and argue that the best measure for this is CDD) from (b) the actual judicial interpretation. Rephrased in the terms of behavioural reflection: algorithmic-driven assessment can explore a larger coverage of the network, but further layers exist beyond that, requiring human experts to stay in the decision-making loop.

Let us have a further look at CDD. Suppose a norm aims to protect certain groups of people (identified by means of protected attributes), and suppose a certain decision process produces a positive or negative outcome (according to a certain value structure), dividing people whose data is under scrutiny in two classes, *advantaged* and *disadvantaged*. The authors propose that a *prima facie* assessment of discrimination can be expressed if $A_R < D_R$ for any R, where R are conditions used to divide the population into sub-populations, A_R is the proportion of people of the sub-population with protected attributes put in the advantaged class, D_R is the proportion with protected attributes put in the disadvantaged class.

But how to decide upon the possible R to take into account? Following Kamiran [15], these conditions should be *explanatory*, i.e. they should hypothetically explain the outcome even in the absence of discrimination against the protected class. For instance, a reason for different salaries between men and women might be different working hours. Indeed, as argued by Pearl [19], the only way out of Simpson's paradox (opposite conclusions using different granularity of observation) is to deal with *causation*. However, questions about "what caused what" have also a strong connection with the idea of *responsibility*. This suggests that other elements may need to be added to the picture in order to evaluate the "reverberations" of the agents' actions throughout the system.

4   Function and Types of Responsibility

Human communities exhibit ascription of responsibility as a spontaneous, seemingly universal behavior, but variants of this construct can be found even in technical and abstract domains. Here we propose a simplified outline, with no pretension of being exhaustive, or covering all the available views on those matters. Our aim is to highlight some specific characteristics that are generally overlooked in technical settings.

4.1   Function of Responsibility

From a systematic standpoint, responsibility attribution is functional to the *localization* of *failures* in constructions whose components are deemed to be *autonomous*. This construct applies not only to social systems, but to any type of system (natural, artificial, etc.), as it is prerequisite to properly implement remedy/repair function. Software engineers, for instance, are suggested to follow e.g. the *single-responsibility* design principle (one module encapsulates one functionality)—one of the SOLID principles for object-oriented programming, and also referred to in *agile development* [17]—because it helps to localize bugs.

4.2   Epistemic Responsibility

Practical failures directly map to failures of *expectations*, namely with respect to the mechanisms attributed to the system or its components. On a conceptual level, those mechanisms, and the events fed to them as inputs, contribute to our understanding of the system as a whole, and for this reason they can be given *epistemic responsibility*. Also in this case, lack of predictive power exhibited at the occurrence of a failure triggers remedy/repair functions, i.e. an investigation on whether the provided input is correct, or the search of better mechanisms to be assigned, or constructed if not available. In stratified systems such *retrodictive, explanatory* construction might be a recursive process, targeting defective lower-level components. However, a similar analysis can be also executed in absence of failure, to understand on which basis the system works. Indeed, *explainable AI* techniques leverage concepts as e.g. *Shapley value* (SHAP [16])—the payoff a player can expect in a coalition game for its contribution to the outcome—to interpret the contribution of a certain feature in producing a certain conclusion.

4.3   Causal Responsibility

Returning on the practical dimension, *causal responsibility* is meant to identify which ones, amongst the components involved in a chain of events, *actually caused* (or prevented) a certain outcome. Several properties have been identified (not without discussion) in the literature related to actual causation, as e.g. *counterfactuality* (the outcome would not have occurred if that agent had behaved otherwise), and *sufficiency* (agent behaviour was the ultimate determinant of the outcome). In general, however, some degree of responsibility is also assigned to *concurrent contributions*, i.e. to enabling conditions that allowed a sufficient event to occur; a problem associated to this extension is to identify the relative contribution of causes (see e.g. the experiments in [26]).

4.4 Moral Responsibility

Moral responsibility builds upon causal responsibility (although in some circumstances it might overdetermine it), but it also presupposes a preferential structure (or of an underlying value structure) about outcomes in the world. Although certain contributions in the analytical literature (e.g. [10]) neglect this aspect, *blame* or *praise* would not make sense for morally irrelevant outcomes.

Empirical studies (e.g. [22], for a unifying computational model see e.g. [27]) suggest that moral responsibility:

– may hold for actions merely initiating potential causes of an outcome;
– grows with the impact of the outcome;
– is diminished e.g. if the action is not under the (expected) control of the agent, or the outcome is (justifiably) not foreseeable from her standpoint.

4.5 Agentive Responsibility

Rather than facing the question of what makes an agent a moral agent, we can more conservatively identify, considering the previous concepts, three requirements for assessing *agentive responsibility*:

1. The agent has the *ability to control* its behaviour;
2. It has the *ability to foresee* the associated outcomes;
3. It has the *ability to assess* their impact according to a preferential/value structure.

None of these three abilities can be absolute. In general, they can be attributed to any (direct and indirect) participants of an interaction, depending on their characteristics and role in the processing network. For instance, a dedicated module (cf. an expert) is expected to have better controllability and foreseeability than a general purpose module (cf. a layman person). Furthermore, they are all context dependent—and the definition of context may not be consistent across observers. Note that foreseeability and assessment of impact play a central role in formulating *risk*.

4.6 Accountability, Liability

If responsibility is concerned primarily by actions (or activities), *accountability* is generally seen as concerned by providing reasons and justifying those actions (or their omission). Additionally, the occurrence of unmet shared expectations might entail consequences, especially in the presence of a (semi-)formalized system of norms: *liability* refers to potential duties (e.g. paying damages) associated to those failures, or to other special contexts.

5   Operationalizing Responsible Computation

Several contributions in the field of *ethical AI* have presented a number of principles for the design and deployment of artificial devices. Consider for instance the ART principles [5]: *accountability*: motivations for the decision-making (values, norms, etc.) needs to be explicit; *responsibility*: the chain of (human) control (designer, manufacturer, operator, etc.) needs to be clear; *transparency*: actions need to be explained in terms of algorithms and data, and it should be possible to inspect them. Or the two requirements for *meaningful* (human) *control* [23]: *tracing*: the system needs to be able to trace back the outcome of its operations to specific directives given by humans during design or operational phases; *tracking*: the system needs to respond to (moral) reasons deemed relevant by directives given by humans guiding the system and to relevant facts in the environment in which the system operates. Or still, the seven requirements (human oversight, technical robustness, privacy and data governance, transparency, fairness, well-being, and accountability) identified by the expert group appointed by the European Commission [14].

At the moment, however, there is no framework bridging those higher-level principles to the abstraction level of technical solutions as e.g. algorithmic fairness and differential privacy. Impediments can be identified both on a societal dimension (explicit power allocations are conflictual in nature) and from an operational point of view (e.g. policies are expressed at different levels of abstraction, are dynamic, etc.). Additionally, those higher-level proposals tend to look at technological artefacts as essentially monolithical and the computational domain as separated from the human domain.

5.1   Responsible Networking

Interestingly, a recent paper by Hesselman et al. on the concept of *responsible Internet* [12] takes an orthogonal view over this matter, both in terms of operationalization, and of decentralization. The authors do not focus on the processing of data for decision-making, but on its transmission across the network (cf. the data-flow view of Sect. 2): "the Internet has only one high-level task, which is to securely and reliably provide end-to-end communications". This task needs to be solved on a decentralized architecture with distributed ownership and control.

The paper revisits and slightly modifies the ART principles [5], operationalizing them on the dimensions of data and infrastructure. For instance, *data transparency* holds if the system is able to describe how network operators transport and process a certain data-flow, whereas *infrastructure transparency* concerns instead the properties and relationships between network operators (location, software, servers, etc.); *data accountability* holds if network operators explain the processing of specific data flows, e.g. their routing decisions or incidents during transmission; *infrastructure accountability* means that network operators explain their infrastructural design decisions. Instead of responsibility, however, Hesselman et al. prefer to refer to *controllability*, to focus more on the ability of users to specify how network operators should handle their data (generally by means of *path control*), and to the ability of infrastructure maintainers to set constraints over network operators.[3] Note that for implementing accountability, norms on which decisions are based need to be made explicit.

The authors also sketch an architecture of how a responsible Internet could work, consisting of three main components: a *network inspection plane* (NIP), enabling users to query the infrastructure for details about its internal operations in terms of network operators; a *network control plane* (NCP), enabling users to specify their expectations on the data which is transmitted by network operators, based on network descriptions; a *policy framework* (POL), enabling infrastructure maintainers to specify policies and have network operators abiding to certain norms, by means of auditing or other enforcement techniques.

5.2   From Operational to Agentive Responsibility?

How does this more operational view on responsibility relate to the properties of responsibility sketched in the previous section?

Accountability and transparency are instrumental to the ascription of responsibility in the moment of failure; they refer to two distinct standpoints over the investigated component, respectively at *functional/extra-functional* levels (accountability), and *non-functional* or implementation level (transparency). The choice of the concept of "controllability" rather than "responsibility" highlights the requirement of setting up the control structure that enables licit outcomes, and prevents illicit outcomes to occur.

As we saw in the previous sections, however, (computational) agentive responsibility is not only a matter of controllability, but also of foreseeability, and of the ability of the agent of assessing foreseen outcomes in terms of a given preferential/value structure. Even if (part of) the preferential/value structure (of the user, infrastructure maintainer, etc.) can be considered to be part of the input exploiting controllability, the picture implicitly misses the contextual domain knowledge necessary for the agent to make a proper judgement, and that users will seldom have. To correct this, each agent (e.g. a network operator) should in principle autonomously assess its own and other agents' conduct, informed by (i) user policies and norms, (ii) known and potentially relevant scenarios (together with some information about their relative occurrence), attempting to form a properly grounded *risk assessment*.[4] In this view, solutions for algorithmic fairness or differential privacy would be controlled instrumentally to reduce dynamically identified risks.

An important comment on this point: in many aspects the term "risk" has already a prominent role in governance technology. However, as several authors

---

[3] Additionally, they introduce the *usability* principle: the working of the system needs to be expressed in a way that enables further analysis (a practical requirement impacting both transparency and accountability).

[4] Similar considerations apply looking beyond the technological boundaries, cf. Helberger et al. [11] with the concept of "*cooperative responsibility*". In principle, observability should be spread more widely over e.g. civil society actors and not merely individuals and regulators.

observed (e.g. Rouvroy [21], Dillon [6]), the alignment of risk analysis with competitive value extraction contributes to a very particular policy platform which is not neutral. These critics do not make risk a necessarily illegitimate category, but point to ways to further elaborate the importance of context, including specific contextual features to acknowledge policy concerns going beyond value extraction. The account proposed here takes indeed this direction.

Conclusion

The paper results from an effort to organize insights coming from different disciplines and domains related to the topic of *responsible computing*. The bottom line of our investigation is that, in contrast to the most common view taken today in technical approaches, issues like privacy and fairness refer to context-dependent and plural norms (where norm is used as in normative, and as in normality, cf. the concept of *normware* [25]), that cannot be directly translated to optimization tasks. Not all bias is unfair, it depends on how it is used and for what. Not all disclosure is illicit; in fact, some might be beneficial to the data subject and to society. To protect against misuses and improvident disclosures, and thus to achieve responsible computing, computation needs to be looked at in distributed terms (including the associated human activities), and computational agents need to be furnished with some degree of autonomy to be able to assess independently, on the basis of (plural) directives given by humans and (plural) knowledge constructed from system practices, whether a certain requested processing is indeed justified. Interestingly, the "distributed responsibility" sketched here is also hinted to in modern legislation as the GDPR, as for instance in Art. 28, according to which the data processor is not any more a mere executor, but has responsibility that the processing requested by the data-controller is complying with the rules.

References

1. Boer, A., van Engers, T.: An agent-based legal knowledge acquisition methodology for agile public administration. In: Proceedings of the 13th International Conference on Artificial Intelligence and Law (ICAIL 2011), pp. 171–180. ACM, New York (2011)
2. Capra, L., Blair, G.S., Mascolo, C., Emmerich, W., Grace, P.: Exploiting reflection in mobile computing middleware. ACM SIGMOBILE Mob. Comput. Commun. Rev. **6**(4), 34–44 (2002)
3. Corbett-Davies, S., Goel, S.: The measure and mismeasure of fairness: a critical review of fair machine learning (2018)
4. De Goede, M., Bosma, E., Pallister-Wilkins, P.: Secrecy and Methods in Security Research: A Guide to Qualitative Fieldwork. Routledge, New York (2019)
5. Dignum, V.: Responsible autonomy. In: Proceedings of International Joint Conference on Artificial Intelligence (IJCAI), pp. 4698–4704 (2017)
6. Dillon, M.: Underwriting security. Sec. Dialogue **39**(2–3), 309–332 (2008)

7. Dwork, C.: Differential privacy: a survey of results. In: Agrawal, M., Du, D., Duan, Z., Li, A. (eds.) TAMC 2008. LNCS, vol. 4978, pp. 1–19. Springer, Heidelberg (2008). https://doi.org/10.1007/978-3-540-79228-4_1
8. Friedler, S.A., Choudhary, S., Scheidegger, C., Hamilton, E.P., Venkatasubramanian, S., Roth, D.: A comparative study of fairness-enhancing interventions in machine learning. In: FAT* 2019 (2019)
9. Gaier, A., Ha, D.: Weight agnostic neural networks. Adv. Neural Inf. Process. Syst. **32**(NeurIPS), 1–19 (2019)
10. Halpern, J.Y.: Cause, responsibility and blame: a structural-model approach. Law, Prob. Risk **14**(2), 91–118 (2015)
11. Helberger, N., Pierson, J., Poell, T.: Governing online platforms: from contested to cooperative responsibility. Inf. Soc. **34**(1), 1–14 (2018)
12. Hesselman, C., et al.: A responsible internet to increase trust in the digital world. J. Netw. Syst. Manag. **28**(4), 882–922 (2020)
13. Hewitt, C.: What is computation? Actor model versus turing's model. In: A Computable Universe: Understanding and Exploring Nature as Computation, pp. 159–186 (2012)
14. High-Level Expert Group on Artificial Intelligence (AI HLEG): Ethics Guidelines for Trustworthy AI (2019)
15. Kamiran, F., Žliobaitė, I., Calders, T.: Quantifying explainable discrimination and removing illegal discrimination in automated decision making. Knowl. Inf. Syst. **35**(3), 613–644 (2013)
16. Lundberg, S.M., Lee, S.I.: A unified approach to interpreting model predictions. In: Proceedings of Advances in Neural Information Processing Systems (NIPS), pp. 4766–4775 (2017)
17. Martin, R., Rabaey, J., Chandrakasan, A., Nikolic, B.: Agile Software Development: Principles, Patterns, and Practices. Pearson Education, New Jersey (2003)
18. Nissenbaum, H.: Privacy In Context: Technology Policy And The Integrity Of Social Life. Stanford University Press, Stanford (2009)
19. Pearl, J.: Understanding Simpson's Paradox. Am. Stat. **68**(1), 8–13 (2014)
20. Rieder, B., Gordon, G., Sileno, G.: Mapping value(s) in AI: the case of YouTube. In: AoIR 2020: The 21th Annual Conference of the Association of Internet Researchers (2020)
21. Rouvroy, A.: The end(s) of critique. Priv. Due Process. Comput. Turn **39**, 143–167 (2008)
22. Saillenfest, A., Dessalles, J.L.: Role of Kolmogorov complexity on interest in moral dilemma stories. In: Proceedings of the 34th Annual Conference of the Cognitive Science Society, pp. 947–952 (2012)
23. Santoni de Sio, F., van den Hoven, J.: Meaningful human control over autonomous systems: a philosophical account. Front. Rob. AI **5**, 15 (2018)
24. Sileno, G., Boer, A., van Engers, T.: Reading agendas between the lines, an exercise. Artif. Intell. Law **25**(1), 89–106 (2017)
25. Sileno, G., Boer, A., van Engers, T.: The role of Normware in trustworthy and explainable AI. In: 1st XAILA Workshop on Explainable AI and Law, in conjunction with JURIX 2018 (2018)
26. Sileno, G., Dessalles, J.L.: Qualifying Causes as Pertinent. In: Proceedings of the 40th conference of the Cognitive Science Society (CogSci 2018), vol. 1, no. 2 (2018)
27. Sileno, G., Saillenfest, A., Dessalles, J.L.: A Computational Model of Moral and Legal Responsibility via Simplicity Theory. JURIX 2017 FAIA 302, pp. 171–176, Amsterdam (2017)

28. Wachter, S., Mittelstadt, B., Russell, C.: Why fairness cannot be automated: bridging the gap between EU non-discrimination law and AI. In: SSRN Electronic Journal, pp. 1–72 (2020)
29. Wilkinson, M.D., et al.: The FAIR guiding principles for scientific data management and stewardship. Sci. Data **3**, 1–9 (2016)

Law as Web of linked Data and the Rule of Law

The Rule of Law and Compliance: Legal Quadrant and Conceptual Clustering

Pompeu Casanovas[1,2(✉)] ⓘ, Mustafa Hashmi[1,3,4] ⓘ, and Louis de Koker[1,5] ⓘ

[1] La Trobe LawTech, La Trobe Law School, La Trobe University,
Melbourne, Australia
{P.CasanovasRomeu,M.hashmi,L.DeKoker}@latrobe.edu.au
[2] Institute of Law and Technology, Autonomous University of Barcelona,
Bellaterra, Spain
[3] Data61, CSIRO, Canberra, Australia
[4] Federation University, Brisbane, Australia
[5] University of the Western Cape, Bellville, South Africa

Abstract. We present in this paper: (i) a regulatory quadrant to describe the rule of law; (ii) a cluster of concepts to describe instruments and processes of the law; (iii) the methodology followed to select the technical papers concerning regulatory compliance; and (iv) an initial mapping to frame the selected papers about legal compliance that we used in our final survey. The result is a conceptual clustering that is useful to analyse and differentiate Compliance by Design (CbD) and Compliance through Design (CtD).

Keywords: Legal theory · Regulatory compliance · Legal compliance · Legal conceptual analysis

1 Introduction

This paper maps and positions a cluster of concepts that are used to describe instruments and processes of the law. It first presents a quadrant reflecting the four basic concepts at play in the societal implementation of the rule of law, and the relationship among them. It then presents a comprehensive clustering of these concepts and eight sub–sets of concepts that can be distinguished. This clustering was required to map and frame selected papers concerning regulatory, business, and legal compliance in a range of different datasets. These formed the framework for an extended survey on legal compliance that we carried out between 2017 and 2019. Its preliminary results have been presented in Casanovas et al. [11], and Hashmi et al. [22]. A first comprehensive explanation of the components of the legal quadrant can be found in [45]. A complete survey on business and regulatory compliance can be found in Hashmi et al. [23]. This paper presents a more refined presentation of the concepts that were used to deepen the analysis. The final results of the survey on legal compliance will be published shortly.

© Springer Nature Switzerland AG 2021
V. Rodríguez-Doncel et al. (Eds.): AICOL-XI 2018/AICOL-XII 2020/XAILA 2020, LNAI 13048, pp. 215–229, 2021.
https://doi.org/10.1007/978-3-030-89811-3_15

This paper is organized as follows: Section 2 describes the legal quadrant, shaped as a conceptual compass of the sources of the rule of law from a legal governance perspective. The different phases of the survey and the analytical research approach are presented in Sect. 3. Section 4 focuses on the outcomes, i.e. the clustering of legal concepts. Section 5 is devoted to conclusions and future work.

2   Legal Quadrant

Conceptual dimensions of the rule of law have been analysed by legal theorists, jurists, and social scientists, both from Civil and Common Law cultures (e.g. Tamanaha [48]). It can be broadly defined as the restriction of the arbitrary exercise of power. According to Tamanaha [ibid.] we can distinguish a formal definition of the rule of law—a law set forth in advance, public, general, clear, stable and certain, and applied to everyone according to its terms—and a more substantive one "embracing fundamental rights, democracy, and/or criteria of justice". The rule of law is recognised as an important concept to be maintained and developed in national, international and transnational law, as it is deemed to combine legality with justice and ethical values such as human dignity and fairness [41,49]. The promotion of the rule of law at the national and international levels is therefore a key target of the United Nation's 2030 Sustainable Development Goals [3] and its promotion features in the work of many multilateral bodies. The rule of law, for example, lies at the heart of many of law and development policies of the World Bank [14].

Legal theory—*Scientia Juris*—links this political philosophical concept with the notion of valid norms expressed in legal sources. The classification of sources of law has been the focus of extensive work from the early Middle Ages rule of law (actually, rule by law), stemming from the Roman tradition in the 12th century [35] to the modern legal theory of the Nordic school, from Torstein Eckhoff to Alf Ross [42,43]. After the dominance of a sovereignty approach in the 20th century, which informed the assumption that sources of law should be hierarchically ordered, Peczenik [43] observed at the turn of the millennium that globalisation had brought new sources that are not as precise as legislation or case–based law. These include United Nations resolutions, human rights, commercial custom, foreign precedents, arbitration, mediation, recommendations of "more or less" authoritative organisations, soft law, "more or less" globalised doctrine, etc. We could easily add other examples coming from industry, technology, and corporate management—technical protocols, W3C recommendations and standards, ISO/IEC standards, corporate governance principles, risk management guidelines, etc.

It is worth mentioning that there is a significant body of research already done on the automated modelling of the sources of law, i.e. automated recognition of norms according to their sentence structures using parsers and NLP [32,33]. In the last ten years, many legal parsers and classifiers have been constructed to assist in the storage, management, retrieval and cross-referencing of multilingual legal instruments and provisions [25]. We have also witnessed the evolution of legal ontology building, enhancing the interoperability of documents and their normative content [10,38], and the construction of knowledge graphs [24]. These are examples of what is known as "law as data" or the "linked legal data landscape" among semantic scholars [16,46].

Our work is consistent with this approach. We identified four basic components for the societal implementation of the rule of law and the relationship among them: hard law, soft law, policies, and ethics. We considered the sources, domains, and relationships with respect to citizens (interconnectedness of norms or rules). Rather than discrete categories or lists of requirements, defining their regulatory dimension and setting the boundaries of their conceptual properties are a matter of degree and conditions of values and principles, dealing with the pragmatic dimension of the rule of law [9], that is to say, its *legal governance* through technical and AI means [40]. This expression embraces the general factual and normative dimensions of the concept, but it mainly focuses on the social conditions and technical requirements that are required to operate in, and in relation to, the Internet of Things (IoT) and the Web of (Linked) Data (WoLD) (i) to embed protections into systems, apps and platforms, and (ii) to empower stakeholders (citizens, consumers, organisations, communities ...), (iii) to protect and enhance their individual and collective rights. Legal governance is related to the creation and emergence of (socio-)legal ecosystems within machine–human and machine–machine interfaces [45]. The epistemic assumptions of this theoretical position have been described from a meso–level [45] and middle–out [40] approach, with social and political roots in cognitive science [13], and economic institutional [39] and evolutionary analysis [15].

We considered the implementation of the rule of law along two different but related dimensions at the empirical level: (i) institutional power, and (ii) social dialogue (negotiation, compromise, mediation, agreement). We also considered law and regulations focusing on power and how it is handled and eventually shared. Even at the micro level, this includes a proportional and gradual system of sanctions. There is a wide range of sanctions, from mere disincentives to criminal punishment. We identified values to be assigned to them according to the level of binding force of norms and their acceptance by stakeholders. The intuitive approach to first separate binding from non–binding norms according to the nature of the objectives and procedures has been employed by many previous descriptions, for instance, Brous, Janssen and Vilminko-Heikkinen [7], Mondorf and Wimmer [7], the EU Better Regulations scheme for interoperability [1][1] and the new EU (2017) interoperability framework promoting seamless services and data flows for European public administrations.

Figure 1 plots our regulatory quadrant for the rule of law. It shows how the validity of norms (i.e. their 'legality') emerges from four different types of regulatory frames, with some distinctive properties. Properties are understood here as correlating dynamic patterns. This is a scheme, a conceptual compass

---

[1] Under review (2021) https://ec.europa.eu/isa2/shaping-future-interoperability-policy_en

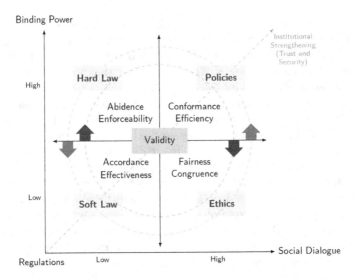

Fig. 1. Legal quadrant for the rule of law (adopted from [22,45])

to be used for a first cluster of norms, according to their type and degree of compliance: *abidance* (for hard law), *conformance* (for policies), *accordance* (for soft law), and *congruence* (or congruity) for ethics. According to the degree of abstraction at the implementation level, these four categories can blur into overlapping concepts. For example, compliance with some company policies can be viewed by company officials and employees as more imperative and mandatory in practice than compliance with some statutes.

Hard law refers to legally binding obligations, either in the national or international arena, with legal rules that can be enforced in court processes. Soft law, on the contrary, is not legally binding and compliance is therefore optional. Soft law consists of rules, best practices and principles that facilitate the governance of networks, social organisations, companies and institutions, leaving room for dialogue, negotiation, common accord and general convergence among relevant actors. Although soft law is not legally binding sometimes in practice the non–legal consequences of non-compliance with a particular standard or soft law rule are so severe that actors have little option but to comply with it. Soft and hard law are not discrete categories but should rather be viewed as categories of regulatory instruments on a continuum. Appropriate use of these instruments enables different powers and authorities to produce global regulatory frameworks–regulations across borders among citizens, organisations, and the different states – as well as national regulatory frameworks within each state.

3   Analytical Research Approach

In this section we outline the multiphase analytical research approach adopted for creating a clustering of core legal concepts. Given the complex and extensive nature of the legal domain and the diversity of legal sources, we carried out this research in several phases, each one consisting of multiple tasks, namely: (i) data collection;(ii) data synthesis and analysis; and (iii) results reporting, as shown in Fig. 2.

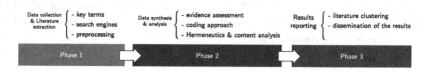

Fig. 2. Multiphase analytical research methodology

3.1   Phase 1: Data Collection and Literature Extraction

We collected literature from a range of primary and secondary sources such as research articles, industry and project reports, and technical articles on regulatory and legal compliance. We loosely followed the data collection, acquisition protocol, and hermeneutic circle from [6], depicted in Fig. 3. We identified and extracted literature from two types of literature sources: (a) technical literature covering the technical aspects of regulatory compliance such as tools and techniques for modelling, extraction and automated verification of legal norms, and (b) conceptual legal literature covering the social and legal aspects of regulatory compliance such as concepts, statutory documents etc.

To search the relevant literature we first compiled a list of core concepts and terms from the interdisciplinary domains of computer science and law focusing on regulatory and legal compliance. This included the alternative spelling of related terms to legal compliance e.g., "legal effectiveness", "law enforcement", "hard law", "legal efficacy", "softlaw", "legal efficacy", "legal norms". For technical articles we used key terms such as "compliance life-cycle", "compliance frameworks", "rule modelling languages", etc. During this process, we found several quasi–synonymous terms interchangeably used in the literature such as "conformance" and "compliance", "backward compliance" and "auditing", and "retrospective" and "design–time" compliance. We also included them in the list. The compiled terms were then combined using various search building blocks, boolean operators, logical operators, wildcards and truncation. This resulted in fine-grained search terms, for example, Compliance***/3 (legal or regulat***) AND "Judicial" ADJn "Systems" AND "Adversarial". The literature search process started with querying prominent databases. For the technical literature from computer science, we queried premium and open–source scholarly databases such as Springer Link, ACM Digital Library, Web of Science (WoS), Ebsco Host, IEEE Xplore, and free search databases DBLP, arXiv, etc. Referring to legal and conceptual sources we searched both open source and proprietary databases such as HeinOnline, austlii, ProQuest, LexisNexis, and Westlaw (US,UK,AUS), etc. The query process yielded a rich corpus of literature, and initially we collected 600+ articles on broader concepts on legal and regulatory compliance. To search literature about specific terms/concepts, we applied various techniques such as lemmatisation, proximity operator, logical and negation operators, and other filtering techniques to narrow down search terms. In addition, we applied backward search verification techniques to extract existing literature surveys and other relevant works not extracted in the first round. With the backward searcher, and after removing duplicates, we extracted a rich collection of more than 900 articles comprising both primary and secondary sources on regulatory compliance.

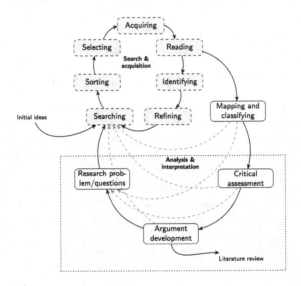

Fig. 3. Hermeneutic Circle (adapted from [6])

3.2   Data Analysis and Synthesis

In this phase the collected literature was preprocessed for further interpretations of the concepts, coding, and further analysis in the later stages. The preprocessing was carried out in two distinct phases. In the first phase, we scrutinised the credibility and quality of the collected literature. We followed the recommendations from [6,29,34], and used the quality assurance criteria from [23] to assess the relevance of collected articles. Any article not meeting the quality assessment criteria was removed from the list of articles suitable for further analysis. This rigorous process addressed concerns about the scholarly quality of the collected literature and ensured high–quality and credible interpretations, mapping and classification, and inferencing from the collection.

Parallel to this we created a list of core concepts and provided their definitions. For this purpose, we used the content analysis approach to qualitative research [19,26]. Both types of content analysis—conceptual and relational

Fig. 4. Relationship nodes for clustered themes

analysis—have been used to find the frequency and relationships, co-occurrences, and explicit concepts in legal and scientific literature. Through the conceptual analysis, a range of concepts occurring in the literature were identified and their frequency was recorded. Then each identified concept was defined, and its features were interpreted and discussed. Essentially, similar to any other domains, most concepts continuously emerge and evolve in the legal compliance domain. Very little agreement, however, exists on proper definitions of some of the basic concepts. In the subsequent steps, the relationships, dependencies and other associations were analysed. Defining, and interpreting the core concepts, confirming their existence in the literature, and studying their relationships was an important preliminary step for the usefulness of the analysis.

To gain a more detailed understanding and insights, we then coded the literature. We derived a codification protocol to meet the objectives of the analysis. In the coding process, we used a sample of the most frequently used concepts—and we created 327 nodes across four clusters of distinct themes according to the hard law, ethics, policies and soft law quadrant (see Fig. 1). Within each cluster we maintained the level of coding depth into no more than five levels of hierarchy to manage the complexity of the analysis. Figure 4 shows the relationship nodes for the coded concepts.

Along these lines, we also created 157 additional relationship nodes, expanding the analysis to 484 nodes. The relationship nodes helped us to form a more detailed understanding of several characteristics and inter/intra relationships among concepts. The coding process resulted in a matrix of nodes reflecting the interactions of various concepts and dependencies between them.

3.3 Results Reporting

In this last phase of the analysis, the coded concepts were further analysed to form deeper insights. We applied Pearson's coefficient correlation [28] and

Jaccard's [27] statistical techniques[2] to investigate the relationships between the concepts, and across inter/intra clustered themes. The former developed intersection matrix and boolean search techniques were used to evaluate the correlation and dependencies between the themes across all clusters. Finally, the analysis results were documented for further dissemination [22].

4 Discussion: Clustering of Legal Concepts

Alternative analyses more focused on validity as a supervenient property of legislation [18], as a property actually belonging to hard law from an internal (non–external) approach [20], or as a historical construct [30] are also possible. In our case, as already stated, we confined ourselves to construct a compass that could be used to build conceptual clusters primarily focused on the conditions for legal compliance as a type of emergent empirical and cognitive pattern of social behaviour in the WoLD and the IoT. We focused on the causal conditions to set both (socio-)legal ecosystems and the notion of ecological validity of rules, patterns, and norms in such ecosystems. As this is an empirical view, we searched for causal relations among concepts.

Legal causation has a long history in jurisprudence [21,36]. In the Common law tradition, it is used to assign legal responsibility to actors who cause harms to others. It is therefore closely associated with legal fields such as criminal law and the laws of torts [36]. This is also compatible with the Civil law tradition where its application can be found in criminal law and law of damages or delicts. The legal rules in those fields of law are dogmatic applications of causation, reflecting how its application is expressed in norms and how these are interpreted and eventually applied. This line of reasoning has been recently followed by Liepina et al. [31] to set the requirements for establishing defeasible logic factual causation arguments in legal cases. Our usage is somewhat different and rather closer to the scientific meaning of causality—the relationship of influence between two events, processes, states or objects. For the notion of (socio-)legal ecosystem, we needed to find some evidence that such a relation effectively occurred among concepts, i.e. that there is a statistical relation that reflects some kind of empirical linkage among them.

How behaviour relates to norms can be considered in many different ways. In the AI&Law literature this kind of plain or direct causality has been already addressed from an institutional ontology point of view—e.g. differentiating constitutive from regulative rules [47]. In the normative Multi-Agent Systems and Socio-Technical Systems communities, reflexivity and autonomy have been

---

[2] In addition, we applied the Sørenson similarity coefficient [5] to enhance our analysis, and validate the similarity and strenght of the relationship between the concepts. Essentially, with the Sørenson coefficient we compared other two coefficients (Jaccard and Pearson). However, the Sørenson coefficient is somewhat similar to Jaccard's coefficient and takes the same number of values but it only counts true positives. Hence, we immediately discarded the Sørenson coefficient from the analysis. See [22] for further details.

addressed to set "self–governing socio–technical systems where reflexivity with respect to ecological and environmental impact is enabled through algorithms for deliberation, introspection and self–organization" [44]. The authors integrate four dimensions of reflexivity to achieve algorithmic governance for socio–technical–ecological systems, i.e. systems conceiving "society, technology, and environment as co–constituted and co–emergent entities" [2].

This is a framework to link meta-models, in the same way that the WIT framework applies to socio-cognitive-technical systems (SCTS) W(world), I(institutions), T(technology), i.e. a system that exists in the real world, and "that is composed by two ('first class') entities: a social space and the agents who act within that space" [37]. A metamodel "consists of a collection of languages, data structures and operations that serve to represent the agents and the social space of a given SCTS with an appropriate level of detail and accuracy".

Although also compatible with such an approach, a compass is not a general framework nor a metamodel in the same sense but a practical scheme or template to be used to shape the context of implementation for legal regulatory models. The essential point is that rules, norms, values and principles—in fact all elements of a regulatory system that may include in addition behavioural patterns—are not only assumed but defined as dynamic constitutive parts of a social and complex reality as components of (socio-)legal ecosystems. The compass simply reflects their elements onto a manageable scheme. The assertion that "all novel technologies are constructed by combining assemblies and components that already exist" [4] applies to evolving legal concepts as well. It is worth mentioning here that we do not need to refine at this stage the "count as" relationship to assess how a certain conduct acquires meaning in institutional terms [47], nor the reflexivity within institutional settings [44]. These are important aspects that can come later. From a theoretical point of view, we have adopted a lower level of abstraction and we restricted ourselves to what "already exists" in the legal field.

We were interested in concepts as bricks, as instruments used to build up the field of law, crossing and overlapping several domains (such as jurisprudence, national and international dogmatics, cultural analysis, legal studies, history, political science, practical philosophy, administration, policy studies, ethics...). We came up with a non-exhaustive list, after several internal and external discussions among the members of the team (stemming from jurisprudence, dogmatics, political science, economy, socio-legal studies, linguistics, and cognitive and computer sciences). We decided to apply a flexible hermeneutical and dialogical procedure at the beginning to reach consensus about the main concepts. However, as we have shown in the previous section, the selection and analysis of the articles and the clustering and validation of the approach followed a strict statistical methodology to surpass the possible bias of the initial collection. We commenced the work on the basis of the legal quadrant that we have been using for some years now [8] to investigate its statistical validity: (i) we produced a conceptual clustering from the articles included in the survey, and (ii) we populated the concepts of the survey with data extracted from the database.

Moreover, (iii) we carried out the analyses to show the degree of relationship (or lack of relationship) between concepts referring to legal compliance, and (iv) to establish the level of association between them.

The result is a categorial scaffolding, simple enough to flesh out in a reusable and scalable way the legal system's affordances and functionalities within a variety of environments and scenarios. For instance, within the same D2D CRC project, for the implementation of the Australian Spent Convictions Scheme, we could develop at the same time (i) the construction of semantic formal rules to process and solve automatically the most common cases, (ii) the NLP techniques to classify the Australian case-based law on this subject, and (iii) the theoretical model to understand the meaning, functionality, and consistency of the Spent Convictions system as such [12]. Thus, law as rules, law as data, and law as knowledge were made compatible. They are but different dimensions of the same problem which was, in the case of the D2D CRC project, how to facilitate the effective exchange of the spent conviction data at the federal or Commonwealth level [17].

Figure 5 shows an excerpt of the clustering of the core legal concepts, distributed according to the legal quadrant. We identified four sets and eight paired subsets related to hard law, soft law, ethics, and policies. The pairs were created having in mind the effective projection of concepts for their implementation, i.e. their use for legal governance. Therefore, the subsets contain (i) general common concepts as elements of hard and soft law, ethics and policies, and (ii) concepts as elements of normative, dialogical, behavioural and organisational systems.

The cluster and statistical analysis have shown the difference between business, regulatory, and legal compliance in the existing literature. Many concepts of public law (including ethics, constitutional rights, and human rights) have not been mentioned in the technical literature on compliance of the last ten years. Until recently, only the aspects of market and especially labour law, finance, insurance and the possibility to be fined by the agencies, were included. This has been changing from 2019 onwards, as the entry into force of the European Data Protection Regulation (GDPR) in 2018 has been followed by a new wave of studies on "compliance with legal requirements" (legal compliance) related to privacy and data protection. But the results of the application of different metrics are clear: regulatory compliance technology, as methodology and policy, has been focused on the market, on corporate management, on monitoring and control, rather than on rights or on concepts pertaining to the public sphere. And likewise, technical studies for legal services have captured only a small set of legal concepts such as contract, obligation, or consumption rather than those related to the constitutional and human rights, both political and socio–economic.

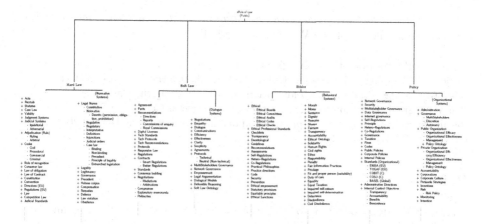

Fig. 5. Excerpt of the clustering of core legal concepts

5 Conclusions

Our survey on legal compliance has shed some light on the different conceptual usages of this notion in the legal and technological fields. Compliance *by* design and *through* design should be treated separately, as legal compliance and business compliance do not always refer to the same concepts and requirements. There is especially a gap between the way how management and business studies refer only to certain areas and legal fields, neglecting many others, and the way in which jurisprudence and legal doctrine refer to the concepts of compliance, obedience, conformity, accordance, abidance and enforcement. As a consequence: (i) the existing studies on legal compliance do not cover the entire legal field (including ethics and policies), (ii) business and regulatory compliance cannot be projected as such to the legal field, and (iii) more work has to be done to link business, regulatory and legal compliance.

We have discussed in this paper our use of a legal quadrant (or legal compass) to classify the legal sources and to understand the regulatory instruments for the interpretation of our results. This approach has been validated statistically, as we were able to show that the degree of association and correlation (one–way and symmetrical relationship) between the main concepts analysed (through 900 articles) are in a statistically acceptable range.

In the immediate future (i) we will refine our conceptual and statistical approach to regulatory, business and legal compliance, (ii) we will exploit our data to provide further relevant result—for instance, we can show that international human rights provide impetus to legal pluralism rather than monism, (iii) we will give more analytical depth to interpretation, reaching the level of conceptual attributes, i.e. beyond correlative and associative relationships, (iv) we will provide more insights to build theoretical concepts—e.g. offering a better definition of what *legal governance, ecological validity,* and *meta rule* of law mean within the new environment of the WoLD and the IoT.

Acknowledgements. The work presented in this paper has been funded under Project C, Compliance by Design (CbD) and Compliance through Design (CtD) Solutions to Support Automated Information Sharing, within the Australian Law and Policy Program, Data to Decisions Cooperative Research Centre (D2D CRC) (2019). Results have been also partially produced under the auspices of the EU H2020 Projects LYNX and SPIRIT.

References

1. TOGAF: An introduction to the European interoperability reference architecture (EIRA©), version 2.1.0 (2017)
2. Ahlborg, H., Ruiz-Mercado, I., Molander, S., Masera, O.: Bringing technology into social-ecological systems research-motivations for a socio-technical-ecological systems approach. Sustainability **11**(7) (2019). https://doi.org/10.3390/su11072009. https://www.mdpi.com/2071-1050/11/7/2009
3. Arajärvi, N.: The rule of law in the 2030 agenda. Hague J. Rule Law **10**(1), 187–217 (2018). https://doi.org/10.1007/s40803-017-0068-8
4. Arthur, W.B., Polak, W.: The evolution of technology within a simple computer model. Complexity **11**(5), 23–31 (2006). https://doi.org/10.1002/cplx.20130
5. Binanto, I., Warnars, H.L.H.S., Abbas, B.S., Heryadi, Y., Sianipar, N.F., Lukas, Perez Sanchez, H.E.: Comparison of similarity coefficients on morphological rodent tuber. In: 2018 Indonesian Association for Pattern Recognition International Conference (INAPR), pp. 104–107 (2018). https://doi.org/10.1109/INAPR.2018.8627050
6. Boell, S.K., Cecez-Kecmanovic, D.: Literature reviews and the hermeneutic circle. Aust. Acad. Res. Libr. **41**(2), 129–144 (2010). https://doi.org/10.1080/00048623.2010.10721450
7. Mondorf, A., Wimmer, M.A.: Requirements for an architecture framework for pan-European e-government services. In: Scholl, H.J., et al. (eds.) EGOV 2016. LNCS, vol. 9820, pp. 135–150. Springer, Cham (2016). https://doi.org/10.1007/978-3-319-44421-5_11
8. Casanovas, P.: A note on validity in law and regulatory systems (position paper). Quaderns de filosofia i ciència **42**, 29–40 (2012). https://ddd.uab.cat/pub/artpub/2012/119545/quafilcie_a2012n42p29iENG.pdf
9. Casanovas, P., Rodríguez-Doncel, V., González-Conejero, J.: The role of pragmatics in the web of data. In: Poggi, F., Capone, A. (eds.) Pragmatics and Law. PPPP, vol. 10, pp. 293–330. Springer, Cham (2017). https://doi.org/10.1007/978-3-319-44601-1_12
10. Casanovas, P.: Open source intelligence, open social intelligence and privacy by design. In: Proceedings of the European Conference on Social Intelligence (ECSI-2014), Barcelona, Spain, 3–5 November 2014, pp. 174–185 (2014). http://ceur-ws.org/Vol-1283/paper_24.pdf

11. Casanovas, P., González-Conejero, J., de Koker, L.: Legal compliance by design (LCbD) and through design (LCtD): preliminary survey. In: Proceedings of the 1st Workshop on Technologies for Regulatory Compliance Co-located with the 30th International Conference on Legal Knowledge and Information Systems (JURIX 2017), Luxembourg, 13 December 2017, pp. 33–49 (2017). http://ceur-ws.org/Vol-2049/05paper.pdf

12. Casanovas, P., et al.: Summary legal and technical report on spent convictions, July 2019. https://doi.org/10.5281/zenodo.3271525. DC25008: Compliance by Design (CbD) and Compliance through Design (CtD) solutions to support automated information sharing (2018–19). Law and Policy. Project C. Spent Convictions Use Case. Australian Government funded Data to Decisions Cooperative Research Centre (2018–2019), end-user: Australian Criminal Intelligence Commission

13. Castelfranchi, C.: The micro-macro constitution of power. Protosociology **18**, 208–265 (2003)

14. Dann, P.: The Law of Development Cooperation: A Comparative Analysis of the World Bank, the EU and Germany. Cambridge International Trade and Economic Law, Cambridge University Press (2013). https://doi.org/10.1017/CBO9781139097130

15. Dopfer, K., Foster, J., Potts, J.: Micro-meso-macro. J. Evol. Econ. **14**(3), 263–279 (2004)

16. Filtz, E., Kirrane, S., Polleres, A.: The linked legal data landscape. linking legal data across different countries. Artif. Intell. Law (2021, to appear). http://www.polleres.net/publications/filt-etal-2021AILAW.pdf

17. Governatori, G., Romeu, P.C., de Koker, L.: On the formal representation of the Australian spent conviction scheme. In: Gutiérrez-Basulto, V., Kliegr, T., Soylu, A., Giese, M., Roman, D. (eds.) RuleML+RR 2020. LNCS, vol. 12173, pp. 177–185. Springer, Cham (2020). https://doi.org/10.1007/978-3-030-57977-7_14

18. Grabowski, A.: Juristic Concept of the Validity of Statutory Law. A Critique of Contemporary Legal Nonpositivism. Springer, Cham (2013). https://doi.org/10.1007/978-3-642-27688-0

19. Graneheim, U., Lundman, B.: Qualitative content analysis in nursing research: concepts, procedures and measures to achieve trustworthiness. Nurse Education Today **24**(2), 105–112 (2004). https://doi.org/10.1016/j.nedt.2003.10.001. http://www.sciencedirect.com/science/article/pii/S0260691703001515

20. Hage, J.: What Is legal validity? Lessons from soft law. In: Westerman, P., Hage, J., Kirste, S., Mackor, A.R. (eds.) Legal Validity and Soft Law. LPL, vol. 122, pp. 19–45. Springer, Cham (2018). https://doi.org/10.1007/978-3-319-77522-7_2

21. Hart, H., Honoré, T.: Causation in the Law. Clarendon Press (1985). https://books.google.com.au/books?id=k1gQAQAAMAAJ

22. Hashmi, M., Casanovas, P., de Koker, L.: Legal compliance through design: preliminary results. In: Rodríguez-Doncel, V., Casanovas, P., González-Conejero, J., Montiel-Ponsoda, E. (eds.) Proceedings of the 2nd Workshop on Technologies for Regulatory Compliance Co-located with the 31st International Conference on Legal Knowledge and Information Systems (JURIX 2018), Groningen, The Netherlands, December 12, 2018. CEUR Workshop Proceedings, vol. 2309, pp. 59–72. CEUR-WS.org (2018). http://ceur-ws.org/Vol-2309/06.pdf

23. Hashmi, M., Governatori, G., Lam, H.P., Wynn, M.T.: Are we done with business process compliance: state of the art and challenges ahead. Knowl. Inf. Syst. **57**(1), 79–133 (2018). https://doi.org/10.1007/s10115-017-1142-1

24. Hogan, A., et al.: Knowledge graphs (2021)

25. Howe, J.S.T., Khang, L.H., Chai, I.E.: Legal area classification: a comparative study of text classifiers on Singapore supreme court judgments (2019)
26. Hsieh, H.F., Shannon, S.E.: Three approaches to qualitative content analysis. Qual. Health Res. **15**(9), 1277–1288 (2005). https://doi.org/10.1177/1049732305276687. pMID: 16204405
27. Jaccard, P.: Distribution of alpine flora in the Dranses basin and some neighboring regions. Bull. Soc. Vaudoise Sci. Nat. **37**, 241–272 (1901). https://ci.nii.ac.jp/naid/10027880482/en/
28. Kirch, W. (ed.): Pearson's Correlation Coefficient, pp. 1090–1091. Springer, Dordrecht (2008). https://doi.org/10.1007/978-1-4020-5614-7_2569
29. Kitchenham, B., Brereton, P.: A systematic review of systematic review process research in software engineering. Inf. Softw. Technol. **55**(12), 2049–2075 (2013). https://doi.org/10.1016/j.infsof.2013.07.010
30. Köpcke-Tinturé, M.: Legal Validity: The Fabric of Justice. Hart Publishing, Oxford (2019)
31. Liepina, R., Sartor, G., Wyner, A.: Causal models of legal cases. In: Pagallo, U., Palmirani, M., Casanovas, P., Sartor, G., Villata, S. (eds.) AICOL 2015–2017. LNCS (LNAI), vol. 10791, pp. 172–186. Springer, Cham (2018). https://doi.org/10.1007/978-3-030-00178-0_11
32. de Maat, E., Winkels, R.: A next step towards automated modelling of sources of law. In: Proceedings of the 12th International Conference on Artificial Intelligence and Law, pp. 31–39. ICAIL 2009, Association for Computing Machinery, New York (2009). https://doi.org/10.1145/1568234.1568239
33. de Maat, E., Winkels, R.: Automated classification of norms in sources of law. In: Francesconi, E., Montemagni, S., Peters, W., Tiscornia, D. (eds.) Semantic Processing of Legal Texts. LNCS (LNAI), vol. 6036, pp. 170–191. Springer, Heidelberg (2010). https://doi.org/10.1007/978-3-642-12837-0_10
34. Moher, D., Liberati, A., Tetzlaff, J., Altman, D.G.: Preferred reporting items for systematic reviews and meta-analyses: the prisma statement. PLoS Med. **6**(7), 1–6 (2009). https://doi.org/10.1371/journal.pmed1000097
35. Møller, J.: Medieval origins of the rule of law: the Gregorian reforms as critical juncture? Hague J. Rule Law **9**(2), 265–282 (2017)
36. Moore, M.: Causation of law. The Stanford Encyclopedia of Philosophy (Winter 2019 Edition), Zalta, E.N. (ed.) (2019). https://plato.stanford.edu/archives/win2019/entries/causation-law/
37. Noriega, P., Verhagen, H., d'Inverno, M., Padget, J.: A manifesto for conscientious design of hybrid online social systems. In: Cranefield, S., Mahmoud, S., Padget, J., Rocha, A.P. (eds.) COIN-2016. LNCS (LNAI), vol. 10315, pp. 60–78. Springer, Cham (2017). https://doi.org/10.1007/978-3-319-66595-5_4
38. de Oliveira Rodrigues, C.M., de Freitas, F.L.G., Barreiros, E.F.S., de Azevedo, R.R., de Almeida Filho, A.: Legal ontologies over time: a systematic mapping study. Expert Syst. Appl. **130**, 12–30 (2019). https://doi.org/10.1016/j.eswa.2019.04.009. http://www.sciencedirect.com/science/article/pii/S0957417419302398
39. Ostrom, E.: Understanding Institutional Diversity. Book Collections on Project MUSE, Princeton University Press (2009). https://books.google.com.au/books?id=LbeJaji_AfEC
40. Pagallo, U., Casanovas, P., Madelin, R.: The middle-out approach: assessing models of legal governance in data protection, artificial intelligence, and the web of data. Theory Pract. Legislation **7**(1), 1–25 (2019). https://doi.org/10.1080/20508840.2019.1664543

41. Palombella, G.: The Rule of Law as an Institutional Ideal, pp. 1–37. Brill, Leiden (2010). https://doi.org/10.1163/ej.9789004181694.i-215.6
42. Pattaro, E., Rottleuthner, H., Shiner, R.A., Peczenik, A., Sartor, G.: A Treatise of Legal Philosophy and General Jurisprudence. Springer, Dordrecht (2005). https://doi.org/10.1007/1-4020-3505-5
43. Peczenik, A.: Scientia Iuris - an unsolved philosophical problem. Ethical Theory Moral Pract. 3(3), 273–302 (2000). https://doi.org/10.1023/A:1009948025411
44. Pitt, Jeremy, J.D., Ober, J.: Algorithmic reflexive governance for socio-techno-ecological systems. IEEE Technol. Soc. Mag. 39(2), 52–59 (2020)
45. Poblet, M., Casanovas, P., Rodríguez-Doncel, V.: Legal Linked Data Ecosystems and the Rule of Law, pp. 87–126. Springer, Cham (2019). https://doi.org/10.1007/978-3-030-13363-4_5
46. Rodríguez-Doncel, V., Casanovas, P., González-Conejero, J. (eds.): Proceedings of the 1st Workshop on Technologies for Regulatory Compliance Co-located with the 30th International Conference on Legal Knowledge and Information Systems (JURIX 2017), Luxembourg, 13 December 2017, CEUR Workshop Proceedings, vol. 2049. CEUR-WS.org (2018). http://ceur-ws.org/Vol-2049
47. Sileno, G., Boer, A., van Engers, T.: Revisiting constitutive rules. In: Pagallo, U., Palmirani, M., Casanovas, P., Sartor, G., Villata, S. (eds.) AICOL 2015-2017. LNCS (LNAI), vol. 10791, pp. 39–55. Springer, Cham (2018). https://doi.org/10.1007/978-3-030-00178-0_3
48. Tamanaha, B.Z.: On the Rule of Law: History, Politics, Theory. Cambridge University Press, Cambridge (2004)
49. Waldron, J.: The rule of international law. Harvard J. Law Public Policy 30, 15–30 (2006)

The Web of Data's Role in Legal Ecosystems to Address Violent Extremism Fuelled by Hate Speech in Social Media

Andre Oboler[1,2(✉)] [iD] and Pompeu Casanovas[1,3] [iD]

[1] LawTech Research Group, Law School, La Trobe University, Melbourne, Australia
{a.oboler,p.casanovasromeu}@latrobe.edu.au, ceo@ohpi.org.au,
pompeu.casanovas@uab.cat
[2] Online Hate Prevention Institute (OHPI), Sydney, Australia
[3] Autonomous University of Barcelona (IDT), Barcelona, Spain

Abstract. It is usually said that technical solutions should operate ethically, in compliance with the law and subject to good governance principles. In this position paper we face the problem of behavioural compliance and law enforcement in the case of hate speech and extremism online. Law enforcement and behavioural compliance are ways of coping with the objective of stopping the spread of hate and radicalisation online. We contend that a combination of regulatory instruments, incentives, training, proactive self-awareness and education can be effective to create legal ecosystems to improve the present situation.

Keywords: Hate speech · Terrorism · Rule of law · Semantics · NLP · Legal governance

1 Introduction

Violence is a pervasive phenomenon in contemporary global societies. It has been fostered by the expansion of the Internet, social media networks and the fast development of the web of data. Violent language reflected in bias attitudes is the first step in the pyramids of hate, escalation of conflicts, and radicalization of individuals. Even in the most extreme case of inhumanity, Rabbi Abraham Joshua Heschel noted how "the Holocaust did not begin with the building of crematoria, with tanks and guns. It began with uttering evil words, with defamation, with language and propaganda" [1]. This is the opinion of most linguists in the 20th and 21st c., e.g. [2]. It has also been expressed judicially, for example in Canada [3]. According to some recent studies, media and the way in which minority groups are targeted are fuelling this phenomenon. Dichotomic, binary categories, and the rise of white supremacy with its practice of depicting non-white cultures as "alien" ("othering"), play a major role in reinforcing negative, weak, or fearful images of migrants and refugees and spreading xenophobia [4]. This language can be seen in recent terrorist manifestos, for example, the 2019 attacks in Christchurch, New Zealand [5], and in Halle, Germany [6].

© Springer Nature Switzerland AG 2021
V. Rodríguez-Doncel et al. (Eds.): AICOL-XI 2018/AICOL-XII 2020/XAILA 2020, LNAI 13048, pp. 230–246, 2021.
https://doi.org/10.1007/978-3-030-89811-3_16

However, detecting, tracking, and monitoring these particular uses of language on the web has turned out to be a difficult task, as it implies a meta-cognitive operation of annotating, classifying and clustering terms and expressions from a previous interpretation of their context of usage. Hate speech and extremist speech can partially be fear speech as well. But what is hate or extremism and what is fear disguised by such speech? [7] Violence attracts, fascinates and repeal, as shown by the 'beautiful' war images displayed on newspapers and on the media [8].

In this paper, we contend that (i) it is much better to take a proactive ethical stance than adopting a passive *laissez-faire* approach, (ii) there is an effective possibility of making errors of judgment (false positives and negatives), (iii) technology offers at present some means to overcome or at least reduce these risks (although not completely), (iv) the rise of online hate speech is an indicator of cultural change that is just starting to be taken seriously, as it should, (v) there is no simple solution to stop this based on traditional legal instruments (i.e. enactment of rules and enforcement of laws), (vi) hence, some regulatory imagination is needed, stemming from a combination of hard and soft law, smart regulations, multi-stakeholder governance, policies and ethics.

2 Definition

The first problem is the meaning of the expression. We can identify four stages: (i) before World War II and in the inter-war period hate speech was defined as 'race hate' or 'group libel', (ii) in the second half of the past century, definitions become more inclusive and sensitive to victimisation processes, e.g. Human Rights Watch defined it as 'any form of expression regarded as offensive to racial, ethnic and religious groups and other discrete minorities, and to women' (iii) in the 21st century, even this meaning that included all kind of sexual and political biases has been broadened to cover all kinds of oppression (religious, cultural, political or technological—i.e. based on the lack of knowledge or technological skills) [9], this is an unfolding area, for example, a Parliamentary committee in Victoria, Australia, has between 2019 and 2021 been investigating expanding the state's *Racial and Religious Tolerance Act 2001,* which makes racial and religious vilification unlawful, in order to also cover attributes such as gender, sexual orientation, gender identity, sex characteristics and disability [10], (iv) an emerging phase in 2021 in which the link between hate speech is being more widely recognised as an aspect of radicalisation into terrorism. This can be seen, for example, in comments in the United States by Representative Elissa Slotkin, chairwoman of the House Homeland Security Subcommittee on Intelligence and Counterterrorism, who said one of the questions she is considering in her role is "how to hold social media companies accountable for their role in facilitating the spread of domestic terrorism?" [11], and even more explicitly the terms of reference of a Parliamentary inquiry into extremist movements and radicalism in Australia which include investigating "further steps the Commonwealth could take to disrupt and deter hate speech and establish thresholds to regulate the use of symbols and insignia associated with terrorism and extremisms" [12].

The idea behind prohibiting hate speech is that human rights and its political side, civil rights, are deemed to *empower* people; hence, all sorts of humiliation imply a loss of dignity that constitutes in itself a form of *disempowerment*, i.e. an aggression that can be qualified as a form of violence. This is even more where online spaces have a visible extremist presence, which is at least an implied, and often explicit, threat of violence. On the Internet violence finds its own 'connectomes', producing a permanent and structural harm that can be easily amplified for political and economic reasons [13]. Words create worlds, that is, they shape the very fabric of our environment. In "linked democracy" scenarios, this particular threat should be avoided and considered the first step to tyranny, thus, a negative condition for the construction of the global (linked) space [14].

This approach represents a turning point that shifts the way in which the jurisprudence and legal philosophy of the 20th c. described the problem as a constituent of political democracies. The USA is the only Western democracy to exclude any kind of legal punishment against extreme forms of language intended to foster hatred in the public space.[1] Free speech, the First Amendment provision, prevails. This has been reaffirmed, unanimously by the US Supreme Court as recently as 2017 [16], however, the American Bar Association has noted how the controversy over hate speech has now been renewed in light of the Black Lives Matter Movement and the MeToo movement [17]. Against hate speech bans one of the more persuasive arguments was advanced by Ronald Dworkin [18], who pointed out that law enforcement would deny subjects an adequate opportunity for dissent. Freedom of speech 'guarantees and preserves liberalism's commitment to equality by offering everyone an opportunity to speak, whereas any other policy, such as state regulation, would fail to offer this equal opportunity' [19]. This egalitarian liberalism has recently been contested by Jeremy Waldron, stemming from the perspective of the construction of a public space based on dignity, a human constituent that cannot be politically bartered nor negotiated [20, 21].

Europe has taken a very different approach and is often held up in counterpoint to the United States. In-light of the Holocaust, in Germany and other countries, laws against Holocaust denial and the glorification of Nazism have long been deemed a necessary and valid curtailments of free speech. The European Union first enacted law to combat racism and xenophobia in 1996 [22], followed by the current Framework Decision in

---

[1] The *International Convention on the Elimination of All Forms of Racial Discrimination* entered into force on January 4th 1969 [15]. It has been ratified by 88 states. The Convention also requires its parties to outlaw hate speech and criminalize membership in racist organizations. USA ratified the Convention, but upon ratification, it stated the following reservations: "1. That the Constitution and laws of the United States contain extensive protections of individual freedom of speech, expression and association. Accordingly, the United States does not accept any obligation under this Convention, in particular under articles 4 and 7, to restrict those rights, through the adoption of legislation or any other measures, to the extent that they are protected by the Constitution and laws of the United States. 2. That the Constitution and laws of the United States establish extensive protections against discrimination, reaching significant areas of non-governmental activity. Individual privacy and freedom from governmental interference in private conduct, however, are also recognized as among the fundamental values which shape our free and democratic society. [...] 3. That with reference to article 22 of the Convention, before any dispute to which the United States is a party may be submitted to the jurisdiction of the International Court of Justice under this article, the specific consent of the United States is required in each case." Cfr. https://treaties.un.org/Pages/ViewDetails.aspx?src=TREATY&mtdsg_no=IV-2&chapter=4&lang=en#EndDec.

2008 [23], which is supported by the Implementation Report of 2014 which provides further clarifications around hate speech [24]. The *Additional Protocol to the Convention on Cybercrime* sought criminalization of online racist and xenophobia in 2013 [25], with the Council of the European Union invited ratification by member states [26]. Focusing on online hate particularly in social media, there was first an effort to reach voluntary agreements with technology companies to curtain hate speech in 2016 [27], but after this failed to sufficiently address the problem, a Communication with guidelines and principles followed in 2017 [28], and a Commission Recommendation in 2018 [29]. Individual countries implement the European Framework through national laws and Germany, for example, implemented a law with heavy fines for companies that don't remove clear hate speech within 24 h. Recently there have also been high profile police raids on those who post hate speech online. For example, in November 2020 European police launched simultaneous raised in seven European countries on those who had posted online hatred and incitement to violence with 96 suspects questions in Germany [30].

3 Technology: Fostering Dignity and Preventing Extremism

From a technological point of view the nature of the argument, fostering dignity, has been perceived as a real need. [31] 'explores the limitations of unilateral national content legislation and the difficulties inherent in multilateral efforts to regulate the Internet':

> The exponential growth in the Internet as a means of communication has been emulated by an increase in far-right and extremist web sites and hate based activity in cyberspace. The anonymity and mobility afforded by the Internet has made harassment and expressions of hate effortless in a landscape that is abstract and often beyond the realms of traditional law enforcement [31].

As mainstream platforms moved to de-platform white nationalism [32], then QAnon [33–35], an exodus of users to alternative social media space occurred [36]. A number of the newer platforms are located in countries whose values may be at odds with democratic norms.

Australia represents a particularly interesting case study in the unilateral approach to regulating online content. First the Australian government unilaterally regulated with respect to certain forms of terrorist content, specifically video captured by a terrorist during an attack, which it pushed in legislation in the wake of the Christchurch terrorist attack with penalties for technology platforms despite their stiff opposition [37]. A second effort at unilateral regulation, this time in relation to the use of Australian news content by technology platforms, led to Facebook taking the nuclear option and banning all Australian news from its platform [38]. The ban, to promote Facebook's corporate interests, was implemented far more rapidly than any previous action to prevent hate speech.

Hate speech, in the realms of social media, is a kind of writing that disparages and is likely to cause harm or danger to the victim. It is a 'bias-motivated, hostile, malicious speech aimed at a person or a group of people because of some of their actual or perceived innate characteristics' [9]. It is a kind of speech that demonstrates a clear intention to be hurtful, to incite harm, or to promote hatred. The environment of social media and the interactive Web 2.0 provides a particularly fertile ground for creation, sharing and exchange of hate messages against a perceived enemy group. These sentiments are expressed at news review sites, Internet forums, discussion groups as well as in micro-blogging and social networking sites [39]. Extremist content relates to groups that are identified as sharing an extremist ideology, usually one linked to individuals or groups of people who have engaged in violence while presenting that ideology as their justification. The rationale for removing or banning extremist ideologies is that often violent attacks motivated by an ideology, that is terrorist attacks, inspire further attacks. This can be seen in the three attacks that followed the Christchurch attack [6].

[40] addresses the problem of hate speech detection in online user comments. Hate speech is defined as 'an abusive speech targeting specific group characteristics, such as ethnicity, religion, gender, sexual orientation, gender identity, sex characteristics or disability'. Where this content can lead users down a path of extremism, it poses a significant safety risk to society.

Automated detection, clustering, monitoring and managing, and tracking on real time are the most common problems. Several approaches have been proposed so far, mostly leaning on NLP, AI and semantics: (i) classifiers can be used to detect the presence of hate speech, using sentiment analysis and subjectivity detection in pre-defined areas (e.g. race, gender, religion) [41], (ii) lexicons can be created and also used for this purpose, (iii) practical projections to real-world discourses can then be applied [42], (iv) distributed low-dimensional representations of hate comments can be identified using neural language models that can then be fed as inputs to a classification algorithm [40], (v) machine learning [43], (vi) annotated datasets, impact of extra-linguistic features in conjunction with character n-grams for hate speech detection [42, 43], (vii) qualitative and discourse analysis [44]. The table below, by Silva et al. [45], displays the top ten expressions in Twitter and Wisper (Table 1):

Table 1. Top ten expressions in Twitter and Wisper [45].

Twitter	% posts	Whisper	% posts
I hate	70.5	I hate	66.4
I can't stand	7.7	I don't like	9.1
I don't like	7.2	I can't stand	7.4
I really hate	4.9	I really hate	3.1
I fucking hate	1,8	I fucking hate	3.0
I'm sick of	0.8	I'm sick of	1.4
I cannot stand	0.7	I'm so sick of	1.0
I fuckin hate	0.6	I just hate	0.9
I just hate	0.6	I really don't like	0.8
I'm so sick of	0.6	I secretly hate	0.7

Some social media platforms are now using their own artificial intelligence approaches to remove hate speech proactively before users report it. On Facebook the pro-active removal rate has risen from 23.6% at the end of 2017, when it was first introduced, to 97.1% by the end of 2020 [37], while on Instagram it has risen from 44.6% at its introduction at the end of 2019 to 95.1% by the end of 2020. The pro-active rate is the fraction of content removed by the AI before users report it as a percentage of all hate speech that was removed (i.e. the total of AI proactive action and responses by Facebook staff to user reports) (Fig, 1).

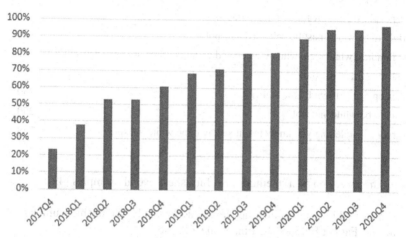

Fig. 1. Facebook's proactive removal of hate speech, graph by the authors based on Facebook data at [37]

Other platforms are far less proactive. Research analysis of one white supremacist account on Twitter found that of the 11,000 tweets which could be accessed, there were only 12 unique posts (ignoring variations in the URLs used to store multiple copies of the same image) [47]. Many of the words and phrases used are well known white supremacy slogans, yet the content remained on Twitter for years, and through an automation tool tweets were reposted each day. The data in Table 2 strongly suggests the absence of a well configured AI system at Twitter.

Another failure could be seen in relation to a terrorist manifesto that included clear examples of incitement to kill people from a range of groups. Copies of the manifesto were not only visible in Google search results, they appeared on webpages that also hosted Google Ads. In this case the result was government advertising along-side a terrorist manifesto [48]. This highlights the need for the widespread use of technology to identify hate speech, fear speech and extremism.

Table 2. Content of 11,000 Tweets from one Twitter account, data source: [38]

Tweet	Count
It's okay to be white #auspol" + an image reading "It's okay to be white"	340
God loves white people #auspol	303
White people have interests #auspol	303
White people are awesome! #auspol	251
I love white people #auspol	251
White people are our greatest strength! #auspol	251
The right to be white must be defended #auspol	251
White families, white men, white women and white children are okay #auspol	251
White lives matter. #auspol	250
White people are alright by me #auspol	250
It's okay to be white #auspol	250
Hi everyone, just letting you know that it's okay to be white #auspol	249

A recent survey on NLP methods also furnishes several examples [49]:

(1) Go fucking kill yourself and die already useless ugly pile of shit scumbag.
(2) The Jew Faggot Behind The Financial Collapse
(3) Hope one of those bitches falls over and breaks her leg

While the set of features examined by [49] in the different works present a great diversity, the classification methods mainly focus on supervised learning, surface-level features to classify, and generic features, such as bag of words or embeddings. According to the authors, character-level approaches work better than token-level approaches, and lexical resources, such as list of slurs, may help classification, but usually only in combination with other types of features. A benchmark or annotated dataset would be needed, as inferences, suppositions and associative tropes are difficult to detect and could benefit from a semantic approach considering the contexts and possible scenarios.

An interesting approach is taken when annotations and descriptions are grounded on a crowdsourced basis. Oboler [50] identified ten years ago the main elements of antisemitic discourse in social media—what he called "antisemitism 2.0"—as follows: (i) The content denies its antisemitic nature; (ii) it promotes antisemitic tropes, (iii) it claims its message is a legitimate view people should be free to hold (no different from choosing to support a particular sports team), (iv) the content is designed to go viral by making sharing the content both technically easy and socially acceptable in social media, (v) the audience is not the dedicated antisemites but rather the susceptible public. These elements have since been found to hold true for a range of other types of hate and the concept generalised to Hate 2.0 [51].

The next stage has been the creation of social and collective bonds, seeking for awareness and participation [50, 52]:

> Based on the recommendations of the Global Forum, the Online Hate Prevention Institute (OHPI) in Australia developed FightAgainstHate.com, a cloud-based tool, later turned into a software-as-a-service, for reporting, monitoring, and measuring the response to online antisemitism as well as other forms of online hate. Using the tool the public can report various types of online hate speech and assign both a category and sub-category to the hate they report [53]. The system is designed to calibrate members of the public against experts and other members of the crowd who have already been assessed as being of similar quality to the experts for a given type of content. The system can be built into a larger eco system for real-time data-sharing between the public, civil society groups and other stakeholders, allowing the data to also serve as input to further AI based tools [54].

4 Regulatory Models: Socio-legal Ecosystems

How should hate speech be effectively regulated? How can compliance with universal values such as peace and tolerance be achieved? As mentioned, to do this we need some regulatory imagination and a combination of hard and soft law, smart regulations, multi-stakeholder governance, policies and ethics.

Reducing hate speech is a matter of changing social practice. Pierre Bourdieu explains how social practice is formed with the equation "[(habitus)(capital)] + field = practice" [55]. In this construct, different types of stakeholders bring their different dispositions and values (i.e. habitus) as well as different forms of capital (e.g. different economic, cultural, social and symbolic capital) to the different social areas (i.e. fields). Bourdieu also introduced the notions of 'symbolic violence' and 'symbolic power' in his work, showing their relevance in the construction of non-egalitarian social bonds. The introduction of policies and ethics can help shape the habitus, while hard law, soft law, and smart regulations alter the field. Banks [31] suggests that "a broad coalition of government, business and citizenry is likely to be most effective in reducing the harm caused by hate speech". This builds on the fact that the different stakeholders bring not only different forms of capital to the problem, but different dispositions that focus on different aspects of the harm. Working together provides both more tools and greater coverage of the problem, leading to a more complete change to social practice.

While this broad approach is a reasonable goal, it is not always easily achievable. Some governments use hate speech for other political reasons—e.g., to prosecute citizens participating in demonstrations or to divide society to better secure political power.

We think that what is required to address this is a set of regulatory tools to create *socio-legal ecosystems*, e.g., patterns of behaviour able to show resilience, i.e. *leaning on behavioural rather than normative compliance* [13, 56]. Even though, this is not simple. In a previous version of this work [57], we focused on behavioural compliance only. We have revisited this point with the notion of *ecological* compliance, which is broader, and it takes into account the empirical dimension of regulatory compliance, bringing together the descriptive (empirical) and normative (prescriptive) aspects of behaviour.

The normative side of compliance—i.e., the models against which a determined set of behaviours can be tested—cannot be ignored because it is a substantial part of the cognitive skills and affordances of the agents (be they human or artificial). Ecological compliance can be defined as the set of technical requirements and social conditions that a system must consider to create a sustainable socio (legal) ecosystem, which is the main objective of an effective hate speech policy.

Behavioural compliance has been investigated in organisations, companies, and administrations. Several studies highlight the importance of social bonding, social influence, and cognitive processing [58, 59]. Deterrence does not suffice [60]. Social bonds largely influence attitudes toward compliance and foster the adoption of personal codes of conduct. However, social bonds that work against racism are not spontaneous. Waseem [44] concludes:

> We find that amateur annotators are more likely than expert annotators to label items as hate speech, and that systems trained on expert annotations outperform systems trained on amateur annotations.

Thus, expert knowledge, guidance (and political will), matter [61]. To make effective the protections of the rule of law in the age of linked data, a combination of sanctions, training, and educative efforts should be put in place. Therefore, ethics should play a new regulatory role on the web of data [62]. We prefer the expression "legal governance" rather than "law", following the work already done by the AI4People Group on good AI governance [63]. This is a new cultural turn not (or not only) for coercive measures, but for *relational law and justice* on the web of data and the internet of things [64].

This leaves the question of how governments should response to technology platforms that refuse to cooperate or follow local laws. Facebook's removal of news in Australia shows that large corporations are willing to go head-to-head with governments if they believe it serves their interests. Ultimately governments can seek to regulate expectations of platforms, something seen in a draft Australian eSafety Bill [65], and platforms can be forced to either comply or exit the market space. Platforms may be willing to pull out of some markets to protect their interests, but cooperation between governments and international agreements make self-exclusion from markets to avoid regulation unviable for social media platform. With the stakes rising this high, the threat of platform interference in democratic elections, to serve their own corporate ends, may be edging closer [66].

5   Ecological Compliance

We would like to raise some more questions to shed some light on this debate. Ecological compliance is more difficult to achieve than regulatory, normative and (just) behavioural compliance, for more conditions apply to the available regulatory means and instruments. Enforcement can only be a component, along with agreement, conformance, and acceptance of values, principles, and rules. Hence, the acquiescence and cooperation of the subjects must be represented as a necessary condition for the regulatory pattern to occur.

Therefore, the tension between free speech and hate speech limitations cannot be solved in one single dimension. At the epistemic level we should introduce (i) the complexity entailed by collective interactions and decision-making, (ii) the different levels of abstraction in which these concepts are used, (iii) the micro- and macro- societal layers in which the implementation of regulations operates.

Gould observes that 'hate speech is fuzzed in the abstract but more apparent when confronted in person' [67]. He carried out an interesting empirical analysis, showing that despite the judicial hurdles based on the first amendment the concept has pervaded American society. We are not facing a discrete category, but a continuum in which semantic and pragmatic elements are entangled to produce social adhesion and bonds. This would be an example of *societal regulation:*

Hate speech regulation has permeated other elite institutions like the media and has trickled down to influence mass opinion and common understandings of institutional norms. [So] extra-judicial law and the power of legal meaning-making [...] informal law or mass constitutionalism is as powerful as the formal constitution, providing vehicles to change that exists without the intervention of courts [67].

Delgado and Stefancic [68] observe that, at least in USA, there is a tendency to frame the debate in "legal" terms, i.e. as one of procedure rather than substance. On the contrary, defenders of setting hate speech limitations: (i) ponder the importance of social power, and recognize the connection between general, nontargeted hate speech and the rise of destructive social movements, (ii) point out that hate speech often targets individuals who, by reason of his or her race or physical appearance, have been the object of similar attacks many times before.

The pressure from society is building and social media platforms have responded to varying degrees of success. The attack on the Capitol in January 2021 by QAnon and Trump followers brought a new imperative to US law makers, as have expectations of society following mass movements like Black Lives Matter and Me Too. The harm in hate speech and extremism is felt on a widespread basis and even in the halls of power. Norms, laws and community expectations are out of sync, and efforts to recalibrate are needed.

Another problem on the research front is the reliability of annotations, as "the presence of hate speech should perhaps not be considered a binary yes-or-no decision, and raters need more detailed instructions for the annotation" [69]. Researchers working on a German hate speech corpus for the refugee crisis in 2016 noticed that building a classifier (i.e. rating the offensives of tweets on a 6-point Likert scale) entailed discussions not only among raters but researchers, due to personal attitudes.

The difficulty of automated detections should not be underestimated. In the recent First Shared task on Aggression Identification organized with the TRAC workshop at COLING 2018, in which 30 teams finally submitted their system, "performance of the neural networks-based systems as well as the other approaches do not seem to differ much. If the features are carefully selected, then classifiers like SVM and even random forest and logistic regression perform at par with deep neural networks" [60]. The task was to develop a classifier that could discriminate between Overtly Aggressive, Covertly Aggressive, and Non-aggressive texts. The participants were provided with a dataset of 15.000 aggression-annotated posts and comments (in English and Hindi). Systems obtained a weighted F-score between 0.50 and 0.64. This is consistent with similar scores in current research summarised in this paper (Sect. 3). The data from Facebook suggests that with enough data and a large enough sample of human reviews for training, higher accuracy can be achieved, but the scale required may only be possible for a few very large platforms.

Thus, crowdsourced hate speech reporting face two main challenges: (i) cooperation between lay and expert knowledge to annotate the corpus, (ii) the difference between the surface of discourse and the environments and contexts that discourses contribute to create.

What is crucial is differentiating between the individual expression and the course of collective action in which this expression is embedded. This would help to separate *hate speech* from *fear speech*. Figures 2 and 3 show how cooperation between lay people (reporting), experts (evaluating and counselling) and institutions (receivers) can help to solve the puzzle. But even in this case, independent monitoring and evaluation matters, as governments may fail in reducing the volume of abusive content on social media platforms [70]. In addition, some governments may also divert the definition of hate speech, broadening it to target political adversaries. Thus, hate speech regulations should not be understood only from a narrow national perspective, but as a global exercise of implementation of human and democratic rights.

Fig. 2. Types of organised threats. Source: Oboler [72]

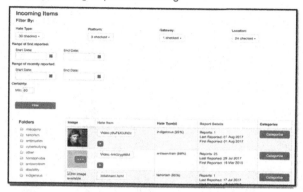

Fig. 3. Facilitation of experts' tasks. Source: Oboler [72]

6 Final Remarks

Concern over online hate and extremism has risen sharply since 2019 when the Christchurch terrorist attack was live streamed through Facebook. The attacker's self-radicalisation online, and his online content including the video of the livestreamed attack and the manifesto he uploaded just before the attack began, provided a global wakeup call [6]. The content was rapidly duplicated and spread across the internet. It led to concern from the public and responses by both governments and technology companies. One response from the technology sector was a greater focus on cooperation to tackle terrorism through GIFCT [73]. Efforts by technology platforms to prevent the use of their services by extremists slowly increased over the following two years, then took a major leap forward in early 2021 following the attack on the US Capitol. The change resulted in sweeping deplatforming efforts across major platforms. Twelve technology companies took the unprecedented step of banning or severely restricting accounts belonging to the President of the United States [74]. The change is qualitative as the threshold for unacceptability has been lowered, different kinds of material are now subject to sanction. The link between hate speech, radicalization and violent extremism is now seen more clearly.

Research in this space is increasing rapidly. Figure 4 shows the number of papers published each year, according to a Google Scholar search, on the topic of "social media" and "online hate". The research comes from a variety of fields and focuses on the problem from many different perspectives. While there is still much work to be done in this space, the research intensity is growing. A focus at the intersection of technology and law is likely to be a vital part of this research into the future.

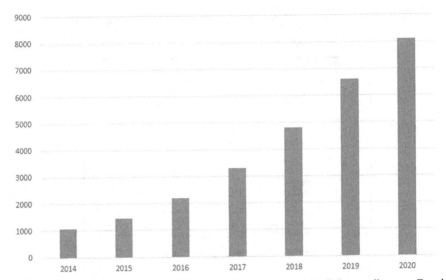

Fig. 4. Papers published per year on "Hate Speech" and "Social Media" according to a Google Scholar search

7 Conclusion

The change we see unfolding, including the exodus to smaller platforms by those with more radical views, highlights the urgent need for a more rounded use of the web of data to tackle the threat of violent extremism. In recent years we have learned the importance of focusing not only on direct threats of terrorism and incitement to violence, but also more broadly on hate speech which fuels radicalization.

Removing the content is not enough to counter the threat, we have seen the use of coded language and shifting symbols, and ultimately, we have seen entire communities relocate to alternative platforms. Legal remedies against terrorism are also only a part of the problem. A more complete and complex response is needed, one that helps to shift online culture. One driven by the data and with greater cooperation between technology platforms, governments, civil society organisations and the public.

Civic technologies, CivicTech, should be aligned with a proactive attitude towards regulations, understood as sustainable behavioural patterns through regulatory models. With the right ecosystem in place those at the early stages of travel down a path to radicalization can, through their use of mainstream platforms, be identified and deradicalized. The attitudes of the community, which can develop into intolerance for minorities of various types, can be shifted. New threats can be identified. New responses can be developed.

A combination of regulatory instruments, incentives, training, proactive self-awareness and education can be effective in shifting cultural practices so as to counter the problem of online hate and incitement to extremism.

Acknowledgements. This research was partially funded by the Data to Decisions Cooperative Research Centre (D2D CRC, Australia) (http://www.d2dcrc.com.au/), and Meta-Rule of Law

(DER2016-78108-P, Spain). Views expressed herein are however not necessarily representative of the views held by the funders. A former and shorter version of this work was presented at TERECOM@Jurix 2018, pp. 135–134 http://ceur-ws.org/Vol-2309/11.pd.

References

1. Eisen, A.: The Spiritual Audacity of Abraham Joshua Heschel. On Being (2012). https://onb eing.org/programs/arnold-eisen-the-spiritual-audacity-of-abraham-joshua-heschel/
2. Klemperer, V.: LTI Lingua Tertii Imperii. A Philologist's Notebook [1975]. Continuum, London (2006)
3. Andrews, R.v., Smith: 65 O.R. (2d) 161 (Ont.CA) at 179; R v. Keegstra [1990] 3 SCR 697, 699 (1989)
4. Naffi, N.: The Trump effect in Canada: A 600 per cent increase in online hate speech, November 2, 9.37am AEDT The Conversation (2017). https://theconversation.com/the-trump-effect-in-canada-a-600-per-cent-increase-in-online-hate-speech-86026
5. Oboler, A.: New Zealand Terrorist Attack: Analysis from OHPI. Online Hate Prevention Institute, Melbourne (2019). https://ohpi.org.au/new-zealand-terrorist-attack/
6. Oboler, A., Allington, W., Scolyer-Gray, P.: Hate and Violent Extremism from an Online Sub-Culture: The Yom Kippur Terrorist Attack in Halle, pp. 15, 16, 22, 33, 34, 40. Online Hate Prevention Institute, Germany, Melbourne (2019)
7. Naffi, N.: Ceci n'est pas un discours haineux, 23/04/2017 08:21 EDT | Actualisé 23/04/2017 08:21 EDT. https://quebec.huffingtonpost.ca/nadia-naffi/islamophobie-discours-haineux_b_16151548.html
8. Shields, D.: War Is Beautiful. The New York Times Pictorial Guide to the Glamour of Armed Conflict. Powerhouse Books, Nova York (2015)
9. Siegel, M.L.: Hate speech, civil rights, and the Internet: the jurisdictional and human rights nightmare. Alb. LJ Sci. Tech. 9, 375–398 (1998)
10. Elphick, L.: Victoria's new anti-vilification bill strikes the right balance in targeting online abuse. The Conversation, 11 September 2019. https://theconversation.com/victorias-new-anti-vilification-bill-strikes-the-right-balance-in-targeting-online-abuse-123014
11. Oswald, R.: A month after Capitol riot, a look at domestic terrorism laws, 4 February. https://www.rollcall.com/2021/02/04/a-month-after-capitol-riot-a-look-at-domestic-terrorism-laws/
12. Terms of Reference - Inquiry into Anti-Vilification Protections, Victorian Parliament, 12 September 2019. https://www.parliament.vic.gov.au/images/stories/committees/lsic-LA/Inq uiry_into_Anti-Vilification_Protections_/Terms_of_Reference_-_Inquiry_into_Anti-Vilifi cation_Protections.pdf
13. Poblet, M., Casanovas, P., Rodríguez-Doncel, V.: Linked Democracy. Springer Briefs. Springer, Cham (2019). https://doi.org/10.1007/978-3-030-13363-4
14. Casanovas, P., Mendelson, D., Poblet, M.: A linked democracy approach for regulating public health data. Heal. Technol. 7(4), 519–537 (2017). https://doi.org/10.1007/s12553-017-0191-5
15. United Nations: International Convention on the Elimination of All Forms of Racial Discrimination Adopted and opened for signature and ratification by General Assembly resolution 2106 (XX) of 21 December 1965. https://www.ohchr.org/en/professionalinterest/pages/cerd.aspx
16. Matal v. Tam: 137 S.Ct. 1744 (2017). 1750, 1764, 1769
17. Wermiel, S.J.: The ongoing challenge to define free speech. Hum. Rights Mag. 43(4) (2018). https://www.americanbar.org/groups/crsj/publications/human_rights_magazine_home/the-ongoing-challenge-to-define-free-speech/the-ongoing-challenge-to-define-free-speech/

18. Dworkin, R.: Freedom's Law. Oxford University Press, Oxford (1996)
19. Levin, A.: Pornography, hate speech, and their challenge to Dworkin's Egalitarian liberalism. Public Aff. Q. **23**(4), 357–373 (2009)
20. Waldron, J.: The Harm in Hate Speech. Cambridge University Press, Cambridge (2012)
21. Oboler, A., Allington, W., Scolyer-Gray, P.: Hate and Violent Extremism from an Online Sub-Culture: The Yom Kippur Terrorist Attack in Halle, Germany, Melbourne, pp. 1–2. Online Hate Prevention Institute (2019)
22. EU 96/443/JHA: Joint Action of 15 July 1996 adopted by the Council on the basis of Article K.3 of the Treaty on European Union, concerning action to combat racism and xenophobia
23. EU Council Framework Decision 2008/913/JHA of 28 November 2008 on combating certain forms and expressions of racism and xenophobia by means of criminal law
24. Report from the Commission to the European Parliament and the Council on the implementation of Council Framework Decision 2008/913/JHA on combating certain forms and expressions of racism and xenophobia by means of criminal law, COM(2014)27 final, 3–7
25. Council conclusions on combating hate crime in the European Union – Justice and Home Affairs Council meeting, Brüssel, 5–6 December (2013). http://www.consilium.europa.eu/uedocs/cms_data/docs/pressdata/en/jha/139949.pdf
26. Council of Europe: Additional Protocol to the Convention on Cybercrime, concerning the criminalisation of acts of a racist and xenophobic nature committed through computer systems CETS No 189, 20 October 2013
27. European Commission: European Commission and IT Companies announce Code of Conduct on illegal online hate speech, ec.europa.eu, 31 May 2016. http://europa.eu/rapid/press-release_IP-16-1937_en.htm
28. European Commission: COM(2017) 555 final, 28.9.2017, Communication From The Commission To The European Parliament, The Council, The European Economic And Social Committee And The Committee Of The Regions - Tackling Illegal Content Online
29. European Commission Recommendation 2018/334 of 1.3.2018 on measures to effectively tackle illegal content online (C(2018) 1177)
30. European police in coordinated raids against online hate speech. Reuters, 3 November 2020. https://www.reuters.com/article/us-europe-crime-internet-idUSKBN27J1C3
31. Banks, J.: Regulating hate speech online. Int. Rev. Law Comput. Technol. **24**(3), 233–239 (2010). https://doi.org/10.1080/13600869.2010.522323
32. Ingber, S.: Facebook Bans White Nationalism and Separatism Content From Its Platforms, NPR, 27 March 2019. https://www.npr.org/2019/03/27/707258353/facebook-bans-white-nationalism-and-separatism-content-from-its-platforms
33. Ohlheiser, A., Shapira, I.: Gab, the white supremacist sanctuary linked to the Pittsburgh suspect, goes offline (for now), Washington Post, 29 October 2018. https://www.washingtonpost.com/technology/2018/10/28/how-gab-became-white-supremacist-sanctuary-before-it-was-linked-pittsburgh-suspect/
34. Dick, S.: Pro-trump supporters switch to gab after twitter bans, parler removal. New Dly. (2021). https://thenewdaily.com.au/news/world/us-news/2021/01/13/trump-social-media-gab/
35. Collins, B., Zadrozny, B.: Facebook bans QAnon across its platforms. NBC News (2020). https://www.nbcnews.com/tech/tech-news/facebook-bans-qanon-across-its-platforms-n1242339
36. Sharples, S.: Donald Trump's Twitter suspension sees conservative social media site Gab claim thousands of new users, News.com.au, 11 January 2021. https://www.news.com.au/finance/business/technology/donald-trumps-twitter-suspension-sees-conservative-social-media-site-gab-claim-thousands-of-new-users/news-story/8753861b38c1981fb89a5916c5432bd9

37. Cave, D.: Australia passes law to punish social media companies for violent posts. N. Y. Times (2019). https://www.nytimes.com/2019/04/03/world/australia/social-media-law.html
38. Flynn, K.: Facebook bans news in Australia as fight with government escalates. CNN, 18 February 2021. https://edition.cnn.com/2021/02/17/media/facebook-australia-news-ban/index.html
39. Gitari, N.D., Zuping, Z., Damien, H., Long, J.: A lexicon-based approach for hate speech detection. Int. J. Multimedia Ubiquitous Eng. 10(4), 215–230 (2015)
40. Djuric, N., Zhou, J., Morris, R., Grbovic, M., Radosavljevic, V., Bhamidipati, N.: Hate speech detection with comment embeddings. In: Proceedings of the 24th International Conference on World Wide Web, pp. 29–30. ACM, May 2015
41. Gambäck, B., Sikdar, U.K.: Using convolutional neural networks to classify hate-speech. In: Proceedings of the First Workshop on Abusive Language Online, pp. 85–90 (2017)
42. Waseem, Z., Hovy, D.: Hateful symbols or hateful people? Predictive features for hate speech detection on twitter. In: Proceedings of NAACL-HLT 2016, San Diego, California, 12–17 June, pp. 88–93. Association for Computational Linguistics (2016)
43. Burnap, P., Williams, M.L.: Cyber hate speech on twitter: an application of machine classification and statistical modeling for policy and decision making. Policy Internet 7(2), 223–242 (2015)
44. Waseem, Z.: Are you a racist or am I seeing things? Annotator influence on hate speech detection on Twitter. In: Proceedings of the First Workshop on NLP and Computational Social Science, pp. 138–142 (2016)
45. Silva, L.A., Mondal, M., Correa, D., Benevenuto, F., Weber, I.: Analyzing the targets of hate in online social media. In: ICWSM, pp. 687–690 (2016)
46. Facebook: Community Standards Enforcement Report. https://transparency.facebook.com/community-standards-enforcement#hate-speech
47. Oboler, A.: Extremism Online: The Automation of White Supremacy. Online Hate Prevention Institute (2020). https://ohpi.org.au/extremism-online-the-automation-of-white-supremacy/
48. Oboler, A.: Advertising supports hosting of terrorist manifesto. Online Hate Prevention Institute (2020). https://ohpi.org.au/advertising-supports-hosting-of-terrorist-manifesto/
49. Erjavec, K., Kovačič, M.P.: You don't understand, this is a new war! Analysis of hate speech in news web sites' comments. Mass Commun. Soc. 15(6), pp. 899–920 (2012)
50. Oboler: Online Antisemitism 2.0. "Social Antisemitism" on the "Social Web", JCPA, 1 April. (Prereleased in February 2008)
51. Oboler, A.: Aboriginal Memes and Online Hate, p. 19. Online Hate Prevention Institute (2012). https://ohpi.org.au/aboriginal-memes-and-online-hate/
52. Schmidt, A., Wiegand, M.: A survey on hate speech detection using natural language processing. In: Proceedings of the Fifth International Workshop on Natural Language Processing for Social Media, Valencia, Spain, 3–7 April 2017, pp. 1–10. Association for Computational Linguistics (2017)
53. Oboler, A., Connelly, K.: Hate speech: a quality of service challenge. In: IEEE Conference on e-Learning, e-Management and e-Services (IC3e), pp. 117–121 (2014)
54. Oboler, A., Connelly, K.: Building SMARTER communities of resistance and solidarity. Cosmopol. Civ. Soc. Interdiscip. J. 10(2), 91–110 (2018). https://doi.org/10.5130/ccs.v10i2.6035
55. Bourdieu, P.: Distinction: A social Critique of the Judgement of Taste [1979]. Harvard University Press, Cambridge (1996)
56. Gunderson, L., Cosens, B.: Case studies in adaptation and transformation of ecosystems, legal systems, and governance systems. In: Cosens, B., Gunderson, L. (eds.) Practical Panarchy for Adaptive Water Governance, pp. 19–31. Springer, Cham (2018). https://doi.org/10.1007/978-3-319-72472-0_2

57. Casanovas, P., Oboler, A.: Behavioural compliance and law enforcement in online hate speech. In: TERECOM@Jurix 2018, pp. 135–134 (2018). http://ceur-ws.org/Vol-2309/11.pd

58. Ifinedo, P.: Information systems security policy compliance: an empirical study of the effects of socialisation, influence, and cognition. Inf. Manage. 51(1), 69–79 (2014)

59. Vroom, C., Von Solms, R.: Towards information security behavioural compliance. Comput. Secur. 23(3), 191–198 (2004)

60. Ogbonna, E., Harris, L.C.: Managing organizational culture: compliance or genuine change? Br. J. Manage. 9(4), 273–288 (1998)

61. Oboler, A.: Technology and regulation must work in concert to combat hate speech on line, March 12 2018 6.09pm AEDT (2018). https://theconversation.com/technology-and-regulation-must-work-in-concert-to-combat-hate-speech-online-93072

62. Casanovas, P.: Semantic web regulatory models: why ethics matter. Philos. Technol. 28(1), 33–55 (2015)

63. Pagallo, U., et al.: AI4People-On Good AI Governance: 14 Priority Actions, a SMART Model of Governance, and a Regulatory Toolbox (2019). https://www.eismd.eu/wp-content/uploads/2019/11/AI4Peoples-Report-on-Good-AI-Governance_compressed1.pdf

64. Casanovas, P., Poblet, M.: Concepts and fields of relational justice. In: Casanovas, P., Sartor, G., Casellas, N., Rubino, R. (eds.) Computable Models of the Law. LNCS, vol. 4884, pp. 323–339. Springer, Heidelberg. https://doi.org/10.1007/978-3-540-85569-9_21

65. Online Safety Bill 2020 – Exposure Draft. https://www.communications.gov.au/have-your-say/consultation-bill-new-online-safety-act

66. Oboler, A., Welsh, K., Cruz, L.: The danger of big data: social media as computational social science. First Monday 17(7) (2012). https://doi.org/10.5210/fm.v17i7.3993

67. Gould, J.B.: Speak No Evil: The Triumph of Hate Speech Regulation. University of Chicago Press (2010)

68. Delgado, R., Stefancic, J.: Four observations about hate speech. Wake Forest L. Rev. 44, 353–370 (2009)

69. Ross, B., Rist, M., Carbonell, G., Cabrera, B., Kurowsky, N., Wojatzki, M.: Measuring the reliability of hate speech annotations: the case of the European refugee crisis. arXiv preprint arXiv:1701.08118. (2017)

70. Kumar, R., Ojha, A.K., Malmasi, S., Zampieri, M.: Benchmarking aggression identification in social media. In: Proceedings of the First Workshop on Trolling, Aggression and Cyberbullying (TRAC-2018), Santa Fe, USA, 25 August, pp. 1–11 (2018)

71. Oboler, A.: Technology and regulation must work in concert to combat hate speech online, March 12, 2018 6.09pm AEDT. https://theconversation.com/technology-and-regulation-must-work-in-concert-to-combat-hate-speech-online-93072

72. Oboler, A.: Building peace by fighting online hate. Yitzhak Rabin Memorial Lecture, 4 November 2018, Slides (2018)

73. About GIFCT: Global Internet Forum to Counter Terrorism. https://gifct.org/about/

74. Crichton, D.: The deplatforming of president trump. Tech Crunch (2021). https://techcrunch.com/2021/01/09/the-deplatforming-of-a-president/

SPIRIT: Semantic and Systemic Interoperability for Identity Resolution in Intelligence Analysis

Costas Davarakis[1]($\boxtimes$), Eva Blomqvist[2] (iD), Marco Tiemann[3] (iD),
and Pompeu Casanovas[4,5] (iD)

[1] Singular Logic, SPIRIT, Athens, Greece
kdavarakis@singularlogic.eu
[2] Department of Computer and Information Science, Linköping University, Linköping, Sweden
eva.blomqvist@liu.se
[3] Innova Integra Ltd., Reading, United Kingdom
marco.tiemann@innovaintegra.com
[4] La Trobe Law School, La Trobe University, Melbourne, Australia
p.casanovasromeu@latrobe.edu.au, pompeu.casanovas@uab.cat
[5] Institute of Law and Technology, Autonomous University of Barcelona, Barcelona, Spain

Abstract. This paper introduces the SPIRIT H2020 Project. The SPIRIT identity resolution service has been designed to learn about identity patterns, to build up a social graph related to them, and thereby facilitate LEA's investigation work. The paper will briefly discuss the main task of identity resolution, the privacy controller system, the SPIRIT prototype that will realise the solution, and the ontology to embed privacy into the system. It also discusses a specific technical and legal challenge—i.e., semantic interoperability when integrating SPIRIT data—and its coordination at the agency level with human decision making—systemic interoperability. This paper takes into account the SPIRIT testing prototype and the first revision version (proof of concept prototype).

Keywords: Identity resolution · Social graph · Semantic interoperability · Privacy

1 Introduction

This paper presents some results of the EU H2020 Project SPIRIT.[1] The project aims at developing a semantically rich sense-making capability to provide Law Enforcement Agencies (LEA) with resolved identities and their provenance trail as substantiated evidence to be presented in court [1]. However, associative searches over all sources that correlate information from textual and multimedia data might involve the use of large quantities of personal data retrieved from social media and other open-source origins (OSINT). Some privacy protection must therefore be put in place [2]. The toolkit must

---

[1] SPIRIT. Scalable privacy preserving intelligence analysis for resolving identities. https://cordis.europa.eu/project/id/786993.

© Springer Nature Switzerland AG 2021
V. Rodríguez-Doncel et al. (Eds.): AICOL-XI 2018/AICOL-XII 2020/XAILA 2020, LNAI 13048, pp. 247–259, 2021.
https://doi.org/10.1007/978-3-030-89811-3_17

also comply with existing data protection and security protocols and be capable of receiving automated products from existing in-house Police systems to (semi) automatically produce a list of potential identities who may belong to a single person. The remainder of the paper is distributed into five sections: (i) SPIRIT system components, (ii) Semantic interoperability, (iii) Privacy controller system, (iv) Integration and systemic interoperability (v) and final section of conclusions and future work.

2   SPIRIT System Components

SPIRIT tools are developed in an implementation loop. Figure 1 presents the overview of the SPIRIT system components developed and integrated prototype:

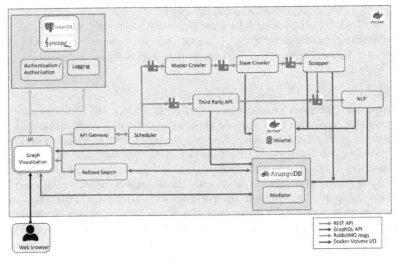

Fig. 1. Spirit system components overview for Y1 prototype

A set of standard components have been deployed as a base of the SPIRIT platform: (i) Docker (an independent container platform to seamlessly build, share and run applications in a way that developers can manage their infrastructure *and* applications)[2], (ii) Apache Syncope (user's authentication and authorisation, a very important function in SPIRIT)[3], (iii) PostgreSQL users DB[4] (open source object-relational database system that uses and extends the SQL language combined with many features that safely store and scale data workloads), (iv) ArangoDB content DB (open-source native multi-model database for graph, document, key/value and search needs - in SPIRIT used to store a content as a property graph)[5], (v) RabbitMQ (asynchronous message broker supporting multiple messaging protocols, message queuing etc.)[6]. SPIRIT partners provided a

---

[2] https://www.docker.com/.

[3] https://syncope.apache.org/.

[4] https://www.postgresql.org/.

[5] https://www.arangodb.com/.

[6] https://www.rabbitmq.com/.

second set of components related to services: (i) UI service, (ii) Refined Search service, (iii) API Gateway service, (iv) Scheduler service, (v) Crawler services (master crawler service, slave crawler service), (vi) Scraper service, (vii) Third Party API service, (viii) NLP service, (ix) Face Detection service, (x) Face Matching service. Integration tools are the following: (i) Rest services (access to SPIRIT platform with specific services), (ii) RabbitMQ messages. It is worthwhile specifically mentioning (i) the authentication and authorisation service, (ii) the activity logging service, (iii) the so-called mediator service, which translates a given query or data modification request, sent through a GraphQL[7] API, into transactions of operations to be executed by ArangoDB, (iv) the API gateway (single point of entry into the SPIRIT microservices system), (v) the schedular service (from which the user is able to trigger a fully automated content intake job)., (vi) the crawler service, (vii) the scraper service (it extracts textual content and images from the crawled web pages, and saves them in the shared volume and their metadata in the Content Database). These services—plus the Third Party API and NLP and refined search services— support some important internal controls related to the accountability, transparency and traceability of the system. They activate the inner handling of monitoring and controlling identity queries [3], in compliance with the ethical and legal security requirements, and the SPIRIT guidelines and incidental and residual risks policies [2, 4].

3 Semantic Interoperability

SPIRIT has only started to work on ontologies and semantic interoperability, but this is an important function. The first step was to identify potential ontology use cases. The work involved the use of experiences from previous projects (VALCRI[8]) and surveying examples of data integration and their challenges in SPIRIT. Semantic interoperability means to determine the unambiguous meaning of the data, in order to be able to exchange data knowing when it actually means the same thing. A prerequisite for automating semantic interoperability is being able to formally describe meaning (i.e., semantics)— ontologies are one way of semantically describing data. Several standard languages exist for expressing ontologies, and associated data, such as the W3C standards OWL[9] and RDF[10], which are used in the SPIRIT project. To provide LEAs with semantics-based data integration and search capabilities, and to provide a basis for semantics-based graph analysis and sense making, the semantics of the data in the social graph needs to be made explicit.

It should be highlighted that having the mediator component as the only entry point for accessing the SPIRIT content database (i) makes it possible to log all database accesses in a uniform manner, c.f. the privacy controller described in the next section; (ii) the mediator provides various abstraction levels, or views, over the social graph, e.g., such as ontology-based representations of the semantics of the data [5]. For identity resolution it may also be the case that data should be linked to external police databases. To cover these data integration requirements, the database schema needs to also reflect

---

[7] https://graphql.org/.

[8] http://valcri.org/.

[9] https://www.w3.org/OWL/.

[10] https://www.w3.org/RDF/.

the structure of such policing data. This raises the problem of what kind of ontologies, and what level of abstraction and granularity, are needed to abstract over both internal and external data. A concrete example is how the different notions of a "person", i.e., both the user's view and the views of components internal to the system, could be represented. In a policing dataset used by SPIRIT (anonymised by West Midlands Police, originating from the VALCRI project) a person is represented by the concept "nominal"—but a nominal can be either an organisation OR a person, e.g., an organisation can also be the victim of a crime and this needs to be represented. In SPIRIT internal data we are mainly targeting information about individual physical persons. However, in an ontology we can express that the notion of nominal subsumes both the concepts 'organisation' and 'person', and thus integrate data from the policing dataset at the appropriate level of abstraction.

Another example, related to the creation of views over data concerns the structure of the social graph generated. In the internal data structure of the data storage, the evidence supporting the identification of a person, e.g., as a face in an image, or through a mention in text, is quite complex, involving several nodes and edges in the graph database representing provenance information and internal tracking of the software processes used to produce the resulting connection. However, from a user perspective, the interesting part may be simply if a person was identified or not (considering some confidence threshold selected) in a specific piece of information analysed, e.g., an image or text etc. Figure 2 illustrates how an ontology-based view over a more complex graph data structure can be used to simplify the data structure and create a more intuitive view over the social graph for the end-user visualisation. Such graph abstractions also have the potential to be used for hiding details in the graph, e.g., for privacy reasons. However, the latter use case has not been explored in SPIRIT so far.

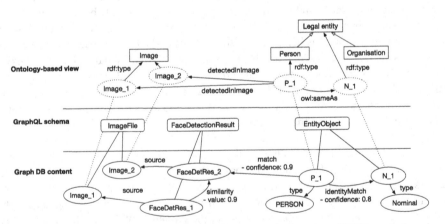

Fig. 2. Illustration of a potential graph abstraction, as an ontology-based RDF graph, based on an ontology view over the underlying property-graph database.

At the bottom of Fig. 2 is a property graph representation of data, similar to the way SPIRIT represents data internally, with an excerpt of a corresponding GraphQL schema in the middle (properties omitted for readability). The nominal node (N_1), to the bottom right, could be viewed as imported from an external data source, and then matched to the internal SPIRIT data through the identity resolution service. Several edge annotations are used in the property graph to represent the confidence level of edges. At the top is an ontology-based view of the data, where some of the actual nodes in the internal property graph representation (dashed lines) are viewed as instances of ontology concepts instead of GraphQL schema entities. This view could be represented as an RDF graph, hence, the usage of RDF and OWL-specific relations (properties) in the figure. Further, direct links, such as the "detectedInImage"-property have been inferred based on the data graph at the bottom. The latter can be done using tailored rules encoded in GraphQL queries over the property graph, or by executing more generic queries for retrieving an RDF graph mirroring the property graph at the bottom, and then reasoning over the retrieved graph, e.g. using OWL property chains to drive the new property instances, such as those using "detectedInImage".

In SPIRIT we will not materialise the RDF graphs, similar to the one illustrated above, but will rather use the approach usually denoted Ontology-based Data Access (OBDA), which has recently been extended beyond abstractions over relational databases [6]. This means that the ontology expressing the view over data will only be used to express mappings to the underlying data structure, while the data remains in its original form in the storage system. The system can then provide an additional API (or even directly a SPARQL endpoint), based on the view over data specified by the ontology, for querying the data in accordance with the ontology. A query received is then mapped, based on a set of mapping rules, into one or more GraphQL queries using the internal data representation (GraphQL schema), and the result of these queries are further combined to generate the result of the ontology-based query. In the case of SPIRIT, providing a generic SPARQL endpoint does not provide sufficient security and privacy control option, but rather we will merely extend the current GraphQL API to allow for ontology-based queries in addition to directly expressing the queries over the internal GraphQL schema, as illustrated in Fig. 3. The intention is to allow both for system components to use the rather low-level data representation provided by the GraphQL schema, mainly for

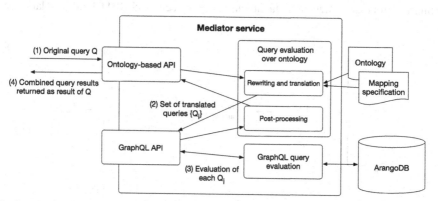

Fig. 3. Illustration of the workflow of the mediator service extended with an OBDA capability, as an additional API of the service.

SPIRIT-internal processing of data, as well as allowing for components and functionalities closer to the system end-user to use the more abstract high-level data representation expressed by the ontology, at the same time.

4 Privacy Controller System

The Privacy Controller System (PCS) is a specific and functionally separated system component in SPIRIT that addresses the tasks of.

a) gathering and logging user and system activities carried out when using the SPIRIT Platform,
b) analysing the gathered data to identify activities that may raise issues in terms of privacy, ethics or data protection requirements for data processing in SPIRIT,
c) presenting the gathered and analysed data to staff tasked with carrying out ethics monitoring of SPIRIT Platform usage, and
d) automatically triggering system activities to protect privacy, ethical or data protection rights where this may be necessary.

Within SPIRIT, the system evaluates available log data in terms of two main questions: a) To which extent is it necessary for an activity log entry to be reviewed by a relevant human operator? b) Does the output generated by the SPIRIT platform need to be modified prior to presenting it to a particular user? [7] Both questions are relevant, because they may refer the system to an external (technical, ethical or institutional) human controller. Figure 4 depicts the processing workflow of the PCS that enables the fulfilling the above purposes. The PCS integrates with the overall SPIRIT Platform in order to gather a range of input data that are necessary in order to provide a thorough activity log account of user and system actions. As part of this integration, the PCS gathers user actions directly at the user interface level, gathers processing data as they are communicated through the SPIRIT Platform backend, and retrieves processing data and primarily processing metadata that are added to the central SPIRIT Platform database.

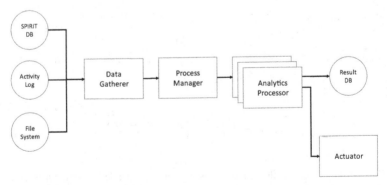

Fig. 4. Overview of privacy controller processing chain [7]

The Privacy Controller System transforms acquired data into a uniform data representation and then evaluates the incoming data in order to generate or update composite data points (such as counts of failed login attempts or searches executed in a specific investigation), assign data points with a criticality score, evaluate whether data points should be flagged to ethics review personnel or whether data points trigger any specified actions in the system.

Several types of Analytics Processor systems can be enabled for the Privacy Controller System. Since no statistical data sets are available for use with the SPIRIT Privacy Controller System, a set of evaluation rules has been defined to identify various types of events ranging from brute force credential hacking attempts to disproportionate use of specific types of searches or analytics processes within a single investigation or by a specific system user. The system also facilitates complex rule processing, scripted rule sequences and retrieval of supplemental data during rule evaluation. Manually specified rules are generally human-readable and can be explained to human end users, which can be beneficial for staff reviewing log and alert data generated by the Privacy Controller System.

Individual rules in the Privacy Controller Subsystem can be enabled, disabled and customised for individual system deployments, so that they can express specific monitoring and compliance requirements and goals of individual law enforcement agencies. Rules can also be seen as operational expressions of goals specified in leaf nodes of data protection ontologies, and triggering a rule can in principle be used for event-driven ontology updating; similarly, a data protection ontology can be used in order to organise and aggregate observed events for summarisation and organised presentation to users.

As next steps we will integrate such a data protection ontology into the system, as a general vocabulary to which the rules mentioned above can be attached. Several ontologies for this purpose have already been developed, such as the model proposed by Bartolini&Muthuri [8], PrOnto [9] and the GDPRov [10], for instance. However, we have chosen to reuse the most recent an integrative development in this area, which is the Data Privacy Vocabulary[11] (DPV), which is an ontology being developed and maintained by the W3C Data Privacy Vocabularies and Controls Community Group. This ontology is based on several of the earlier efforts in the area, and hence aim to cover similar concepts and use cases, including reasoning and rule-based checking of compliance. The ontology is being actively maintained by the community group, and already has a specialised extension (DPV-GDPR) covering GDPR[12], which we will specifically use in SPIRIT. Further, we will create another extension of DPV-GDPR, more specifically targeted at the SPIRIT use case, by subclassing the concepts in the existing ontology. Since the ontology itself is modular, and so will our extension, it will also be possible to create exchangeable ontology components that represent specific rules and regulations within a specific jurisdiction where SPIRIT will be use, i.e. catering for the difference in legal systems and regulations that LEAs around Europe are subject to. The latter makes the SPIRIT Privacy Controller System configurable to different legal contexts.

---

[11] https://dpvcg.github.io/dpv/.

[12] https://github.com/dpvcg/dpv-gdpr.

5 Integration and Systemic Interoperability as Regulatory Devices

This paper introduces the work on identity analysis and resolution carried out by the H2020 SPIRIT project. Identity resolution is a well-trodden path in computer science [11]. Compared to other related works based on attribute-based approaches and social links [12], or fuzzy compositional models [13], SPIRIT OSINT work can define habitual patterns as a social graph related to entity patterns that tend to reappear on the web. A System Integration describe the components and interactions, as well as the interoperability and scalability design. The integration and functional validation process define the roles and the workflow of integration (Fig. 5). For this presentation, we have put aside speaker identification and matching, face analysis entity extraction and matching, and acoustic scene classification.

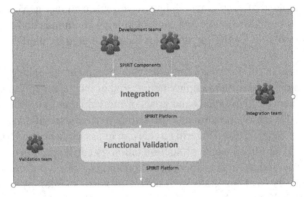

Fig. 5. Overview of the integration and validation workflow.

SPIRIT is a platform for law enforcement in which LEAs, private companies and universities cooperate. However, the terms of such a cooperation are strict. Data sharing with third parties constitute a well-known issue, for it increases the risk of breaches. It has already happened with the development of low-cost health mobile apps and devices equipped with a variety of sensors [14]. The risk of non-compliance with legal and ethical requirements increases when third parties are involved [15]. Thus, when the analysis has to be performed by third parties, privacy and compliance become a relevant issue too. As Vicenç Torra states from the very beginning of his book, 'similar problems arise when other actors not directly related to the data analysis enter into the scene (e.g. *software developers who need to test and develop procedures on data that they are not allowed to see*' [16] [our emphasis]. This is the case with the SPIRIT project. Following ethical requirements, developers of the system are not working with real data coming from LEA's repositories. This means that protections have to comply with several types of different requirements, dealing with:

a) the protection of citizens whose identity will be processed,
b) data protection as it has been conceptualised by Directive (EU) 2016/680 [17] and the Regulation (EU) 2016/679 that came into force on May 25th 2018 [18],
c) the monitoring and control of police behaviour in their daily investigatory roles.

Thus, three different agents—police, citizens and the researchers themselves—that require separate incidental risks internal policies and a set of redress mechanisms in case of failure, data breaches, wrong identification, or false positives bringing about social and personal consequences [2, 4]. This creates a complex environment for integration purposes, in which human and artificial agency coexist according to different regulatory patterns and the diverse conditions and roles played by the human agents.

Protections and policies can be partially embedded into the system in different ways, both by means of ontologies and the Privacy Controller System (PCS). This integration has a regulatory effect, close to what is called 'systemic interoperability'. Semantic interoperability refers to the ability of computer systems to unambiguously exchange data with an explicit, shared meaning. In information systems processing, systemic interoperability goes beyond semantic interoperability and refers to the ability of complex *systems* to interact. It focuses onto the coordination of practices and organisational structures, *in between* human behaviour and artificial systems [19]. This is what will be produced—or we are intending to produce—both in LEA's investigatory workplaces and through the process of building the security platform for identity resolution.

The EU guidelines defines interoperability as 'the ability of organisations to interact towards mutually beneficial goals, involving the sharing of information and knowledge between these organisations, through the business processes they support, by means of the exchange of data between their ICT systems' [20]. We are taking into account the four layers defined by the EU policy for integrated public service governance to reach 'interoperability by design' [20]:

a) *legal* ('about ensuring that organisations operating under different legal frameworks, policies and strategies are able to work together'),
b) *organisational* ('the way in which public administrations align their business processes, responsibilities and expectations to achieve commonly agreed and mutually beneficial goals'),
c) *semantic* ('ensures that the precise format and meaning of exchanged data and information is preserved and understood throughout exchanges between parties'),
d) *technical* ('applications and infrastructures linking systems and services', including 'interface specifications, interconnection services, data integration services, data presentation and exchange, and secure communication protocols').

This policy develops the articles and recitals of the INSPIRE Directive [21] assessing that 'Member States should ensure that any data or information needed for the purposes of achieving interoperability are available on conditions that do not restrict their use for that purpose' (R16) and 'for sharing spatial data between the various levels of public authority in the Community' (R17). Article 7 provides that "interoperability' means the possibility for spatial data sets to be combined, and for services to interact, without repetitive manual intervention, in such a way that the result is coherent, and the added value of the data sets and services is enhanced'.

Key interoperability enablers (KIE) have been promoted and facilitated by the ISA and ISA[2] programs [22][13]. KIE can be defined as the interoperability solutions (e.g., services and tools, standards, and specifications) that are necessary for the efficient and effective delivery of public services across administrations. This holds for private companies and its relationship with public administrations as well, as 'the reuse of shared services may be provided by the public sector, the private sector or in public-private partnership (PPP) models' [20]. SPIRIT fits into this legal scheme. It also falls under the Directive concerning LEAs on data protection and 'personal data breaches', defined as 'a breach of security leading to the accidental or unlawful destruction, loss, alteration, unauthorised disclosure of, or access to, personal data transmitted, stored or otherwise processed' [17].

Interoperability also has a public dimension related to Schengen area and policies. For at least fifteen years now, since the 2005 *Communication on interoperability and synergy among EU databases in the area of Justice and Home Affairs* [23], scholars have raised the issue about its consequences. In opposition to the Communication—which asserted that 'Interoperability" is a technical rather than a legal or political concept' and that 'it is disconnected from the question of whether the data exchange is legally or politically possible or required' [23]—they have stressed its political nature [24]. Successive EU proposals regarding migration, cross-border relationships, and identity—such as the implementation of the European Search Portal, a Shared Biometric Matching Service, a Common Identity Repository, and a Multiple-Identity Detector [25]—have recently raised identical concerns [26]; thus, reinforcing the need to adopt the least intrusive measures in effective police cooperation, and to respect the principle of proportionality and purpose limitation.

The Spirit privacy controller system, ontologies, and internal policies address these issues through the integration of modules and the construction of a specific regulatory model to monitor it. It is a 'cumulative structure', for semantic interoperability can only be achieved when standards for syntactic and technical interoperability have successfully been implemented [27]. But, most of all, it encompasses a systemic and holistic perspective, as the proposal is implementing the responsible AI guidelines [28] and the AI4People regulatory toolbox for good legal governance [29].

6   Conclusions and Future Work

To our knowledge, this is the first time that this integrative solution has been put in place in a security platform. It is certainly the result of the collective thinking of the whole team that has created an *intermediate organisational space* to bring together ontologies, a privacy controller system, and a hybrid (machine-human) policy set of principles and procedures. This intermediate space facilitates at the same time the wide *interpretation* of policies making them compatible with LEA's investigatory objectives and the regulatory compliance with a set of ethical requirements. Cooperation from the LEA's side, setting what they have called a protective *firewall* between researchers and police investigators

---

[13] ISA[2] Programme (2016–2020) supported 'the development of digital solutions that enable public administrations, businesses and citizens in Europe to benefit from interoperable cross-border and cross-sector public services.' https://ec.europa.eu/isa2/isa2_en.

and a *red ethical line* (or 'red thread implementation approach')—DCI Geoff Robinson, Thames Valley Police—has been essential. However, there still is room for improvement.

First, some meta-ethical work must be tackled to avoid the tensions between (i) privacy vs. transparency, (ii) accuracy vs. explainability, (iii) and accuracy vs. fairness pointed out at [30].

Second, as recently noticed [31] the advantage of description logics (OWL2) in ontology building 'is that all the main policy-reasoning tasks are decidable (and tractable if policies can be expressed with OWL2 profiles), while compliance checking is undecidable in rule languages, or at least intractable—in the absence of recursion—because it can be reduced to data log query containment'. We still have to check whether high-level data protection ontologies and semantic rules—as developed for business compliance [32]—can be made compatible for our specific security policy purposes. We have started working on this (non-trivial) problem.

Third, the SPIRIT system will reduce false positives on identity by combining results from several differing approaches using the crawler, a multi-layer perceptron, self-organising map, a NLP algorithm and a text matching algorithm. But this remains still to be tested in the immediate future. The AI ethics strategy in SPIRIT has also been developed with the aim of addressing the risks of misuse, stigmatisation and bias posed by the face recognition and AI techniques underlying the SPIRIT Tool.

The important conclusion is that no single component nor language is sufficient by itself to build up a satisfactory regulatory model. All of them must be separately selected and built, and then coordinated and integrated to reach acceptable and effective results.

Acknowledgments. SPIRIT. *Scalable Privacy Preserving Intelligence Analysis for Resolving Identities.* European Commission. Contract 786993. 01/08/2018-31/07/2021.

References

1. Blomqvist, E., Davarakis, C.: Intelligence analysis and semantic interoperability for identity resolution (abstract). TWSDetection, Toulouse, 27–28 February 2020. http://ceur-ws.org/Vol-2606/12invited.pdf
2. Casanovas, P., Morris, N., González-Conejero, J., Teodoro, E., Adderley, R.: Minimisation of incidental findings, and residual risks for security compliance: the SPIRIT Project. In: TERECOM@JURIX 2018, pp. 85–96. CEUR 2309 (2018). https://ceur-ws.org/Vol-2309/09.pdf
3. Adderley, R., Adderley, S., Rovatsou, R., Kazemian, H., Raffaelli, M., Ferrara, F.: Blomqvist: SPIRIT. Deliverable n° D6.2. Resource Flow Pattern Recognition Algorithms (WP6), 30 November 2019
4. Casanovas, P., Morris, N., Teodoro, E., González-Conejero, J., Adderley, R.: SPIRIT. Deliverable n. D9.2. Incidental Findings Policy (WP9), 31 October 2018. https://doi.org/10.5281/zenodo.3815050
5. Hartig, O., Blomqvist, E., Capshaw, R., Raffaelli, M., Adderley, R., Kazemian, H.: SPIRIT. Deliverable n. ° D5.1. Graph Infrastructure and Analysis (Report and software) (WP5), 30 November 2019
6. Botoeva, E., Calvanese, D., Cogrel, B., Corman, J., Xiao, G.: Ontology-based data access – beyond relational sources. Intelligenza Artificiale **13**(1), 21–36 (2019). IOS Press

7. Tiemann, M., Badii, L., Faulkner, R.: SPIRIT. Deliverable n. 9.7 Privacy Controller for Modelling and Filtering Software and Report (a). (WP2), 30 October 2019

8. Bartolini, C., Muthuri, R.: Reconciling data protection rights and obligations: an ontology of the forthcoming EU regulation. In: Workshop on Language and Semantic Technology for Legal Domain, p. 8 (2015). https://orbilu.uni.lu/bitstream/10993/21969/1/main.pdf

9. Palmirani, M., Martoni, M., Rossi, A., Bartolini, C., Robaldo, L.: PrOnto: privacy ontology for legal reasoning. In: Kő, A., Francesconi, E. (eds.) EGOVIS 2018. LNCS, vol. 11032, pp. 139–152. Springer, Cham (2018). https://doi.org/10.1007/978-3-319-98349-3_11

10. Pandit, H.J., Lewis, D.: Modelling provenance for GDPR compliance using linked open data vocabularies. In Proceedings of the 5th Workshop on Society, Privacy and the Semantic Web - Policy and Technology (PrivOn2017) (PrivOn) (2017). http://ceur-ws.org/Vol-1951/PrivOn2017_paper_6.pdf

11. Jonas, J.: Identity resolution: 23 years of practical experience and observations at scale. In: 2006 ACM SIGMOD International Conference on Management of Data, p. 718 (2006)

12. Bartunov, S., Korshunov, A., Park, S.T., Ryu, W., Lee, H.: Joint link-attribute user identity resolution in online social networks. In: The 6th SNA-KDD Workshop 2012 (SNA-KDD 2012), 12 August. ACM (2012)

13. Fu, X., Boongoen, T., Shen, Q.: Evidence directed generation of plausible crime scenarios with identity resolution. Appl. Artif. Intell. **24**(4), pp. 253–276 (2010)

14. Papageorgiou, A., Strigkos, M., Politou, E., Alepis, E., Solanas, A., Patsakis, C.: Security and privacy analysis of mobile health applications: the alarming state of practice. IEEE Access **6**, 9390–9403 (2018)

15. Zimmeck, S., et al.: Maps: scaling privacy compliance analysis to a million apps. Proc. Priv. Enhanc. Technol. **3**, 66–86 (2019)

16. Torra, V.: Data privacy: Foundations, New Developments and the Big Data Challenge. Springer, Cham (2017). https://doi.org/10.1007/978-3-319-57358-8

17. EU 2016: Directive (EU) 2016/680 of the European Parliament and of the Council of 27 April 2016 on the protection of natural persons with regard to the processing of personal data by competent authorities for the purposes of the prevention, investigation, detection or prosecution of criminal offences or the execution of criminal penalties, and on the free movement of such data, and repealing Council Framework Decision 2008/977/JHA. https://eur-lex.europa.eu/legal-content/EN/TXT/?uri=CELEX%3A32016L0680

18. EU 2016: Regulation (EU) 2016/679 of the European Parliament and of the Council of 27 April 2016 on the protection of natural persons with regard to the processing of personal data and on the free movement of such data, and repealing Directive 95/46/EC (General Data Protection Regulation). https://eur-lex.europa.eu/eli/reg/2016/679/oj

19. Casanovas, P., Mendelson, D., Poblet, M.: A linked democracy approach for regulating public health data. Heal. Technol. **7**(4), 519–537 (2017)

20. EU 2017: New European interoperability framework promoting seamless services and data flows for European public administrations. Publications Office of the European Union, Luxembourg (2017). https://doi.org/10.2799/78681

21. EU 2007: DIRECTIVE 2007/2/EC of the European Parliament and of the Council of 14 March 2007 establishing an Infrastructure for Spatial Information in the European Community (INSPIRE). https://eur-lex.europa.eu/legal-content/EN/TXT/PDF/?uri=CELEX:32007L0002&from=EN

22. EU 2017: Communication from the Commission to the European Parliament, the Council, the European Economic and Social Committee and the Committee of The Regions. European Interoperability Framework – Implementation Strategy. Brussels, 23.3.2017 COM(2017) 134 final. https://eur-lex.europa.eu/resource.html?uri=cellar:2c2f2554-0faf-11e7-8a35-01a a75ed71a1.0017.02/DOC_1&format=PDF

23. EU 2005: Communication from the Commission to the Council and the European Parliament on improved effectiveness, enhanced interoperability and synergies among European databases in the area of Justice and Home Affairs, Brussels, 24.11.2005 COM(2005) 597 final
24. De Hert, P., Gutwirth, S.: Interoperability of police databases within the EU: an accountable political choice? Int. Rev. Law Comput. Technol. **20**(1–2), 21–35 (2006)
25. EU 2017: Proposal for a Regulation of the European Parliament and of the Council on establishing a framework for interoperability between EU information systems (police and judicial cooperation, asylum and migration). Brussels, 12.12.2017 COM(2017) 794 final 2017/0352 (COD)
26. Quintel, T.: Interoperability of EU Databases and Access to Personal Data by National Police Authorities under Article 20 of the Commission Proposals. Eur. Data Prot. L. Rev. **4**, 470–482 (2018)
27. Kubicek, H., Cimander, R., Scholl, H.J.: Layers of Interoperability. In: Kubicek, H., Cimander, R., Scholl, H.J. (eds.) Organizational Interoperability in E-Government, pp. 85–96. Springer, Heidelberg (2011). https://doi.org/10.1007/978-3-642-22502-4_7
28. Dignum, V.: Responsible Artificial Intelligence: How to Develop and use AI in a Responsible Way. Springer, Cham (2019). https://doi.org/10.1007/978-3-030-30371-6
29. Pagallo, U., Casanovas, P., Madelin, R., Dignum, V., et al.: AI4People. 2019. On Good AI Governance. 14 Priority Actions as SMART Model of Governance, and a Regulatory Toolbox (2019). https://www.eismd.eu/wp-content/uploads/2019/11/AI4Peoples-Report-on-Good-AI-Governance_compressed.pdf
30. Whittlestone, J., Nyrup, R., Alexandrova, A.,Dihal, K., Cave, S.: Ethical and Societal Implications of Algorithms, Data, and Artificial Intelligence: A Roadmap for Research. Nuffield Foundation, London (2019)
31. Bonatti, P.A., Kirrane, S., Petrova, I.M., Sauro, L.: Machine Understandable Policies and GDPR Compliance Checking. arXiv preprint arXiv:2001.08930 (2020)
32. Governatori, G., Hashmi, M., Lam, H.-P., Villata, S., Palmirani, M.: Semantic business process regulatory compliance checking using LegalRuleML. In: Blomqvist, E., Ciancarini, P., Poggi, F., Vitali, F. (eds.) EKAW 2016. LNCS (LNAI), vol. 10024, pp. 746–761. Springer, Cham (2016). https://doi.org/10.1007/978-3-319-49004-5_48

TimeLex: A Suite of Tools for Processing Temporal Information in Legal Texts

María Navas-Loro[(✉)] [iD] and Víctor Rodríguez-Doncel [iD]

Ontology Engineering Group, Universidad Politécnica de Madrid, Madrid, Spain
mnavas@fi.upm.es

Abstract. In this paper we present a suite of tools named TimeLex, that includes different systems able to process temporal information from legal texts. The first tool, called *lawORdate*, helps preprocessing legal references in texts in Spanish that can be misleading when trying to find dates in texts. The second one, *Añotador*, is a temporal tagger (this is, a tool that finds temporal expressions, such as dates or durations) that identifies temporal expressions in texts and provides a standard value for each of them. Finally, a third tool, called *WhenTheFact*, extracts relevant events from judgments, allowing a full processing of the temporal dimension of this kind of texts, and being a first step towards the complete temporal information processing in the legal domain.

Keywords: Temporal expressions · Events · Timeline generation · Legal texts

1 Introduction

Temporal information is a very important dimension in documents. Being able to extract it would enable higher level functionalities, such as event-based summarization or search, pattern detection in cases, and timeline generation, that would facilitate the understanding of legal documents, usually difficult to comprehend by layman users, as well as enhance other NLP tasks over legal documents. Nevertheless, not a lot of research has been done in the legal domain in the field of temporal information.

TimeLex [1] is a suite of tools that aims to cover this gap in the domain, providing approaches to several parts of the temporal information extraction task. In this paper we briefly present the different contributions we have created in order to process this kind of texts from the temporal perspective.

The remaining of this paper is structured as follows. Section 2 presents related work in previous literature. Section 3 introduces *lawORdate*, a preprocessing tool that deals with legal references in order to facilitate latter temporal tagging task. Section 4 presents *Añotador*[1], a temporal tagger designed to find and normalize temporal expressions in legal texts. Section 5 shows a first approach for event extraction, introducing the

---

[1] *Añotador* is a pun: "Año" means "Year" in Spanish, while "Anotador" is the person or tool that performs the task of annotation. *Añotador* is a merge of the two concepts, and would therefore can be understood as "*What annotates years*".

© Springer Nature Switzerland AG 2021
V. Rodríguez-Doncel et al. (Eds.): AICOL-XI 2018/AICOL-XII 2020/XAILA 2020, LNAI 13048, pp. 260–266, 2021.
https://doi.org/10.1007/978-3-030-89811-3_18

tool *WhenTheFact*, which is able to generate a timeline of events from the information extracted from a document from the European Court of Human Rights[2]. Finally, Sect. 6 presents the conclusions and details the next steps in this research, targeting the semantic representation of the temporal annotations for further applications.

2 Related Work

Most effort related to temporal information in the legal domain has been done in relation to normative texts. This is the case of CronoLex [2], that aims to help lawyers by representing the legal norms in Spanish storing information about their life cycle, among others. Also in this direction, Akoma Ntoso [3] allows to represent several types of legal text in a standard way, including temporal information in the metadata.

Regarding the processing of temporal expressions and events, Schilder [4] analyzed the different types of legal documents with regard to temporal information, and divided them in statutes or regulations (where temporal information usually are constraints), transactional documents (including documents for legal transactions like contracts) and case law. In this paper, Schilder deeply studies the two first types of legal documents, but case law narrative structure was considered similar to the narratives in news, and received no dedicated attention.

Again in normative texts, Isemann et al. [5] used Named Entitiy Recognition and temporal processing in order to process the temporal dimension of regulations. This work, on the other hand, also described usual problems found by temporal taggers find in legal texts (not only in normative texts). Among them we can highlight the similar pattern of legal references and dates, that tend to be misleading to temporal taggers (e.g. "Directive 2012/33/EC"), or the distinction between *generic* events and *episodic* events. While the first refer to abstract events, general truths, rules, expectations or laws, *episodic* events are those that actually happened. Finally, also works on transactional documents [6, 7] and reasoning in legal evidence [8] can be found in literature.

3 LawORdate

lawORdate is a tool that cleans legal references with a date form from text documents. It addresses an important problem when processing legal documents from the temporal perspective, since common legal references in Spanish tend to include dates or patterns that can be misleading to temporal taggers. For instance, in the following excerpt:

> ".. creado via el Real Decreto **2093/2008**[1], de **19 de diciembre**[2]. Ha sido actualizado por ultima vez el **_13 de agosto de 2017_**[3]."

Most temporal taggers would find in this excerpt the three expressions in bold. Nevertheless, expression number one is not a date (despite of following a date-ish pattern) and expression number two is a date but does not belong to the narrative of the text (is part of a legal reference), so they should not be tagged. Therefore, the only one that should be tagged is the one underlined.

---

[2] https://www.echr.coe.int/Pages/home.aspx?p=home.

LawORdate is currently available both as a webapp [9] and as a GitHub repository [10], and finds and replaces misleading legal references in the texts, storing the original references. Once the temporal tagging is done, the references are restored in the text. Figure 1 shows the pipeline of use of lawORdate.

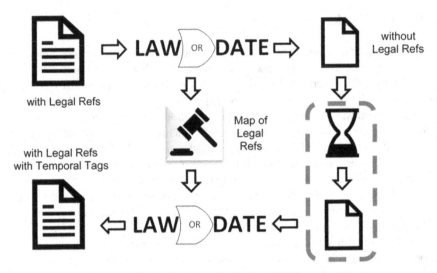

Fig. 1. Pipeline of use of lawORdate.

In the pipeline in Fig. 1, a text with legal references is first sent to the service. Then it finds all the misleading legal references that could affect to the precision of a temporal tagger and replaces them with inoquous expressions, storing the original references for further restoring. The output of this first step is to be used in a temporal tagger (in the demo, *HeidelTime* is offered, but any other can be used). Then, the output of the tagger (in TimeML) is sent back to *lawORdate*, that restores the original legal references. We therefore obtain the original text, but tagged without the interference of any legal references in it.

4 Añotador

Añotador [11] is a temporal tagger for Spanish and English able to find temporal expressions in texts, specially targeted to the legal domain. *Añotador* can detect different types of temporal expressions included in the TimeML standard, namely *dates, times,* sets (this is, expressions that repeat over time such as "*every Thursday*" or "*twice a week*") and *durations,* and some additional temporal expressions developed for the legal domain, such as specific expressions (e.g., "*business days*") and the type *interval. Añotador* outperforms the available state-of-the-art temporal taggers for Spanish [12]. It receives as input the text to annotate and optionally a reference date, called *anchor date* (if no date were introduced, the current date would be considered). With this information, the system is able to both find and normalize temporal expressions, this is, express them as a standard value, usually normalizing with regard to a reference date. If we had for instance the sentence *"I went to the park yesterday"* with "2019-09-20" as reference date, we would consider that the *normalized value* of *'yesterday'* is "2019-09-19". Nevertheless, not every temporal expression is normalized with regard to this initial *anchor date*. Once the temporal expressions in the text are identified using some hand-made rules specifically developed for the Spanish language, we apply a normalization algorithm that takes into account previous dates in the text for normalizing temporal expressions (see Fig. 2).

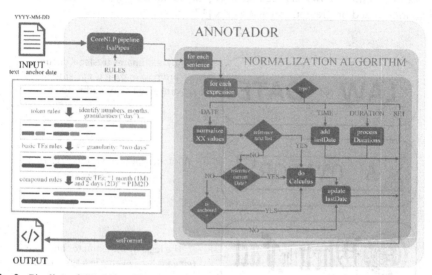

Fig. 2. Pipeline of *Añotador*. The user introduces the text to annotate and optionally a reference date.

Figure 2 shows how the text is first preprocessed using *CoreNLP* [13] and some *IxaPipes* [14] models. Then, different rules apply at different stages in order to detect the temporal expressions in the text. Once we have them, a normalization algorithm is applied in order to find their value. Finally, the system returns the text tagged.

Añotador has been tested against different state-of-the-art temporal taggers, both for legal English and for the Spanish language. Updated results of these evaluations can be found in its website [11].

5 WhenTheFact: Dealing with Events

After being able to identify temporal expressions using *Añotador* and *lawORdate*, the next logical step would be to detect events. Our current work focuses in detecting legal events in judgments, not covering just in the mention of the event (as most temporal taggers do), but also considering all the surrounding information available, such as the parts involved, when and where it happened or the jurisdiction involved. This was already done for a different type of legal document in a previous work, detecting events related to the lifecycle of a contract [15], but while in that case a rule-based approach was successful, taking into account the limited amount of events targeted, for legal judgments the amount of relevant events demands a more flexible approach.

To this aim, we are considering different lines of research in parallel, in order to detect the different types of events we can find in a judgment (e.g., the facts under judgment, that change from case to case, and the legal events, such as applications or decisions, that are court-related and tend to occur in all cases). To test the different approaches, a corpus of legal documents annotated with events (the first publicly available of its kind as far as the authors know) has been built [16], in collaboration with experts from other institutions, based on previous related works [17, 18].

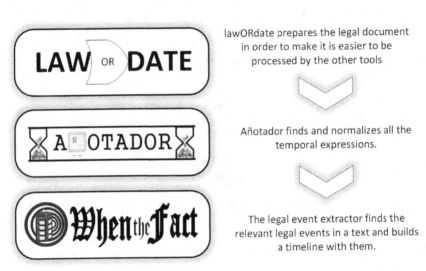

lawORdate prepares the legal document in order to make it is easier to be processed by the other tools

Añotador finds and normalizes all the temporal expressions.

The legal event extractor finds the relevant legal events in a text and builds a timeline with them.

Fig. 3. Different tools available in for temporal processing of legal texts.

In Fig. 3, first, the text is preprocessed by *lawORdate*; then, the temporal expressions can be more accurately found by *Añotador* (once done, *lawORdate* would return the original legal references to the text). Finally, *WhenTheFact* detects the relevant events in the text (current online implementation already includes *Añotador* in order to perform the full processing).

6   Conclusions

In this paper we presented the different tools created for processing temporal information from legal texts. The first service introduced, *lawORdate*, "cleans" the document of misleading legal references; then, *Añotador* is able to tag and normalize the temporal expressions in the text. *WhenTheFact* detects events and builds a timeline from it. The suite therefore covers a full processing from the temporal perspective.

Although *WhenTheFact* is still an ongoing tool, able just to extract events from very specific types of texts whose structure is already known and with room for improvement. Additionally, *WhenTheFact* builds a timeline, but we consider further applications would be useful for the legal domain, such as event-based summarization or pattern recognition. To facilitate these potential applications, represent temporal information in a standard and NLP focused manner would be extremely helpful. For this reason, next steps include the definition of this representation option by gathering different already available ontologies and schemas. Additionally, *WhenTheFact* is currently being expanded to cover more languages and types of documents. All advances in these directions will be reflected in the website of *TimeLex* [1].

References

1. Website of TimeLex. https://mnavasloro.github.io/timelex/
2. Marin, R.H., Hernandez, J.L., Delgado, J.J.I.: Cronolex: a computing representation of law dynamics. In: XXII World congress of Philosophy of Law and Socialphilosophy, Granada, Spain (2005)
3. Palmirani, M., Vitali, F.: Akoma-Ntoso for legal documents. In: Sartor, G., Palmirani, M., Francesconi, E., Biasiotti, M. (eds.) Legislative XML for the Semantic Web. Law, Governance and Technology Series, vol. 4, pp. 75–100. Springer, Dordrecht (2011). https://doi.org/10.1007/978-94-007-1887-6_6
4. Schilder, F., McCulloh, A.: Temporal information extraction from legal documents. In: Dagstuhl Seminar Proceedings. Schloss Dagstuhl-Leibniz-Zentrum für Informatik (2005)
5. Isemann, D., Ahmad, K., Fernando, T., Vogel, C.: Temporal dependence in legal documents. In: Yin, H., et al. (eds.) IDEAL 2013. LNCS, vol. 8206, pp. 497–504. Springer, Heidelberg (2013). https://doi.org/10.1007/978-3-642-41278-3_60
6. Guda, V., Srujana, I., Naik, M.V.: Reasoning in legal text documents with extracted event information. Int. J. Comput. Appl. 28(7), 8–13 (2011)
7. Ramakrishna, K., Guda, V., Padmaja Rani, B., Chakati, V.: A novel model for timed event extraction and temporal reasoning in legal text documents. Int. J. Comput. Sci. Eng. Surv. 2, 39–48 (2011)
8. Vlek, C.S., Prakken, H., Renooij, S., Verheij, B.: Representing and evaluating legal narratives with subscenarios in a Bayesian network. In: 2013 Workshop on Computational Models of Narrative. Schloss Dagstuhl-Leibniz-Zentrum fuer Informatik (2013)
9. LegalWhen demo. http://legalwhen.appspot.com/
10. LawORDate GitHub. https://github.com/mnavasloro/LawORDate
11. Añotador website. https://annotador.oeg.fi.upm.es/
12. Navas-Loro, M., Rodríguez-Doncel, V.: Annotador: a temporal tagger for Spanish. J. Intell. Fuzzy Syst. 39(2), 1979–1991 (2020)
13. Manning, C.D., Surdeanu, M., Bauer, J., Finkel, J.R., Bethard, S., McClosky, D.: The Stanford CoreNLP natural language processing toolkit. In: Proceedings of 52nd Annual Meeting of the Association for Computational Linguistics: System Demonstrations, pp. 55–60 (2014).
14. Agerri, R., Bermudez, J., Rigau, G.: IXA pipeline: efficient and ready to use multilingual NLP tools. In: LREC, vol. 2014, pp. 3823–3828, May 2014
15. Navas-Loro, M., Satoh, K., Rodríguez-Doncel, V.: ContractFrames: bridging the gap between natural language and logics in contract law. In: Kojima, K., Sakamoto, M., Mineshima, K., Satoh, K. (eds.) JSAI-isAI 2018. LNCS, vol. 11717, pp. 101–114. Springer, Cham (2019). https://doi.org/10.1007/978-3-030-31605-1_9
16. Filtz, E., Navas-Loro, M., Santos, C., Polleres, A., Kirrane, S.: Events matter: extraction of events from court decisions. In: Serena, V., Harasta, J., Kremen, P. (eds.) Legal Knowledge and Information Systems – JURIX 2020: The Thirty-third Annual Conference, Brno, Czech Republic, December 9–11 2020, volume 334 of Frontiers in Artificial Intelligence and Applications, pp. 33–42. IOS Press (2020)

17. Navas-Loro, M., Santos, C.: Events in the legal domain: first impressions. In: Proceedings of the 2nd Workshop on Technologies for Regulatory Compliance co-located with the 31st International Conference on Legal Knowledge and Information Systems (JURIX 2018), Groningen, The Netherlands, pp. 45–57, 12 December 2018 (2018)

18. Navas-Loro, M., Filtz, E., Rodríguez-Doncel, V., Polleres, A., Kirrane, S.: TempCourt: evaluation of temporal taggers on a new corpus of court decisions. Knowl. Eng. Rev. **34** (2019)

Data Protection and Privacy Modelling and Reasoning

Inferring the Meaning of Non-personal, Anonymized, and Anonymous Data

Emanuela Podda[1,2,3]($\boxtimes$) (iD) and Monica Palmirani[1]($\boxtimes$) (iD)

[1] ALMA AI, Alma Mater Studiorum, Università di Bologna, Bologna, Italy
emanuela.podda@studio.unibo.it, {emanuela.podda2,
monica.palmirani}@unibo.it
[2] Università di Torino, Turin, Italy
[3] University of Luxembourg, Luxembourg, Luxembourg

Abstract. On the awareness of the dynamism pertaining to data and its processing, this paper investigates the problem of having two mutually exclusive definitions of personal and non-personal data in the legal framework in force. The taxonomic analysis of key terms and their context of application highlights the risk to crystalize the whole system upon which the digital single market is built, suffocating its future development. With this premise, the paper discusses the extent of the two main data processing tools provided by the GDPR, questioning the *ex-ante* categorization of data and its outcome, supporting stakeholders in overcoming this issue.

Keywords: Non-personal data · Anonymization · Pseudonymization

1 Introduction

Everyday people generate data from all spheres of their life, circulating in the IoT environments and *feeding* Big Data Systems (Perera 2015), posing uncountable challenges to data protection and privacy (Sollins 2019). In the last year, the European Commission proposed important initiatives to unlock the re-use of different types of data and create a common European data space[1]. The first pillars were proposed in February 2020 as the Data Strategy[2] and the White Paper on Artificial Intelligence[3], followed by the adoption of a proposed Regulation on Data Governance[4] in November 2020. Lastly, the Artificial

---

[1] https://ec.europa.eu/digital-single-market/en/data-policies-and-legislation-timeline.

[2] A European strategy for data, Brussels, 19.2.2020, COM(2020) 66 final.

[3] The European Commission confirms that data and artificial intelligence (AI) can help find solutions to many of society's problems, from health to farming, from security to manufacturing. However, it also stresses on the risks posed by AI. It stresses on the need to enforce it adequately to address the risks that AI systems create.

[4] https://eur-lex.europa.eu/legal-content/EN/TXT/PDF/?uri=CELEX:52020PC0767&from=EN.

© Springer Nature Switzerland AG 2021
V. Rodríguez-Doncel et al. (Eds.): AICOL-XI 2018/AICOL-XII 2020/XAILA 2020, LNAI 13048, pp. 269–282, 2021.
https://doi.org/10.1007/978-3-030-89811-3_19

Intelligence Act[5]. In this proposal, the European Commission confirms the methodological roadmap based on risk analysis, intermediary services - already proposed in the Digital Services Act[6] - and certifications for the Artificial Intelligence processes and products.

The whole new system relies on principles and rules introduced by the two main Regulations for granting the data flow in the digital single market: the General Data Protection Regulation[7] (GDPR), and the Free Flow Data Regulation[8] (FFDR). While personal data can flow provided that some conditions are respected (e.g., consent, processing, risk evaluation, etc.) non-personal data can freely flow in the digital environment. Thus, the whole legal system is anchored to the dichotomy *personal & non-personal data,* and even its development is strictly dependent on it, facing the risk of suffocating innovation.

Since the entry into force of this legal framework, few areas of improvement have been identified[9], even on the awareness that the context and the infrastructure are rapidly evolving and changing, therefore potentials and risks. The EU Member States will set up a new common digital platform[10], the Data Space, where the international dimension will play a central role[11], creating a level playing field with companies established outside the EU[12]. The ability for private and public sector actors to collect and process data on a large scale will increase: devices, sensors and networks will create not only large volumes of data, but also new types of data like inferred, derived and aggregate data (Abuosba 2015) or synthetic data (Platzer 2021), moving beyond the data dichotomy imposed by the legal framework.

The awareness of the legal vulnerabilities pertaining to this technological evolution, implies that both law and technology must, together, promote and reinforce the beneficent use of Big Data for public good (Lane et al. 2014), but also, people's control of their personal data, their privacy and digital identity (Karen 2019)[13].

---

[5] https://eur-lex.europa.eu/legal-content/EN/TXT/PDF/?uri=CELEX:52021PC0206&from=EN.

[6] https://eur-lex.europa.eu/legal-content/EN/TXT/PDF/?uri=CELEX:52020PC0825&from=en.

[7] Regulation (EU) 2016/679 of the European Parliament and of the Council of 27 April 2016 on the protection of natural persons with regard to the processing of personal data and on the free movement of such data, and repealing Directive 95/46/EC.

[8] Regulation (EU) 2018/1807 of the European Parliament and of the Council of 14 November 2018 on a framework for the free flow of non-personal data in the European Union.

[9] To this extent, refer to the first Report on the evaluation of the GDPR published by the Commission on June 2020 https://ec.europa.eu/info/sites/info/files/1_en_act_part1_v6_1.pdf.

[10] See the program Gaia-X, https://www.data-infra-strucure.eu/GAIAX/Navigation/EN/Home/home.html.

[11] The report stresses on the fact that "the Commission will continue to focus on promoting convergence of data protection rules as a way to ensure safe data flows".

[12] Ibid, 10.

[13] The author affirms that a sustainable IoT Big Data management can be effectively designed only after decomposing the set of drivers and objectives for security/privacy of data as well as innovation into: 1) the regulatory and social policy context; 2) economic and business context; and 3) technology and design context. By identifying these distinct objectives for the design of IoT Big Data management, a more effective design and control is possible.

2 Personal Data, Non-personal Data, Mixed Datasets

The GDPR and the FFDR provide the taxonomy of data. Art. 4(1) of the GDPR specifies that "personal data" means *"any information relating to an identified or identifiable natural person ('data subject') [...]"*.[14] Art. 3 of the FFDR defines "non-personal data" as data other than personal data as defined in point (1) of Article 4 of Regulation (EU) 2016/679.

These definitions are mutually exclusive and strongly chained: the definition of non-personal data is dependent on the definition of personal data. *Ex ante,* they seem not considering the *physiological attitude of data to be treated* thus, not considering the data lifecycle (Wing 2019). *De facto,* the legal framework does not provide any concrete tool able to ensure to check the nature of data during its physiological lifecycle. On the contrary, imposes to data controllers and data processors to keep monitoring the risks linked to such processing.

Data processing, indeed, modifies the status of data, its definition and category. Hence, it is necessary to distinguish between the *static perspective,* based on the reasoning of what can be literally considered as "personal data" and the *dynamic* one, specifically on what kind of *status* modification data can have due to its lifecycle.

In line with these premises, if considering both perspectives, the span of the concepts increases, proportionally implying the risk of overlapping in definitions and sclerotizing the whole system of data flow in the digital single market. Not to mention that, since these definitions are strictly dependant, any vulnerability in one affects the other and *vice-versa.*

These critical points were originally highlighted in the Impact Assessment of the Regulation[15]. Nowadays, after few years of the entry into force, they are largely confirmed and still discussed in the academic debate (Graef et al. 2018; Hu et al. 2017; Finck and Pallas 2020; Leenes 2008; Stalla-Bourdillon and Knight 2017).

The definition of personal data is coming from the centerpiece of EU legislation on data protection, Directive 95/46/EC, adopted in 1995[16] and it has been transposed in the GDPR. Since then, it led to some diversity in the practical application. For example, the issue of objects and items ("things" – referring to IoT systems) linked to individuals, such as IP addresses, unique RFID numbers, digital pictures, geo-location data and telephone numbers, has been dealt differently among Member States[17]. The CJEU played - and keeps playing - an essential role in resolving these diversities, harmonizing the legislation[18].

---

[14] In order to clarify the concept, the WP29 04/2007 on the concept of Personal Data states that the contextual presence of 4 elements connotes personal data: 1) Any information, 2) Relating to, 3) An identified or Identifiable, 4) Natural Person.

[15] Commission Staff Working Paper, Brussels, 25.1.2012, SEC(2012) 72 final, Impact Assessment.

[16] The Directive was also complemented by several instruments providing specific data protection rules in the area of police and judicial cooperation in criminal matters (*ex* third pillar), including Framework Decision 2008/977/JHA.

[17] These diverities are extensively treated in the Impact Assessment.

[18] To this aim, as an example, the judgment in Case C-582/14: Patrick Breyer v Bundesrepublik Deutschland.

The core of the problem leading to legal uncertainty as a major area of divergence in the Member States, and strictly linked to the data processing[19] - is related to the concept of *identifiability*. Specifically, to the circumstances in which data subjects can be said to be "*identifiable*".

The importance of this concept is strengthen by the combined provisions of Recital 26 of GDPR and Recital 8 of the FFDR were it is clearly stated that data processing can modify the nature of data. This problem acquires even more resonance, when literally recalling Art. 2(2) of the FFDR "*In the case of a data set composed of both personal and non-personal data, this Regulation applies to the non-personal data part of the data set. Where personal and non-personal data in a data set are inextricably linked, this Regulation shall not prejudice the application of Regulation (EU) 2016/679.*"

In order to clarify the concept of *inextricability*, the European Commission released a *Practical guidance for businesses on how to process mixed datasets*[20] contextualizing the case and confirming that in most real-life situations, a dataset is however very likely to be composed of both personal and non-personal data (*mixed dataset*), thus it would be challenging and impractical, if not impossible, to split such mixed dataset.

As pointed by some authors (Greaf 2018), this data taxonomy becomes counterproductive to data innovation[21].

Therefore, still nowadays, the meaning and interpretation of identifiability yet represents the main reason why the concept of personal data and its interconnection with non-personal data is widening being still problematic, especially in perspective of the data processing, e.g. anonymization, pseudonymization.

When transposed in the technological environment, this perspective leads to the concept Personally Identifiable Information (hereinafter referred as PII). Referring to

---

[19] Specifically, on the nature of processed data, Data Protection Authorities (hereinafter referred as DPAs) considered encoded or pseudonymised data as identifiable thus, as such, as personal data in relation to the actors who have means (the "key") for re-identifying the data, but not in relation to other persons or entities (e.g. Austria, Germany, Greece, Ireland, Luxembourg, Netherlands, Portugal, UK). In other Member States all data which can be linked to an individual were regarded as "personal", even if the data are processed by someone who has no means for such re-identification (e.g. Denmark, Finland, France, Italy, Spain, Sweden). DPAs in those Member States are "generally less demanding" with regard to the processing of data that are not immediately identifiable, taking into account the likelihood of the data subject being identified as well as the nature of the data.

[20] Guidance on the Regulation on a framework for the free flow of non-personal data in the European Union Brussels, 29.5.2019 COM(2019) 250 final.

[21] A timid tentative of overcoming this problem, it is contained in the proposal of the Data Governance Act where the Commission proposes to create a formal expert group, the European Data Innovator Boards.

the International Standards[22] ISO 27701[23] defines PII as *"any information that (1) can be used to establish a link between the information and the natural person to whom such information relates, or (2) is or can be directly or indirectly linked to a natural person"*.

If considered the amount of data that can be freely gathered in the digital info-sphere and the potential of data mining tools (Clifton 2002) contextualizing these definitions in several datasets, any kind of *value* linked to a person may lead to a PII. Consequently, it could be possible to affirm that in the digital context, affected by the process of datafication (Palmirani and Martoni 2019), an *identity* is any subset of attributed values of an individual person and, therefore, usually there is no such thing as "the identity", but several of them, as many as the number of the values combined with the same *data-subject* (Pfitzmann and Hansen 2010).

The problem presents a broader span if recalling the main premise on the data cycle, thus taking into account that even any PII has a natural lifecycle (Wing 2019; Abuosba 2015). As specifically stated in the ISO standard *"from creation and origination through storage, processing, use and transmission to its eventual destruction or decay. The risks to PII can vary during its lifetime but protection of PII remains important to some extent at all stages. PII protection requirements need to be taken into account as existing and new information systems are managed through their lifecycle"*.

To this extent, it can certainly be said that what defined at the moment of *ex-ante processing* as personal data, cannot necessarily last and be confirmed at the moment of *ex-post processing* as non-personal and *vice-versa*. In this regard, *domino effect* is spilling over the definition of anonymous data. There is indeed no doubt that this category includes data, not linkable by any mean to a data subject[24] but, for which this certainty is not undoubtable as the one namely recalled by Recital 26 of the GDPR referring to *data rendered anonymous in such a way that the data subject is no longer identifiable*.

To what extent a data processing can grant that a data subject is no longer identifiable?

Academics are currently stressing on a more proper evaluation of the differential element between personal and non-personal data (Finck and Pallas 2020), and on the importance of the paramount importance of the legal Principle of Data Minimization to overcome this legal emapasse (Biega et al. 2020).

Others (Stalla-Bourdillon and Knight 2018), also referring to the Breyer case, considers that characterizing the data should be context-dependent.

---

[22] The Commission's policy aims to align European Standards as much as possible with the international standards adopted by the recognized International Standardization Organizations ISO, IEC and ITU. This process is called "primacy of international standardization", meaning that European standards should be based on International standards (COM(2011)-311, point 7). For more info, cfr: https://ec.europa.eu/growth/single-market/european-standards/policy/int ernational-activities_en.

[23] ISO/IEC 27701:2019 (formerly known as ISO/IEC 27552 during the drafting period) is a privacy extension to ISO/IEC 27001. The design goal is to enhance the existing Information Security Management System (ISMS) with additional requirements in order to establish, implement, maintain, and continually improve a Privacy Information Management System (PIMS). The standard outlines a framework for Personally Identifiable Information (PII) Controllers and PII Processors to manage privacy controls to reduce the risk to the privacy rights of individuals.

[24] For example, those referred to businesses, those referred to industrial machinery, stars data like the ones related to Mars, labs data on chemical reactions, etc.

For others (Purtova 2018), the broad notion of personal data is not problematic and even welcome but this will change in future when everything will be personal or will contain personal data, leading to the application of data protection to everything. This will happen because technology is rapidly moving towards perfect identifiability of information, where *datafication* and data analytics will generate a lot of information.

Hence, in order to mitigate the gross risk of re-identification, contextual checks become essential and they should be conceived as complementary to sanitization techniques (Gellert 2018).

3   Anonymization Techniques in the Light of WP29 05/2014 and Its State of the Art

Anonymizing personal data implies a data processing which makes uncertain the attribution of that data to a certain person (data subject), relying on the probability calculation. Stemming from the expansion of data products usually provided by National Statistic, anonymization is considered by the Working Party 29, on the Opinion 05/2014, as a "further processing"[25]. International Standard ISO 29100 considers anonymization as the *"process by which Personally Identifiable Information (PII) is irreversibly altered in such a way that a PII principal cannot longer be identified directly or indirectly, either by the PII controller alone or in collaboration with any other party"*[26].

Differently from pseudonymity[27] which is *generally* (Mourby et al. 2018) distinguished by reversibility[28] (reason why the GDPR considers pseudonymized data still personal data) anonymization therefore should generally imply an *irreversible* alteration of personal data. The European legislation does not provide an explicit regulation on anonymization or an identification on its techniques, neither how the process should be, or could be performed. The legal focus is not on the tool *per se*, rather on its outcome.

---

[25] As such, must comply with the test of compatibility in accordance with the guidelines provided by the Working Party 29 Opinion 03/2013 on purpose limitation and with the de-anonymization risk test as for the Working Party 29 Opinion 05/2014.

[26] International Standard Organization (ISO/IEC) 29100:2011 Information technology – Security techniques – Privacy framework (*Technologies de l'information – Techniques de sécurité – Cadre privé*).

[27] According to Art. 4(5) GDPR 'pseudonymization' means "the processing of personal data in such a manner that the personal data can no longer be attributed to a specific data subject without the use of additional information, provided that such additional information is kept separately and is subject to technical and organizational measures to ensure that the personal data are not attributed to an identified or identifiable natural person".

[28] Pseudonymization is a de-identification process referenced in the GDPR as both security and data protection by design mechanism. There are different levels and scenarios of pseudonymity but as for anonymization process, different levels of security. See in details: https://www.enisa.europa.eu/publications/pseudonymisation-techniques-and-best-practices.

It is solely considered the potential risk linked to this data treatment, thus providing guidance and clarification with the Working Party 29 Opinion 05/2014[29] which, has no binding character, but by some authors (El Emam, Álvarez 2015) it is even considered lacking in some critical topics.

According to the definition provided by the Recital 26 of Directive 95/46/EC, recalled by the Opinion 05/2014, anonymization means *stripping data of sufficient elements such that the data subject can no longer be identified*. Therefore, data must be processed in such a way that it can no longer be possible to identify a natural person by using *"all the means likely reasonably to be used"* by either the controller or a third party. Such processing must be irreversible but, here again, the question to what extent this irreversibility can be granted.

There is no doubt that anonymization has a high degree of uncontrollability, but even that technological development has reached a point, as anticipated, of questioning whether anonymization can still be considered as an irreversible data processing. Moreover, the same approach seems confirmed in Recital 9 of the FFDR *"If technological developments make it possible to turn anonymized data into personal data, such data are to be treated as personal data, and Regulation (EU) 2016/679 is to apply accordingly"*.

Moreover, as said, the WP29 focuses only on the outcome of anonymization strictly related to the risk of de-anonymization, elaborating only on the robustness of few technique based on three criteria:

– *is it still possible to single out an individual[30]*,
– *is it still possible to link records relating to an individual[31]*, and
– *can information be inferred concerning an individual[32]*.

The WP29 recalls the two main anonymization techniques: randomization and generalization. Randomization alters the veracity of data weakening the links between values and objects (data subject), introducing a casual element in the data. This result can be concretely accomplished with few techniques: permutation, noise addition and differential privacy.

---

[29] The Article 29 Working Party (today EDPB – European Data Protection Board) was set up under the Directive 95/46/EC on the protection of individuals with regard to the processing of personal data and on the free movement of such data. It provides the European Commission with independent advice on data protection matters and helps in the development of harmonized policies for data protection in the EU Member States. One of the main tasks of the Article 29 WP was to adopt Opinions without a binding character but fundamental in order to clarify critical data protection issues.

[30] "The possibility to isolate some or all records which identify an individual in the dataset" WP29 Opinion 05/2014 on Anonymization Techniques, WP216, (0829/14/ EN). (2014).

[31] "The ability to link, at least, two records concerning the same data subject or a group of data subjects" WP29 Opinion 05/2014 on Anonymization Techniques, WP216, (0829/14/ EN). (2014).

[32] "The possibility to deduce, with significant probability, the value of an attribute from the values of a set of other attributes" WP29 Opinion 05/2014 on Anonymization Techniques, WP216, (0829/14/ EN). (2014).

According to the Opinion 05/2014, with differential privacy (Dinur and Kobbi 2003; Dwork 2011) singling out, inference and linkability may not represent a risk. However, statistical academics have just underlined its vulnerability (Domingo-Ferrer et al. 2021).

Differently from randomization, generalization dilutes the attributes by modifying the respective scale or order of magnitude and it can be performed using the following techniques: aggregation and K-anonymity (Samarati and Sweeney 1998) (which has been implemented with several algorithms) (Samarati 2001; Le Fevre et al. 2005; Xu 2006), L-diversity (Machanavajjhala et al. 2007) (which seems to be vulnerable and subject to probabilistic inference attacks) and T-closeness (Li et al. 2007) (as a refinement of L-diversity).

Certainly, the state of the art linked to the techniques listed in the Opinion 05/2014 seems confirming that anonymization methods face big challenges with real data and that it cannot longer be considered from a static perspective, but only from the dynamic one, being a dynamic checked process.

The evolution of the academic debate seems confirming the vulnerability of anonymization. Some academics (Ohm 2010; Nissembaum 2011; Sweeney 2001) stress on the unfeasibility of granting a proper and irreversible anonymization and at the same time maintain the data useful, or vice-versa. Others, (Cavoukian 2010; Yakowitz 2011) consider that, despite the awareness of the de-anonymization issue, a compromise between the commercial, social value of sharing data and some risks of identifying people should always be reached, even if producing consequences for personal privacy and data protection.

Moreover, moving beyond the general approach of questioning the concept of anonymization, its values and paradigm, in the last two decades the debate changes perspective. Currently, more and more authors are gathering empirical evidences on the possibility to reverse the process of anonymization, exploring and studying its correlated techniques. The attention is focused on the concrete possibility of de-anonymize data which have undergone a process of anonymization (no matter on which anonymization techniques used) due to the available technology and the technological development. Based on these assumptions, it is implicitly recognized that - within the context of the modern technology and due to uncontrollable technological development - the simple model of anonymization is unrealistic and researchers are currently exploring new models of anonymization.

For these reasons, the new trend is to combine many techniques in a pipeline using a complex monitored process, capable to provide also a dashboard where the human expert is maintained in the loop (Jakob et al. 2020).

In addition, it can be mentioned the model of *"functional anonymization"* which is based on the relationship between data and environment within which the data exists, the so-called "data environment" (Elliot et al. 2016; Elliot and Domingo Ferrer 2018). Researchers provide a formulation for describing the relationship between the data and its environment that links the legal notion of personal data with the statistical notion of disclosure control (Elliot et al. 2018; Hundepool and Willenborg 1996; Sweeney 2001, 2001b; Domingo-Ferrer and Montes 2018).

Assuming that *perfect* anonymization has failed and it is strictly linked to the context, some academics (Rubinstein and Hartzog 2016) remark that while the debate on de-anonymization remains vigorous and productive, *"there is no clear definition for policy"*, arguing that the best way to move data release policy is focusing on the process of minimizing risk of re-identification and sensitive attributes disclosure, rather than trying to prevent harm.

As anticipated, traditional anonymization methods which were originally tailored for the statistical context face big challenges with real data. From a mere legal point of view, the guidance provided by the WP29 in the Opinion 05/2014 needs to be reviewed, in line with the technological development. To confirm it, the fact that recently the European Parliament has recenty adopted a resolution inviting the European Data Protection Board *"to review WP29 05/2014 of 10 April 2014 on Anonymisation Techniques"*[33].

4 ...and Pseudonymization?

The Opinion 05/2014 defines pseudonymization by negation, as *"not a method of anonymization [...]. It merely reduces the linkability of a dataset with the original identity of a data subject, and is accordingly a useful security measure."*

The concept of pseudonymity[34] has a long history and in literature: many writers had a pseudonymous. Nowadays, the term is mostly used with regard to identity and the Internet, and ISO 25237[35] defines pseudonymization as a *"particular type of de-identification that both removes the association with a data subject, and adds an association between a particular set of characteristics relating to the data subject and one or more pseudonyms"*.

The definition provided in the main legal framework in force, is slightly different, and it is contained in art. 4(5) of the GDPR: *"the processing of personal data in such a manner that the personal data can no longer be attributed to a specific data subject without the use of additional information, provided that such additional information is kept separately and is subject to technical and organizational measures to ensure that the personal data are not attributed to an identified or identifiable natural person"*.

Despite the different perspective, despite the fact that the GDPR stresses more on the *"local linkability"* (Hu et al. 2017) there are two common elements in the definitions:

– the removal of the attribution link between the personal data and the data subject
– its replacement with new additional information.

As for anonymization, even for pseudonymization, the GDPR does not define techniques and tools, but provides orientation in terms of context. It places it in two different articles: in art. 25 recalling it as appropriate technical and organizational measure

---

[33] European Parliament resolution of 25 March 2021 (2020/2717(RSP)).

[34] The term pseudonymous stems from the Greek word "ψευδώνυμον *(pseudónymon)*" literally "false name", from ψεῦδος *(pseûdos)*, "lie, falsehood" and ὄνομα *(ónoma)*, "name".

[35] ISO 25237:2017 Health informatics—Pseudonymization. It contains principles and requirements for privacy protection using pseudonymization services for the protection of personal health information.

designed to implement data-protection principles[36], as well as in art. 32 listing it - with encryption - as a security measure that should be implemented by the data controller and the data processor.

These specific collocations explicitly confirm not only that pseudonymization represents a data security measure, but also that the tool can be implemented and adapted to the specific needs and aims of the data controller and the data processor (Drag 2018) in line with the principles of privacy by design (Cavoukian 2010).

The main reference on pseudonymization techniques stemming from the European Institutions, apart from samples recalled in the WP29 Opinions, it is provided by ENISA, the European Agency for Cybersecurity. Listing it among its priorities of the Programming Document 2018–2020, it provides recommendations on shaping technology according to GDPR provisions. Specifically, a complete guidance can be found in three recommendations[37,38,39] thus, as such, not legally binding, confirming the same approach followed by the WP29 Opinion 05/2014 on the Anonymization Techniques.

In ENISA Recommendations, different techniques are described, on the assumption that pseudonymization can relate to a single identifier, but even to more than one. The pseudonymization can be performed with the following techniques: Counter, Random Number Generator (RNG), Cryptographic Hash Function, Message Authentication Code (MAC), and Encryption.

However, not all the pseudonymization techniques are equally effective and the possible practices vary: they can be based on the basic scrambling of identifiers, or to advances cryptographic mechanism. The level of protection may vary accordingly.

In any case, especially for the hash function there is doubt to what extent it represents an efficient pseudonymization technique, especially under certain circumstances such as the case in which the original message has been deleted, thus granting irreversibility. In this case indeed, the hash value might even be considered as anonymized[40], on the basis of the dichotomy *reversible/irreversible* processing.

In term of policy, this decision is of paramount importance to determine the compliance of the rights recognized by the GDPR for certain types of processing (e.g. research,

---

[36] Specifically, art. 25(1) says that "Taking into account the state of the art, the cost of implementation and the nature, scope, context and purposes of processing as well as the risks of varying likelihood and severity for rights and freedoms of natural persons posed by the processing, the controller shall, both at the time of the determination of the means for processing and at the time of the processing itself, implement appropriate technical and organizational measures, such as pseudonymization, which are designed to implement data-protection principles, such as data minimization, in an effective manner and to integrate the necessary safeguards into the processing in order to meet the requirements of this Regulation and protect the rights of data subjects".

[37] ENISA, Recommendations on shaping technology according to GDPR provisions. An overview on data pseudonymization, November 2018.

[38] ENISA, Pseudonymization techniques and best practices. Recommendations on shaping technology according to data protection and privacy provisions, November 2019.

[39] ENISA, Data Pseudonymisation: Advanced Techniques & Use Cases, January 2021.

[40] AEPD, Introduction to the hash function as a personal data pseudonymization technique, October 2019.

traffic data analysis, geolocation, blockchain and others). The last ENISA report on January 2021[41] describes advanced techniques at the state of the art (e.g., zero-knowledge proof), demonstrating that the pseudonymization, like the anonymization, is a dynamic concept depending to the evolution of the technological over time. Additionally, this report remarks also how these techniques are very context dependent and they requires a detailed analysis of all the lifecycle of the data management including custody, key-ring management. In particular the data custodianship (or similar concepts such as data trustees or intermediaries) as a particular agent, trusted intermediaries for supporting confidentiality and protection of data. This may allow to pseudonymize the data and make them available for researchers, or can even be used in the healthcare sector.

The data custodian, or intermediary as defined in the first draft of the Data Governance Act, provides also the service to release synthetic data *"that is not directly related to the identifying data or the pseudonymised data but, still, shows sufficient structural equivalence with the original data set or share essential properties or patterns of those data. Synthetic data is being used instead of real data as training data for algorithms or for validating mathematical models."*

The traditional research debate on pseudonymity tried to clearly define the difference between anonymization and pseudonymization, focusing on the semantic (Pfitzmann and Hansen 2010). After being included in the GDPR as a data processing tool and as a data security measure, a primary focus is given to its risks (Stevens 2017; Bolognini and Bistolfi 2017) and the ambiguity surrounding the concept of pseudonymization in the GDPR (Mourby et al. 2018).

Overall, the state of the art seems confirming that pseudonymization has a greater potential of data protection than anonymization, and the implementation of the different techniques in currently ongoing.

5 Conclusions

The legal uncertainty pertaining to the two mutually exclusive definition of personal and non-personal data is spilling over the two main data processing tools provided by the legal framework in force, and especially the anonymization one.

The current evolution of the techniques in this sector suggests to approach the problem from a dynamic perspective, using a concept of permanent lifecycle checking. This will allow a constant revision of the admissible parameters and techniques, according to the state of the art. In this respect, the proposal for the Data Governance Act seems relying on intermediary certified and trusted services, aiming to different goals: i) a correct implementation of the pseudonymization/anonymization at the best of the state of the art and case-by-case according to the context of application (e.g., health); ii) a constant risk assessment; iii) the peculiar role of data custodian capable to provide a proxy access to other third parties (e.g., research institutions) also through synthetic datasets.

According to these premises, the two mutually exclusive definitions of personal and non-personal data seem obsolete and should be revised in favor of a constant and dynamic process which uses risk analysis, supported by intermediary certified actors. Also relevant for the evolution of anonymization, the concept and role played by "data altruism

---

[41] ENISA, 2021.

organizations[42]" included in the Data Governance Act. Therefore, it could determine a proxy where to anonymize the data using particular conditions and techniques, thanks to the special regime and regulation of this particular processing.

Finally, because some of these anonymization techniques use artificial intelligence artifact, is also relevant the Artificial Intelligence Act which proposes, again, a more detailed risk management approach and the introduction of a European Certification (CE) of the Artificial Intelligence production processes, with related certified actors playing the role of independent intermediators, ensuring the proper application of the regulation according to technological benchmarking.

References

Abuosba, K.: Formalizing big data processing lifecycles: acquisition, serialization, aggregation, analysis, mining, knowledge representation, and information dissemination. In: 2015 International Conference and Workshop on Computing and Communication, IEMCON (2015)

Aggarwal, C.: On k-anonymity and the curse of dimensionality. In: VLDB (2005)

Biega, A.J., Potash, P., Daumé III, H., Diaz, F., Finck, M.: Operationalizing the legal principle of data minimization for personalization, computers and society. In: Proceedings of the 43rd International ACM SIGIR Conference on Research and Development in Information Retrieval (2020)

Bolognini, L., Bistolfi, C.: Pseudonymization and impacts of Big (personal/anonymous) Data processing in the transition from the Directive 95/46/EC to the new EU general data protection regulation. Comput. Law Secur. Rev. 33, 171–181 (2017)

Cavoukian, A.: The 7 Foundational Principles. Identity in the Information Society (2010)

Clifton, C., Kantarcioglu, M., Vaidya, J.: Defining Privacy for Data Mining, in National Science Foundation Workshop on Next Generation Data Mining, Baltimore, MD, pp 126–133, November 2002

Dinur, I., Kobbi, N.: Revealing information while preserving privacy. In: Proceedings of the ACM SIGACT-SIGMOD-SIGART Symposium on Principles of Database Systems (2003)

Domingo-Ferrer, J., Montes, F.: Privacy in statistical databases, PSD. In: International Conference on Privacy in Statistical Databases, UNESCO Chair in Data Privacy, International Conference, PSD 2018, Valencia, Spain, 26–28 September 2018, Proceedings (2018)

Domingo-Ferrer, J., Sánchez, D., Blanco-Justicia, A.: The limits of differential privacy (and its misuse in data release and machine learning) (2011)

Drąg, P., Szymura, M.: Technical and legal aspects of database's security in the light of implementation of General data protection regulation. In: CBU International Conference on Innovation in Science and Education (2018)

Dwork, C.: The promise of differential privacy: a tutorial on algorithmic techniques. In: Proceedings - Annual IEEE Symposium on Foundations of Computer Science, FOCS (2011)

---

[42] Art. 2, point (10) "'data altruism' means the consent by data subjects to process personal data pertaining to them, or permissions of other data holders to allow the use of their non-personal data without seeking a reward, for purposes of general interest, such as scientific research purposes or improving public services", and art. 15 "Register of recognised data altruism organisations. (1) Each competent authority designated pursuant to Article 20 shall keep a register of recognised data altruism organisations. (2) The Commission shall maintain a Union register of recognised data altruism organisations. (3) An entity registered in the register in accordance with Article 16 may refer to itself as a 'data altruism organisation recognised in the Union' in its written and spoken communication."

Elliot, M., Mackey, E., O'Hara, K., Tudor, C.: The anonymization decision-making framework. In: Brussels Privacy Symposium, vol. 1 (2016)

Elliot, M., Domingo Ferrer, J.: The future of statistical disclosure control. Paper published as part of The National Statistician's Quality Review, London, December 2018

Elliot, M., et al.: Functional anonymization: personal data and the data environment. Comput. Law Secur. Rev. **34**(2) (2018)

Finck, M., Pallas, F.: They who must not be identified—distinguishing personal from non-personal data under the GDPR. Int. Data Priv. Law **10**(1) (2020)

Gellert, R.: Understanding the notion of risk in the general data protection regulation. Comput. Law Secur. Rev. **34**(2) (2018)

Graef, I., Gellert, R., Husovec, M.: Towards a holistic regulatory approach for the european data economy: why the illusive notion of non-personal data is counterproductive to data innovation. SSRN Electron. J. (2018)

Hu, R., Stalla-Bourdillon, S., Yang, M., Schiavo, V., Sassone, V.: Bridging policy, regulation and practice? A techno-legal analysis of three types of data in the GDPR (2017)

Hundepool, A., Willenborg, L.: μ- and T-argus: software for statistical disclosure control. In: Third International Seminar on Statistical Confidentiality, Bled (1996)

Jakob, C.E.M., Kohlmayer, F., Meurers, T., Vehreschild, J.J., Prasser, F.: Design and evaluation of a data anonymization pipeline to promote Open Science on COVID-19. Sci. Data **7**, Article no. 435 (2020)

Lane, J., Stodden, V., Bender, S., Nissenbaum, H.: Privacy, Big Data, and the Public Good. Privacy, Big Data, and the Public Good (2014). https://doi.org/10.1017/cbo9781107590205

Leenes, R.: Do you know me? – deconstructing identifiability. Univ. Ott. Law Technol. J. **4**(1&2) (2008)

Li, N., Tiancheng, L., Venkatasubramanian, S.: t-closeness: privacy beyond k-anonymity and l-diversity. In: ICDE (2007)

Le Fevre, K., DeWitt, D.J., Ramakrishnan, R.: Incognito: efficient full-domain k-anonymity. In: SIGMOD Conference (2005)

Machanavajjhala, A., Kifer, D., Kifer, D., Gehrke, J., Gehrke, J., Venkitasubramaniam, M.: L-diversity: privacy beyond k-anonymity. ACM Trans. Knowl. Discov. Data (2007)

Mourby, M., et al.: Are 'pseudonymised' data always personal Data? Implications of the GDPR for administrative data research in the UK. Comput. Law Secur. Rev. **34**(2) (2018)

Ohm, P.: Broken promises of privacy: responding to the surprising failure of anonymization. UCLA Law Rev. **57**(6) (2010)

Palmirani, M., Martoni, M.: Big data, data governance, and new vulnerabilities [big data, governance dei dati e nuove vulnerabilità]. Notizie Di Politeia (2019)

Perera, C., Ranjan, R., Wang, L., Khan, S., Zomaya, A.: Big data privacy in the Internet of Things era. IT Prof. (2015)

Pfitzmann, A., Hansen, M.: A terminology for talking about privacy by data minimization: anonymity, unlinkability, undetectability, unobservability, pseudonymity, and identity management, Technical University Dresden (2010)

Purtova, N.: The law of everything. broad concept of personal data and future of EU data protection law. Law Innov. Technol. **10**(1) (2018)

Rubinstein, I.S., Hartzog, W.: Anonymization and risk. Wash. Law Rev. **91**(2) (2016)

Samarati P., Sweeney, L.: Protecting privacy when disclosing information: k-anonymity and its enforcement through generalization and suppression. Harv. Data Priv. Lab. (1998)

Samarati, P.: Protecting respondents' identities in microdata release. IEEE Trans. Knowl. Data Eng. (2001)

Sollins, K.: IoT big data security and privacy versus innovation. IEEE Internet Things J. (2019)

Stalla-Bourdillon, S., Knight, A.: Anonymous data v. personal data—a false debate: an EU perspective on anonymization, pseudonymization and personal data. Wis. Int. Law J. **34**(2) (2017)

Stevens, L.: The proposed data protection regulation and its potential impact on social sciences research in the UK. Eur. Data Prot. Law Rev. (2017)

Sweeney, L.: Computational disclosure control: a primer on data privacy protection, Ph.D. thesis, Massachusetts Institute of Technology (2001)

Sweeney, L.: Information explosion. In: Zayatz, L., Doyle, P., Theeuwes, J., Lane, J. (eds.) Confidentiality, Disclosure, and Data Access: Theory and Practical Applications for Statistical Agencies, Urban Institute, Washington, DC (2001)

Wing, J.M.: The data life cycle. Harv. Data Sci. Rev. (2019)

Xu, J., Wang, W., Pei, J., Wang, X., Shi, B., Fu, A.W.-C.: Utility-based anonymization using local recoding. In: KDD (2006)

Challenges in the Implementation of Privacy Enhancing Semantic Technologies (PESTs) Supporting GDPR

Rana Saniei$^{(\boxtimes)}$ 🆔

Ontology Engineering Group, Universidad Politécnica de Madrid, Madrid, Spain
r.saniei@upm.es

Abstract. The EU General Data Protection Regulation (GDPR) imposes different requirements for data controllers collecting personal data to protect individuals' privacy. This fact triggered many studies and projects to investigate Privacy Enhancing Technologies (PETs) for the fulfillment of the compliance requirements. In this paper, after reviewing some of the current challenges and gaps in GDPR compliance, we argue the use of Semantic Technologies in PETs in the form of an Intelligent Compliance Agent (ICA) to support data controllers in carrying out a Data Protection Impact Assessment (DPIA). Models and ontologies representing entities involved in the DPIA process can help data controllers determine the risk of their processing activities. Additionally, an inference engine, equipped with a knowledge base of DPIA-related obligations, can effectively assist data controllers in taking specific actions when a legal fact is triggered based on met conditions.

Keywords: Compliance · Semantic web · Rule-based reasoning · Data Protection Impact Assessment · Privacy Enhancing Technologies

1 Introduction

The General Data Protection Regulation (GDPR)[1] went into force on May 25, 2018, with the primary aim of protecting personal data of all European Union citizens, wherever their data is targeted by different organisations. GDPR imposes various obligations on the companies; *data controllers*. Despite the fact that these obligations cover a broad range of requirements, the GDPR's key principles can be summarized as follows: lawfulness, fairness and transparency, purpose and storage limitation, data minimization, accuracy, accountability, and finally, integrity and confidentiality (security). Many principles of GDPR are similar to those in the previous regulations, such as the Data Protection Act 1998 (the 1998 Act) [3]. However, there are a few main changes and novelties. For instance, the principles of individual rights (GDPR, Chapter III), international transfer of personal data (GDPR, Chapter V), accountability, and Data Protection by

---

[1] https://eur-lex.europa.eu/eli/reg/2016/679/oj.

© Springer Nature Switzerland AG 2021
V. Rodríguez-Doncel et al. (Eds.): AICOL-XI 2018/AICOL-XII 2020/XAILA 2020, LNAI 13048, pp. 283–297, 2021.
https://doi.org/10.1007/978-3-030-89811-3_20

Design and Default have not been set out before. Also, for the first time, the GDPR imposes a legal duty on controllers to perform a *Data Protection Impact Assessment* (DPIA).

In the event of a breach, severe financial penalties are imposed to ensure that the cost of compliance will be less than the cost of the violation. However, this is not the only explanation why businesses should comply with GDPR. After several erupted data privacy scandals, in the center of them the 2018 *Cambridge Analytica* data breach, which affected personal data of more than 87 million Facebook users, the general public is becoming more aware of the vast data-driven economies and more concerned about the protection of their personal data collected by major tech companies. As a result, privacy maturity in companies is becoming a way of building trust with customers. Findings from Cisco surveys confirm this claim: In 2019, GDPR-ready companies reported that they experienced shorter sales delays due to *customer's privacy concerns* [2]. One year later, in 2020, Cisco study still represents that most of the compliant organisations experienced positive returns on privacy investments: More than 70% of surveyed companies indicated that they are facing *"significant"* or *"very significant"* benefits in areas such as reducing sales delays, building loyalty and trust with customers, enabling agility and innovation, and mitigating losses from data breaches [14].

From this standpoint, we can see the importance of businesses adhering to the Regulation, which triggered many studies and projects, both in industry and academia, to investigate Privacy Enhancing Technologies (PETs) for the fulfillment of the compliance requirements. The broad definition of PETs includes different technical solutions that support privacy and data protection [16]. In this light and with the main focus on data controllers, we propose the new concept of *"Privacy Enhancing Semantic Technologies"* (PESTs) in this paper. PESTs are PETs taking advantage of Semantic Technologies to support legal compliance, in this case, GDPR. In this research, our main focus is on DPIA. DPIA is a tool for identifying and analyzing risks for individuals posed by an organisation's processing operations or projects. The required steps to mitigate the risks should be selected and applied based on the analysis findings.

The remainder of the paper is structured as follows: Sect. 2 discusses challenges in utilizing PESTs for GDPR compliance. Section 3 reviews related work on supporting GDPR compliance using Semantic Technologies. Section 4 discusses DPIA and how Semantic Technologies can facilitate conducting a DPIA. Section 5 introduces the initial characteristic of a target architecture to respond to the challenges. Finally, Sect. 6 concludes the paper.

2   Challenges in Use of PESTs for GDPR Compliance

2.1   Challenges in Representation of Legal Norms

Specific challenges arise when it comes to modeling legal norms using machine-readable representations. Modeling GDPR text is no exception. Below, using [22,

40], we go over some of the unique features of legal norms that these challenges stem from. Some examples from GDPR are also provided in some cases.

-Different legal interpretations: Legal texts usually include various ambiguous terms as legislators want to make the norms applicable in as many situations as possible. However, even under the same context, this can contribute to varying interpretations of the legal document, which may cause challenges in automatic compliance checking. For example, consider the use of phrase *"without undue delay"* in Article 17(1) (see below) and Article 33 (1, 2), or the use of term *"reasonable"* in Article 12(5-a, 6). Article 17(1) states the data subject's right to erasure: *The data subject shall have the right to obtain from the controller the erasure of personal data concerning him or her without undue delay and the controller shall have the obligation to erase personal data without undue delay where one of the following grounds applies [...].*

Here, the truth-value of "undue delay" in terms of logic depends on the legal interpretation given to it.

-Interconnection: Different pieces of a legislative text (articles, paragraphs, etc.) may have direct links to other parts of the legal text or require external knowledge about the law. This can also put a challenge in the validation of compliance. In many sections of the GDPR, we see pointers to other Articles; for example, one of the cases in which carrying out a DPIA is necessary is explained in Article 35(3-b) with links to two other articles: *"processing on a large scale of special categories of data referred to in Article 9(1), or of personal data relating to criminal convictions and offences referred to in Article 10."*

-Nested deontic operators [40]: Legal texts may contain nested permissions, obligations, and prohibitions. One example of such rules is *"if y is obliged to do action a, then x is obliged to do action c."* Article 17(2) is an example of nested obligation and permission in GDPR [40].

-Defeasibility: Rules are defeasible in the sense that they usually define baseline situations, and then the text may include several exceptions to the baseline. The defeasibility of legal rules can be divided into two following issues: *conflicts* and *exclusionary rules*. Rules may produce incompatible legal effects and cause conflict. Conflicts can have three different forms: first, when a rule is an exception of another; second, when a rule has a different strength; and third, when rules were enacted in different times. For example, GDPR, Article 35(10) specifies some situations for which Article 35(1–7), associated with conducting a DPIA, would not apply. In terms of exclusionary rules, some rules make it possible to explicitly undercut other rules by making them inapplicable.

-Temporal properties: Temporal properties of rules, such as the time of enactment or enforcement, are usually important to validate rules. Thus, capturing the temporal characteristics of a rule is essential. In the case of GDPR, for example, we can see some modifications in the circumstances under which data controllers must appoint a data protection officer (DPO). First, designation of

a DPO applied only to large enterprises with more than 250 employees (Commission's initial proposal), but ultimately, it is mandated to appoint a DPO in certain cases mentioned in Article 37(1), which concerns the kind and scale of the processing activities and not the number of employees.

2.2 Challenges in the Representation of Involved Entities in GDPR

Aside from the difficulties of modeling legal norms in GDPR, there is a lack of appropriate and open source vocabularies and ontologies for describing the GDPR's involved concepts and entities. To fill this gap, several initiatives are being undertaken. One of them, the Data Privacy Vocabulary (DPV)[2], provides a collection of terms for describing and modeling information about personal data processing in accordance with established standards, such as GDPR. Also, as will be mentioned in Sect. 3, some studies have been done to model entities such as *consent* [36], but more standard and open source vocabularies are needed in this field to design and implement different PESTs to support compliance. In particular, with the focus on DPIA in this research, we are interested in answering the following questions:

- Which elements of information need to be considered by a data controller to best evaluate risks (likelihood and severity) in the DPIA and how to represent them using semantic web languages?
- Which elements of information need to be considered by a data controller to check the compliance with the DPIA-related obligations in the GDPR and how to represent them using semantic web languages?
- Which inferences can be made and which rules can be defined to trigger events necessary for a data controller to comply with the obligations related to DPIA in the GDPR?

2.3 Challenges in Personal Data Governance

The Web now has a centralized infrastructure in which organisations, as data controllers, have complete control over personal data typically stored in *data silos*. This fact causes a slew of issues, especially for data subjects, who lose the *sovereignty* they desire over their personal data. On the other hand, while many businesses are driven by the discovery of this form of data, it also places a significant burden on them under privacy regulations such as GDPR, which places a high priority on data security. According to the Cisco survey, 42% of respondents believed that *"meeting data security requirements"* was the top challenge they faced in preparing for the GDPR. Also, the consequences of not fulfilling the GDPR requirements are a major concern for organisations. According to a survey [29] in 2018 on over more than 1100 chef executive officers and IT managers, organisations think that GDPR would not only damage their reputation and lead to loss of revenue, but its huge fines could also threaten their existence.

---

[2] https://dpvcg.github.io/dpv/.

In this context, the advent of decentralized technologies may be an appropriate solution, especially for SMEs. Decentralized technologies like Solid [41] enable data subjects to directly monitor access to their personal data and transfer it between different providers. At the same time, decentralization lightens the burden of compliance for data controllers, albeit it also produces new requirements for handling *communication* and *negotiation* between data subjects and data controllers.

3 Related Work

Several initiatives to promote GDPR compliance in businesses have been launched in recent years. Since this paper focuses on PESTs, we analyze those studies that utilize Semantic Technologies in this section.

Legal ontologies have always been one of the areas of interest for legal professionals and the Semantic Web community. Ontologies provide explicit specifications of conceptualizations [46]. In the same vein, legal ontologies can be seen as tools that add meaning to and represent the semantics of a certain legal area. Thus, they can form the basis of AI legal systems. Thus far, several ontologies have been developed to represent GDPR's main entities. Bartolini et al. [8], propose a coarse grain ontology for personal data protection based on a draft of the GDPR document. The main objective in this preliminary work is to represent the obligations of data controllers under the GDPR, as well as easing the transition of systems and services based on the past existing legislation, Data Protection Directive (DPD), to the new one. In another work, Palmirani et al. propose a legal ontology called PrOnto [32] to represent conceptual cores of the GDPR. The primary goals based on which PrOnto is developed, are supporting legal reasoning and compliance checking by modeling deontic operators. PrOnto contains of five core modules: *data and documents, agents and role, processing purposes and legal basis, data processing and workflow*, and *legal rules and deontic operators*. Pandit *et al.* also present GDPRtEXT [37], a linked data resource of the GDPR, which provides an RDF representation of the GDPR text at the article-paragraph granularity, as well as a vocabulary of terms and concepts relevant to the GDPR. GConsent [36] is another OWL2-DL ontology that models the necessary entities to represent the notion of *consent*, as well as demonstrating its validity according to GDPR conditions. In another work, GDPRoV [33] is proposed with the main goal of highlighting the provenance of consent and data life-cycles within the GDPR domain. GDPRoV is an OWL2, linked open data ontology based on the PROV-O [25] and P-Plan [19]. The main contributions in this work include recognizing important information about the provenance of consent and personal data in terms of compliance documentation, building an ontology to model and represent this provenance information, and finally generating SPARQL queries to traverse the ontology and find answers to compliance-related questions. The process of evaluating the retrieved information, however, is the subject of another research [35] conducted by the same authors. In this work, a validation model based on the Shapes Constraint

Language (SHACL)[3] is proposed to evaluate the correctness and completeness of the retrieved information.

Modeling legal reasoners to automatically perform compliance checking is the subject of several other researches and projects. Arfelt et al. [5] propose a mechanism for monitoring articles in GDPR where entities such as data controller, data processor, and data subject should perform specific actions in response to other actions. The extracted articles are formalized as rules using the metric first-order temporal logic (MFOT) [10]. Then, to observe the actions saved in systems logs and decide if a violation has occurred, a monitoring tool called MonPoly [9] is utilized. MonPoly accepts as input a process log and an MFOT formalization of a rule, then determines parts of the log which violate the formula. In another research [34], the application of Semantic Technologies is investigated in representing and querying compliance information related to the GDPR. In particular, the authors demonstrate the use of SPARQL in the evaluation of compliance. BPR4GDPR[4] [13] is a European project with the main objective of building a set of comprehensive tools to support GDPR compliance. It proposes two main ontologies namely *Information Model Ontology* and *Policy Model Ontology*. The former, which is also called the *Compliance Ontology*, captures the involved entities in the life-cycle of processes, their properties, and their relationships [26]. Using the Compliance Ontology, policies for the process activities are conceptualized around the main quadruples of {*actor, operation, resource, and organisation*} which together describe an *action* [43]. Subsequently, a rule can be defined based on its context, purpose, action, pre-action, and post-action. This rule-based framework in BPR4GDPR has been implemented as a *Policy Model Ontology* and adapted to formalize a set of twenty-four rules extracted directly from the GDPR to form the basic policies within the project. SPECIAL (Scalable Policy-aware Linked Data Architecture For Privacy, Transparency and Compliance) [45] is another European project aiming to build technical solutions for the transparent management of data usage. This project provides the SPECIAL Usage Policy Language (SPL)[5], the SPECIAL Policy Log Vocabulary (SPLog)[6], and an engine for compliance checking and policy verification for companies. SPECIAL suggests an automatic mechanism for compliance checking according to the GDPR based on ODRL [4,18]. In the first step, they model both legal regulations (in the form of permission, prohibition, obligation, and dispensation) and business processing operations (in the form of discrete permissions needed to carry out a business process). For this aim, they suggest an ODRL Regulatory Compliance Profile (ORCP), which extends the ODRL vocabularies via the ODRL profile mechanism. In ORCP, subclasses and properties heavily inspired by licensing usecases are excluded, and to better express deontic concepts, two rule subclasses, *Obligation* and *Dispensation*, are added to the model. To support nested rules, e.g. a permission rule with an exception (as form of

---

[3] https://www.w3.org/TR/shacl/.

[4] Business Process Re-engineering and functional toolkit for GDPR compliance.

[5] https://ai.wu.ac.at/policies/policylanguage/#basic-usage-policies.

[6] https://ai.wu.ac.at/policies/policylog/.

prohibition), the properties permission, prohibition, obligation, and dispensation are added to the *Rule* class. Then, the modelled policies are translated into Answer Set Programming (ASP) [20] rules, which facilitates automated compliance checking and also detects and explains non-compliance. SPECIAL project also illustrates how OWL2 reasoners can be used to verify consent at both the ex-ante and ex-post phase by encoding data usage policy, given consent by data subjects, and parts of the GDPR text [12]. In another work, Palmirani et al. [31] introduce a prototype framework to model legal knowledge for checking GDPR compliance under the Cloud for Europe Project (C4E) [15]. They utilized PrOnto [32] to model the GDPR central conceptual cores under the MIREL project[7]. The LegalRuleML [7] ontology is extended in this work to model deontic operators (such as right, obligation, prohibition, and permission) and the implicit relationships among rights and obligations in the GDPR. Furthermore, PrOnto has a module to support risk management and Data Protection Impact Assessment (DPIA) by defining classes such as *RiskAnalysis*, *Risk*, and *Measure*, as well as their properties and connections. Finally, DAPRECO [17] is an FNR-CORE project aimed to create a knowledge base of the GDPR provisions based on the PrOnto, using the LegalRuleML [7]. The D-KB (DAPRECO Knowledge Base) adds a layer of restrictions in the form of *if-then* statements to the PrOnto. These rules are then represented in the novel reified I/O logic [39], which integrates the I/O logic [27] and the reification-based logic presented in [21]. The reification property makes it possible to deal with three main challenges of representation of norms: nested rules, defeasibility, and legal interpretations. This knowledge base currently contains 271 obligations, 76 permissions, and 619 constitutive rules and is available online[8].

A close inspection of the literature shows the lack of efforts in two main areas: First, lightening the heavy burden of the fulfillment of GDPR obligations, as well as other complementary obligations imposed by each Data Protection Authority (DPA), by developing technical tools which provide a level of automation toward compliance. Second, implementation of vocabularies, ontologies and tools to support the *risk-based* approach toward the GDPR compliance, specifically by assisting data controllers in conducting DPIA. In the following sections, we further explain each of these ideas.

3.1   Rule Interchange Languages

We review some of the more commonly used rule interchange languages for representing legal norms in this section, as it is applicable to parts of our work. These languages include SWRL[9] (Semantic Web Rule Language), SBVR (Semantics of Business Vocabulary and Business Rules) [30], RIF[10] (Rule Interchange Format),

---

[7] https://www.mirelproject.eu/.
[8] https://github.com/dapreco/daprecokb.
[9] https://www.w3.org/Submission/SWRL/.
[10] https://www.w3.org/2005/rules/wg/charter.html.

LKIF (Legal Knowledge Interchange Format)[11], RuelML[12] (Rule Markup Language), and its extension LegalRuleML, and ODRL[13] (Open Digital Right Language). Each of these legal rules has its own advantages and limitations, and currently, none of them meets all of the requirements for encoding legal norms. SBVR, for example, introduces deontic operators to encode obligations and permissions, but because it is based on the first-order logic, it cannot manage conflicts. SWRL, the proposed rule language by W3C for the Semantic Web, is based on a combination of ontologies encoded in OWL-DL and XML format for rules in the Unary/Binary Datalog subset of the RuleML. SWRL provides a syntax to represent Horn rules. Hence it does not support the *isomorphic* representation of legal norms. Isomorphism [11] in the legal domain implies that the rules in the formal model should correspond to the units of natural language text that express the rules in the original legal sources, such as parts of legislation, in a one-to-one correspondence. Also, as SWRL is based on the monotonic semantics of first-order logic, it does not handle conflicts firmly. In addition, SWRL and RIF do not support defeasibility, which is a key requirement in modeling legal norms [22]. LKIF and RuleML, however, seem to be more powerful in this regard. The ability to represent normative effects, such as deontic operators, is another requirement in legal languages which is more supported in SBVR and RuleML. One of the promising recent developments in this field is LegalRuleML, an XML standard based on the RuleML. The key principles of the LegalRuleML include: representing different legal interpretations of a legal rule by multiple semantic annotations, saving all creators of a legal document, providing a method of temporal management for all the temporal aspects of entities that change over time, and building *N:M relationships* between rules and provisions based on IRI. It also has a mechanism to build links between provenance information (such as document creators and textual provisions) and legal documents. LegalRuleML is equipped with mechanisms to handle defeasibility, as well as *penalty* and *reparation* statements. Finally, ODRL, which is a W3C recommendation since February 2018, provides information about permissions, prohibitions and duties related to an asset, as well as constraints and the involved parties. Furthermore, ODRL, in its ODRL Common Vocabulary, introduces *Privacy Policy* subclass, a policy that expresses a rule over an asset containing personal data, which is of particular interest to our work on GDPR. However, since ODRL is solely RDF and not OWL-DL or even full OWL, using regular OWL-based reasoners like Pellet and Hermit to do automatic reasoning over it can be difficult [18].

4   Data Protection Impact Assessment

As mentioned earlier, although various ontologies and models have been created to describe different concepts in GDPR or to verify compliance in general,

---

[11] http://www.estrellaproject.org/lkif-core/.

[12] http://www.ruleml.org/.

[13] https://www.w3.org/TR/odrl-vocab/.

currently, there aren't enough Semantic Web-based tools to support DPIA. DPIA is the process of identifying, assessing, and minimizing the risks involved in the processing activities *likely to result in a high risk to the rights and freedoms of natural persons* (Article 35(1)). Although the phrase "rights and freedoms of natural persons", mentioned in this Article, mainly refers to data protection and privacy rights, it could also include other rights such as freedom of speech, freedom of thought, etc. Recital 75 of GDPR also associates risk to physical, material, or non-material damages to individuals. Conducting a DPIA, which is required in specific cases mentioned in Article 35(3), is essential to comply with at least two obligations under the GDPR: *Accountability* and *Data Protection by Default and Design*.

Risk is the core specification necessary to be analyzed in a DPIA. It is defined as a situation identifying an event, its impacts or consequences, and their levels in terms of severity (the extent of the damage) and likelihood [38]. GDPR along with other guidelines such as Article 29 Working Party of EU data protection Authorities (A29WP) [6], DPAs guideline on DPIA (e.g. Information Commissioner's Office (ICO) [1], CNIL[14], etc.) set out different, albeit consistent, criteria for identifying high-risk processing activities. For example, Article 35(3) GDPR lists three types of risky processing activities: systematic and extensive profiling with significant effects, large scale use of sensitive data, and public monitoring. A29WP guideline further explains this Article by enumerating nine other criteria such as evaluation or scoring, automated decision-making with legal or similarly significant effect, matching or combining datasets, and processing data concerning vulnerable data subjects. Similarly, each DPA suggests a list of processing activities that are, independently or in combination with the criteria mentioned in the A29WP guideline, subject to a data protection impact assessment. Furthermore, for each of the detected high-risk activities, mitigation measures should be identified. These measures include reducing the processing scope, improving security solutions, eliminating specific types of data from processing, having clear data-sharing agreements, etc. Finally, data controllers should reflect the conclusion of the DPIA in the project plan.

In this light, PESTs can help data controllers do a DPIA and assess the risk of their processing activities considering the Regulation and criteria mentioned in other guidelines. In this regard, we aim to support DPIA by:

- Designing vocabularies, models, and ontologies for representing the entities involved in conducting a DPIA in general and assessing the risk (in terms of likelihood and severity), in specific, using RDF and OWL languages.
- Implementing a rule-based system of the DPIA-related obligations listed in GDPR, as well as the guidelines published by the supervisory authorities (in response to Article 35(4)) and other resources; e.g., guidelines of the kind of processing operations subject to the requirement for a DPIA published by A29WP, and relevant ISO standards.
- Developing an intelligent agent-based system to assist data controllers in fulfilling their obligations under the GDPR.

---

[14] https://www.cnil.fr/en/privacy-impact-assessment-pia.

5   An Agent-Based Decentralized Architecture

To address some of the challenges mentioned in the previous sections, I propose the use of a decentralized architecture based on the Semantic Web technology and Solid [41]. Solid is a project proposing a decentralized platform for social Web applications based on RDF and Semantic Web technologies such as WebID [42], Linked Data Platform (LDP) [28], and WebAccessContrrol (WAC) [44] ontology. To shift data sovereignty from companies to users, in Solid personal data is stored in Web-accessible personal online datastores (pods). Pods can be deployed on personal servers by users or public servers offered by providers (like current cloud storage providers).

Two new stakeholders are introduced in the proposed framework: an Intelligent Privacy Agent (IPA) on the data subjects side and an Intelligent Compliance Agent (ICA) on the data controllers side. The intelligence of these agents is due to three key specifications: context-awareness, reasoning, and negotiation. IPA and ICA can be of help to support data subjects and data controllers for better handling their privacy rights and operations necessary for GDPR compliance, respectively. Both of these agents are capable of negotiation with each other regarding personal data sharing conditions, taking into account the privacy preferences of data subjects. Both data controllers' privacy policies and data subjects' preferences are encoded in a machine-readable format using ODRL and DPV. Each rule will be also linked to its relevant chapter(s) and article(s) using GDPRtEXT [37]. By using this framework, data subjects can delegate to the privacy agent their everyday decision-making tasks for giving (or rejecting) consent to different data providers by just declaring their privacy preferences and concerns once and for all. Indeed, this delegation raises some accountability issues, e.g. who is responsible if the agent makes a mistake? First, it is worth mentioning that one of the assumptions in this research is that management of personal information can be lawfully delegated to software agents. In addition, a possible approach to address the accountability issues here is to define different levels of automation for the agents in different situations dealing with various kinds of personal data with different sensitivity levels.

On the other hand, the ICA could favor data controllers by handling *internal* operations that support compliance. As discussed in [23], AI systems, whether rule-based or machine learning-based, are good sources of help for handling compliance operations related to tasks such as giving advice, monitoring, and making assessments. Using Semantic Technologies is particularly important in this context. It allows to represent information in a machine-readable and interoperable format, reason over the generated models, retrieve information by making queries, and validate compliance.

5.1   Legal Reasoning by ICA

Among the AI expert systems, rule-based systems are the most prevalent ones [47]. A rule-based system works based on a model of deductive reasoning by applying a rule of law, legislation, or regulation to a given problem to obtain an answer. The system declares that the response is achieved based on the legal resource's principle established by the authorities. Intelligent legal reasoners or inference engines derive new facts from the existing knowledge base. This process is usually referred to as legal reasoning.

GDPR is a legal document setting out necessary actions to meet different requirements for compliance. In this sense, a set of logical rules related to DPIA can be extracted from its text to specify events and operations (legal effects) triggered when specific conditions are met. Then, a rule-based system acting over this set of rules could be built. To support compliance, the ICA is capable of reasoning over this system to notify data controllers when an operation is required to satisfy a DPIA-related obligation. The first area that the ICA would help controllers in is *whether they should do a DPIA for a processing operation*. For example, if a planned processing activity is of type *systematic monitoring of a publicly accessible area* and *large scale*, then the ICA would notify the data controller that they should conduct a DPIA. Listing 1.1 represents this obligation in ODRL.

Listing 1.1. Data controller's obligation to conduct a DPIA when doing a systematic monitoring of publicly accessible area on a large scale (Article 35(3-c))

```
@prefix rdfs: <http://www.w3.org/2000/01/rdf-schema#> .
@prefix odrl: <http://www.w3.org/ns/odrl/2/> .
@prefix dpv: <http://www.w3.org/ns/dpv#> .
@prefix exns: <http://www.example.com/exns#> .
@prefix gdprtext: <https://w3id.org/GDPRtEXT#> .

exns:conductDataProtectionImpactAssessment
    rdfs:comment ''Data controller A should conduct a DPIA when do a
    systematic monitoring of publicly accessible area on a large scale'';
    rdfs:seeAlso gdprtext:ImpactAssessment ;
    rdfs:seeAlso gdprtext:LargeScaleProcessing ;
    rdfs:seeAlso gdprtext:SystematicMonitoring;
    a odrl: Policy ;
        odrl:obligation [
        a odrl:Obligation ;
        odrl:action _:conduct ;
        odrl:assignee _:ControllerA ;
        odrl:target dpv:DPIA
        odrl:constraint [
            a odrl:Constraint ;
            odrl:leftOperand dpv:processingCategory ;
            odrl:operator odrl:isAllOf ;
            odrl:rightOperand (dpv:LargeScaleProcessing ,
                               exns:SystematicMonitoringOfPublicArea) ;
    ] .
] .
_:ControllerA
    a dvp:DataController , odrl:Party ;
    dpv:hasName ''Controller A'' .

_:conduct
    a :Action , skos:Concept ;
    rdfs:isDefinedBy exm ;
    rdfs:label ''conduct''@en ;
    :includedIn odrl:use ;
    skos:definition ''to conduct a task under GDPR.''@en .
```

If a DPIA is necessary, then the ICA helps data controllers conduct it; for example, it notifies them when prior consultation with the supervisory authority is required (Article 36(1)). However, as mentioned before, ODRL does not help in automatic compliance checking. Thus, here we consider the use of SPINdle [24], a defeasible logic reasoner that can infer defeasible theories. It also supports handling negation and conflicts. In SPINdle, an input theory can be encoded using XML and plain text (with pre-defined syntax). An output theory can also be exported using XML. In addition, ICA and IPA will communicate in two scenarios. First, when data subjects want to tell the data controller about their privacy preferences. Some of the rules in the knowledge base are extracted from these individuals' privacy preferences. For example, if Alice states that she is not happy with transferring her data to the third-parties for advertisement purposes, the ICA adds the following rule accordingly to the knowledge base: "If the purpose of transferring Alice's personal data to a third-party is advertisement, then transmission is prohibited." Second, when the data controller wishes to seek the views of data subjects or their representatives about the expected processing (Article 35(9)) or simply wants to publish a DPIA to create confidence with the affected data subjects.

Because of the open-textured nature of legal texts and documents, which leaves space for various interpretations, it is important to seek the advice of a legal professional to ensure that the extracted rules are sufficient and complete. Also, the ontologies will be evaluated in their ability to address the competency questions for which they were created. In addition, we will use test-cases with different compliance requirements regarding DPIA to evaluate final ontologies. The evaluation will be in terms of adequate representation of relevant data and compliance checking in terms of accuracy and integrity of the inferences. Also, comparative analysis will be carried out between the final ontologies in this work and other similar ontologies in the literature to understand the weaknesses and strengths.

6   Conclusion

In this paper, we first described some of the current challenges in using PESTs for compliance and then reviewed the existing works that support GDPR compliance utilizing Semantic Technologies. Then we explained some initial ideas of how PESTs could help data controllers in the form of an intelligent compliance agent to conduct DPIAs, fulfill DPIA-related obligations, and in general, take a *risk-based* approach in processing personal data. GDPR is not the only EU regulation asking for adopting a risk-based approach. The newly-released proposal for regulating AI[15], is also based greatly upon the concept of *risk* and dedicates a considerable number of sections to define and regulate *high-risk* AI systems. The Commission's weighting of risk management and DPIA methods highlights their importance, as well as the need for complementary technological solutions. As a result, we expect that using PESTs, such as the ones introduced in this paper to handle DPIA and risk management, would make it easier for data controllers to fulfill their obligations.

---

[15] https://ec.europa.eu/commission/presscorner/detail/en/ip_21_1682.

Acknowledgements. This research has been supported by European Union's Horizon 2020 research and innovation programme under the Marie Skłodowska-Curie grant agreement No 813497 (PROTECT).

References

1. Information Commissioner's Office (ICO). Guide to the General Data Protection Regulation (GDPR). https://ico.org.uk/media/for-organisations/guide-to-the-general-data-protection-regulation-gdpr-1-0.pdf
2. CISCO CYBERSECURITY SERIES 2019. Maximizing the value of your data privacy investments, Data Privacy Benchmark Study, January 2019. https://www.cisco.com/c/dam/global/en_hk/products/security/security-reports/2019_cisco_cybersecurityseries_data_privacy_benchmark_study_en.pdf
3. Data Protection Act, Data protection act 1998. In: Retrieved June 5, p. 2007 (1998)
4. Agarwal, S., Steyskal, S., Antunovic, F., Kirrane, S.: Legislative compliance assessment: framework, model and GDPR instantiation. In: Medina, M., Mitrakas, A., Rannenberg, K., Schweighofer, E., Tsouroulas, N. (eds.) APF 2018. LNCS, vol. 11079, pp. 131–149. Springer, Cham (2018). https://doi.org/10.1007/978-3-030-02547-2_8
5. Arfelt, E., Basin, D., Debois, S.: Monitoring the GDPR. In: Sako, K., Schneider, S., Ryan, P.Y.A. (eds.) ESORICS 2019. LNCS, vol. 11735, pp. 681–699. Springer, Cham (2019). https://doi.org/10.1007/978-3-030-29959-0_33
6. Article 29 Working Party, Opinion 03/2013 on purpose limitation (WP 203)
7. Athan, T., Governatori, G., Palmirani, M., Paschke, A., Wyner, A.: LegalRuleML: design principles and foundations. In: Faber, W., Paschke, A. (eds.) Reasoning Web 2015. LNCS, vol. 9203, pp. 151–188. Springer, Cham (2015). https://doi.org/10.1007/978-3-319-21768-0_6
8. Bartolini, C., Muthuri, R.: Reconciling data protection rights and obligations: an ontology of the forthcoming EU regulation (2015)
9. Basin, D.A., Klaedtke, F., Zalinescu, E.: The MonPoly monitoring tool. In: RV-CuBES 3, pp. 19–28 (2017)
10. Basin, D., Klaedtke, F., Müller, S.: Monitoring security policies with metric first-order temporal logic. In: Proceedings of the 15th ACM symposium on Access control models and technologies, pp. 23–34 (2010)
11. Bench-Capon, T.J.M., Coenen, F.P.: Isomorphism and legal knowledge based systems. Artif. Intell. Law 1(1), 65–86 (1992)
12. Bonatti, P.A.: Fast Compliance Checking in an OWL2 Fragment. In: IJCAI, pp. 1746–1752 (2018)
13. BPR4GDPR (Business Process Re-engineering and functional toolkit for GDPR compliance). https://www.bpr4gdpr.eu/
14. Cisco. From Privacy to Profit: Achieving Positive Returns on Privacy Investments. January (2020). https://www.cisco.com/c/dam/global/en_uk/products/collateral/security/2020-data-privacy-cybersecurity-series-jan-2020.pdf
15. Cloud for Europe. https://www.fokus.fraunhofer.de/en/dps/projects/cloudforeurope
16. European Union Agency for Cybersecurity. Privacy Enhancing Technologies. https://www.enisa.europa.eu/topics/data-protection/privacy-enhancing-technologies
17. DAta Protection REgulation COmpliance (DAPRECO). https://www.fnr.lu/projects/data-protection-regulation-compliance/

18. De Vos, M., Kirrane, S., Padget, J., Satoh, K.: ODRL policy modelling and compliance checking. In: Fodor, P., Montali, M., Calvanese, D., Roman, D. (eds.) RuleML+RR 2019. LNCS, vol. 11784, pp. 36–51. Springer, Cham (2019). https://doi.org/10.1007/978-3-030-31095-0_3
19. Garijo, D., Gil, Y.: Augmenting PROV with Plans in P-PLAN: Scientific Processes as Linked Data. In: LISC@ ISWC (2012)
20. Gelfond, M., Lifschitz, V.: Classical negation in logic programs and disjunctive databases. New Generat. Comput. **9**(3–4), 365–385 (1991)
21. Gordon, A.S., Hobbs, J.R.: A formal theory of commonsense psychology: how people think people think. Cambridge University Press, Cambridge (2017)
22. Gordon, T.F., Governatori, G., Rotolo, A.: Rules and norms: requirements for rule interchange languages in the legal domain. In: Governatori, G., Hall, J., Paschke, A. (eds.) RuleML 2009. LNCS, vol. 5858, pp. 282–296. Springer, Heidelberg (2009). https://doi.org/10.1007/978-3-642-04985-9_26
23. Kingston, J.: Using artificial intelligence to support compliance with the general data protection regulation. Artif. Intell. Law **25**(4), 429–443 (2017). https://doi.org/10.1007/s10506-017-9206-9
24. Lam, H.-P., Governatori, G.: The making of SPINdle. In: Governatori, G., Hall, J., Paschke, A. (eds.) RuleML 2009. LNCS, vol. 5858, pp. 315–322. Springer, Heidelberg (2009). https://doi.org/10.1007/978-3-642-04985-9_29
25. Lebo, T., et al.: Prov-o: The prov ontology. In: W3C recommendation 30 (2013)
26. Lioudakis, G., et al.: Compliance Ontology (2019)
27. Makinson, D., Van Der Torre, L.: Input/output logics. J. Philos. Logic **29**(4), 383–408 (2000)
28. Malhotra, A., Arwe, J., Speicher, S.: Linked Data Platform Specification. In: W3C Recommendation (2015)
29. NetApp. NetApp GDPR Survey, Gauging global awareness of business concerns, April (2018). https://www.netapp.com/pdf.html?item=/media/12568-netappgdprsurveyfindings.pdf
30. OMG: Semantics of business vocabulary and business rules (SBVR). https://www.omg.org/spec/SBVR/
31. Palmirani, et al. Legal Ontology for Modelling GDPR Concepts and Norms. In: JURIX, pp. 91–100 (2018)
32. Palmirani, M., Martoni, M., Rossi, A., Bartolini, C., Robaldo, L.: PrOnto: privacy ontology for legal reasoning. In: Kő, A., Francesconi, E. (eds.) EGOVIS 2018. LNCS, vol. 11032, pp. 139–152. Springer, Cham (2018). https://doi.org/10.1007/978-3-319-98349-3_11
33. Pandit, H J., Lewis, D.: Modelling Provenance for GDPR Compliance using Linked Open Data Vocabularies. In: PrivOn@ ISWC (2017)
34. Pandit, H J., O'Sullivan, D., Lewis, D.: Queryable provenance metadata for GDPR compliance. Proc. Comput. Sci. **137**, 262–268 (2018)
35. Pandit, H.J., O'Sullivan, D., Lewis, D.: Exploring GDPR compliance over provenance graphs using SHACL. In: SEMANTICS Posters&Demos (2018)
36. Pandit, H.J., Debruyne, C., O'Sullivan, D., Lewis, D.: GConsent - a consent ontology based on the GDPR. In: Hitzler, P., et al. (eds.) ESWC 2019. LNCS, vol. 11503, pp. 270–282. Springer, Cham (2019). https://doi.org/10.1007/978-3-030-21348-0_18
37. Pandit, H.J., Fatema, K., O'Sullivan, D., Lewis, D.: GDPRtEXT - GDPR as a linked data resource. In: Gangemi, A., et al. (eds.) ESWC 2018. LNCS, vol. 10843, pp. 481–495. Springer, Cham (2018). https://doi.org/10.1007/978-3-319-93417-4_31

38. Article 29 data protection working party. Guidelines on Data Protection Impact Assessment (DPIA) (wp248rev.01). https://ec.europa.eu/newsroom/article29/item-detail.cfm?item_id=611236
39. Robaldo, L., Sun, X.: Reified input/output logic: combining input/ output logic and reification to represent norms coming from existing legislation. J. Logic Comput. **27**(8), 2471–2503 (2017)
40. Robaldo, L., et al.: Formalizing GDPR provisions in reified I/O logic: the DAPRECO knowledge base. J. Logic. Lang. Inf. **29**(4) 401–449 (2020)
41. Sambra, A.V., et al.: Solid: a platform for decentralized social applications based on linked data. In: Technical report, MIT CSAIL & Qatar Computing Research Institute (2016)
42. Sambra, A.V., Story, H., Berners-Lee, T.: WebID Specification (2014)
43. Nikolaos Dellas, S.L.G., Lorenzo Bracciale, U.R.M., Adrián Juan-Verdejo, C.A.S.: Initial Specification of BPR4GDPR architecture (2019)
44. Solid- Web Access Control (WAC). https://github.com/solid/web-access-control-spec
45. SPECIAL (Scalable Policy-aware Linked Data Architecture For Privacy, Transparency and Compliance). https://www.specialprivacy.eu/
46. Studer, R., Benjamins, V.R., Fensel, D.: Knowledge engineering: principles and methods. Data Knowl. Eng. **25**(1–2), 161–197 (1998)
47. Van Engers, T., et al.: Ontologies in the legal domain. In: Chen, H., et al. (eds) Digital Government, pp. 233–261, Springer, Boston (2008)

Publication of Court Records: Circumventing the Privacy-Transparency Trade-Off

Tristan Allard[1][ID], Louis Béziaud[1,2]([✉])[ID], and Sébastien Gambs[2]

[1] Univ Rennes, CNRS, IRISA, Rennes, France
{tristan.allard,louis.beziaud}@irisa.fr
[2] Université du Québec à Montréal, Montreal, Canada
gambs.sebastien@uqam.ca

Abstract. The open data movement is leading to the massive publishing of court records online, increasing the transparency and accessibility of justice, and enabling the advent of legal technologies building on the wealth of legal data available. However, the sensitive nature of legal decisions also raises important privacy issues. Most of the current practices address the resulting privacy/transparency trade-off by combining access control with (manual or semi-manual) text redaction. In this work, we argue that current practices are insufficient for coping with the massive access to legal data, in the sense that restrictive access control policies are detrimental to both openness and to utility while text redaction is unable to provide sound privacy protection. Thus, we advocate for a integrative approach that could benefit from the latest developments in the privacy-preserving data publishing domain. We present a detailed analysis of the problem and of the current approaches, and propose a straw man multimodal architecture paving the way to a full-fledged privacy-preserving legal data publishing system.

Keywords: Privacy · Transparency · Legal data · Anonymization

1 Introduction

The opening of legal decisions to the public is one of the cornerstones of many modern democracies: it allows to audit and make accountable the legal system by ensuring that justice is rendered according to the laws in place. As stated in [9], it can even be considered that *"publicity is the very soul of justice"*. Additionally, in countries following the common law, the access to legal decisions is a necessity as the law in place emerged from the previous decisions of justice courts.

Thus, it is not surprising that the transparency of justice is enshrined in many countries as a fundamental principle, such as the *right to a public hearing*

A version of this work was presented at the Law and Machine Learning workshop at ICML 2020 (no proceeding).

V. Rodríguez-Doncel et al. (Eds.): AICOL-XI 2018/AICOL-XII 2020/XAILA 2020, LNAI 13048, pp. 298–312, 2021.
https://doi.org/10.1007/978-3-030-89811-3_21

provided by the Article 6 of the European Convention on Human Rights, the Section 135(1) of the Courts of Justice Act (Ontario) stating the general principle that *"all court hearings shall be open to the public"* or in Vancouver Sun (Re) *"The open court principle has long been recognized as a cornerstone of the common law"*. The open data movement push for free access to law with for example the Declaration on Free Access to Law [16]. Multiple open government initiatives also consider the need for an open justice [49], such as the "Loi pour une République numérique" in France, the Open Government Partnership, the Open Data Charter and the Canada's Action Plan on Open Government.

Combined with recent advances in machine learning and natural language processing, the (massive) opening of legal data allows for new practices and applications, called legal technologies. Nonetheless, not all legal decisions should directly be published as such due to the privacy risks that might be incurred by victims, witnesses, members of the jury and judges. Privacy issues have been considered and mitigated by legal systems for a long time. For instance, the identities of the individuals involved in sensitive cases, such as cases with minors, are usually *anonymized* by default because they belong to a vulnerable subgroup of the population. In situations in which the risks of reprisal are high (*e.g.*, terrorism or organized crimes cases), judges, lawyers and witnesses might also ask for their identities to be hidden [21, 26]. Finally, the identities of the members of a jury are also usually protected to guarantee that they will not be coerced but also to ensure that the strategy deployed by the lawyers is not tailored based on their background. Legal scholars are aware of the need for privacy when opening sensitive legal reports [8, 13, 25].

In the past, these privacy risks were limited due to the efforts required to access the decisions themselves. For instance, some countries require to go directly to the court itself to be able to access the legal decisions. Even when the information is available online, the access to legal decisions is usually on a one-to-one basis through a public but restricted API rather than enabling a direct download of the whole legal corpus. Typical restriction mechanisms include CAPTCHAs (SOQUIJ[1]), quotas (CanLII[2]), registration requirement as well as policy agreement and limitation of access to research scholars (Caselaw[3]). Furthermore, the fact that a legal decision is public does not mean that it can, legally, be copied and integrated in other systems or services without any restrictions.

A first approach to limit the privacy risks consists in *redacting* the legal decisions before publishing them. Redaction mostly follows predefined rules that list the information that must be removed or generalized and define how [48] (*e.g.*, by replacing the first and last names by initials, by a pseudonym). Redaction is in general semi-manual (and sometimes fully manual) because automatic redaction is error-prone [40]. This makes it extremely costly, not scalable and does not completely remove the risks of errors [48]. For example, 3.9 million decisions are

---

[1] https://soquij.qc.ca.
[2] https://www.canlii.org.
[3] https://case.law.

pronounced in France every year but only 180000 are recorded in governmental databases and less than 15000 are made accessible to the public [22]. Moreover, even a perfect redaction would still offer weak privacy guarantees. A redacted text still contains a non-negligible amount of information, possibly identifying or sensitive, that may be extracted, *e.g.*, from the background of the case or even from the natural language semantics.

Another approach is access control, such as non-publication (*e.g.*, a case involving terrorism was held in secret in Britain [11]), rate limitation or registration requirements. However, access control mechanisms are binary and do not protect against privacy risks for the texts for which the access is granted. Furthermore, restricting massive accesses through blocking strategies also restricts the development of legal technologies that require a massive access to legal data.

In a nutshell, this paper makes the following contributions:

- We state the problem of reconciling transparency with privacy when opening legal data on a large scale (Sect. 2).
- We analyze the limits of the current approaches that are deployed in a widespread manner in real-life (Sect. 3).
- We propose a high-level straw man architecture of a system for publishing legal data massively in a privacy-preserving manner without precluding the traditional open court principles (Sect. 4).

2   Problem Statement

Legal Data. Legal reports are defined as written documents produced by a court about a particular judgment, which is itself a written decision of a court on a particular case. Although the content of a case report varies with courts and countries, it typically consist of elements such as date of hearing, names of judges and parties, facts, issue, etc. [59].

Need for Readability and Accessibility. The access to legal decisions is required both for transparency and practical reasons such as case law, which is the use of past legal decisions to support the decision for future cases. Thus, the judiciary system is built on the assumption that legal decisions are made public and accessible by default (*open-court principle*), so that (1) citizens are able to inspect decisions as a way to audit the legal system and (2) past decisions can be used to interpret laws, and as such must be known from legal practitioners and citizens. It follows that decisions must be made available in a form readable by humans (*i.e.*, natural language). The need for openness, the current practice in terms of open court, and the associated risks are detailed in [13,41]. They conclude that, although there are powerful voices in favor of open court, radical changes in access and dissemination require new privacy constraints, and a public debate on the effect of sharing and using information in records.

Accessibility is also an important issue. In the past, the access to decisions required attending public hearings or reading books called "reporters". Today, web services share millions of decisions and facilitate access to legal records to individuals–law professionals (judges, lawmakers and lawyers), journalists, or citizens. Online publication also enables the large-scale access and processing of records, in particular due to a standardized format.

Need for Massive Accesses (Legal Technologies). The term *legal technologies* encompasses technologies used in the context of justice, such as practice management, analytics and online dispute resolution[4]. These applications often require some form of "understanding" of legal documents, usually performed through natural language processing (NLP) and machine learning (ML) approaches [15,51]. We focus on this category as these applications are based on the analysis of a large number of legal data. One of the main challenges we have faced is that usually companies provide very few technical details about their actual processing and usage of legal documents.

The automatic processing and analysis of legal records have multiple applications, such as computing similarity between cases [38,43,58], predicting legal outcomes [3,32] (*e.g.*, by weighing the strength of the defender arguments and the legal position of a client in a hypothetical or actual lawsuit), identifying influential cases [39,45,55] or important part of laws [44], estimating the risk of recidivism [57], summarizing legal documents [61], extracting entities (*e.g.*, parties, lawyers, law firms, judges, motions, orders, motion type, filer, order type, decision type and judge names) from legal documents [14,52], topic modelling [6,46], concept mapping [10] or inferring patterns [7,35].

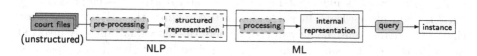

Fig. 1. High-level pipeline of court files processing for Legal Techs

Most of the technologies introduced in the previous section rely on the processing of large database of legal data. However, the unstructured nature of legal data is one of the main challenges of the application of artificial intelligence in law [2]. Consequently, the analysis of a legal text corpus first requires to apply some pre-processing to add structure to the text. Figure 1 represents an abstract processing pipeline for court files, extracted mostly from academic papers[5], and inferred from the current practice of text analysis and descriptions of associated technologies. In the following, we assume that any application involving the use of machine learning (as highlighted by most legal tech companies) is applied to court records. The first NLP step transforms the unstructured data (*i.e.*, natural language) into some structured representation (see below) by pre-processing it.

---

[4] More examples are available at CodeX Techindex at http://techindex.law.stanford. edu which references more than a thousand companies.

[5] The majority of the legal technologies market consists in commercial applications. They do not give information about their inner working and underlying techniques.

Afterwards, the second ML step corresponds to the actual application, which is the training (*i.e.*, processing) of the ML algorithm, whose output is represented by the "internal representation" block. The term instance represents the output of the model given some query (*e.g.*, applicable laws given a set of keywords representing infractions).

The pre-processing can be diverse and depends on the task (*e.g.*, extracting a citation graph between cases). However, most NLP-based applications usually rely on a text model. Many models are based on a bag-of-words (BoW) approach [27]. For example, document-word-frequency decomposes the text into a matrix in which each cell contains the number of times a particular word appears in a document. Other examples include term frequency-inverse document frequency and n-grams [63]. For example, a combination of those techniques are used in [3] to predict decisions from the European Court of Human Rights, and by [33] to identify law articles given a query or to answer to questions given a law article. Another common approach is word embeddings where words are mapped–using e.g. prediction-based or count-based methods–to real-valued vectors along with the context in which they are used [42]. Multiple variations of this structure exist [29,34,36,37,64]. This approach has been used for example in [39] to rank and explain influential aspects of law, or by [44] to predict the most relevant sources of law for any given piece of text using "neural networks and deep learning algorithms".

Need for Privacy. The massive opening of legal decisions for transparency and technological reasons must not hinder the right to privacy as emphasized by current open justice laws. In particular in this setting, the privacy of at least three main actors must be guaranteed: namely the individuals directly involved in decisions (*i.e.*, the parties), the individuals cited by decisions (*e.g.*, experts or witnesses), and the individuals administering the laws (*i.e.*, magistrates).

However, publishing legal decisions while providing sound privacy guarantees is difficult. For instance, authorship attacks [1] may lead to the re-identification of magistrates behind written decisions, or the presence of *quasi-identifiers*[6] within the text decisions may lead to the re-identification of the individuals involved or cited. Famous real-life examples, such as the governor Weld's [56] or Thelma Arnold's re-identification [5], both based on the exploitation of quasi-identifiers, are early demonstrations of the failure of naive privacy-preserving data publishing schemes. Thus despite the fact that legal decisions are written as unstructured text, structured information can be extracted from them, including the formal argument, the decision itself (*e.g.*, "guilty" or "innocent"), as well as arbitrary information about the individuals involved (*e.g.*, gender, age and social relationships).

Pseudonymization schemes simply consist in removing or replacing (e.g. by chainable or non-chainable pseudonyms) directly identifying data (*e.g.*, social

---

[6] A quasi-identifier is a combination of (one or more) attributes that are usually unique in the population, thus indirectly identifying an individual. A typical example is the triple (age, zip code, gender).

security number, first name and last name, address) and keeping unchanged the rest of the information (quasi-identifiers included). These schemes provide a very weak protection level, as acknowledged by privacy legislations (*e.g.*, GDPR), which has led to the development of new approaches for sanitizing personal data in the last two decades (see for instance the survey in [12]). In this paper, we focus on privacy-preserving data publishing schemes providing formal privacy guarantees that hold against several publications (as required by any real-life privacy-preserving data publishing system). These schemes are based on (1) *a formal model* stating the privacy guarantees the scheme as well as one or more *privacy parameters* for tuning the "privacy level" that must be achieved, and (2) *a sanitization algorithm* designed to achieve the chosen model.

A formal model exhibits a set of *composability properties* that defines formally the impact on the overall privacy guarantees of using the scheme on a *log of publications* (also called *disclosures log* in the following). In particular, we will consider the ϵ-differential privacy model [17], defined formally in Definition 1, parametrized by ϵ, and achievable by the Laplace mechanism. Its self-composability properties are stated in Theorem 1 and its overall privacy guarantees are quantified by the evolution of the disclosures log, and in particular by the evolution of the ϵ value along the various differentially-private releases.

Definition 1 (ϵ-differential privacy [17]). *A randomized mechanism $\mathcal{M}$ satisfies ϵ-differential privacy, in which $\epsilon > 0$, if:*

$$\Pr[\mathcal{M}(\mathcal{D}_1) = \mathcal{O}] \leq e^{\epsilon} \cdot \Pr[\mathcal{M}(\mathcal{D}_2) = \mathcal{O}]$$

for any set $\mathcal{O} \in Range(\mathcal{M})$ and any tabular dataset $\mathcal{D}_1$ and $\mathcal{D}_2$ that differs in at most one row (in which each row corresponds to a distinct individual).

In a nutshell, ϵ-differential privacy ensures that the presence (or absence) of data of a single individual has a limited impact on the output of the computation, thus limiting the inference that can be done by an adversary about a particular individual based on the observed output.

Theorem 1 (Sequential and parallel Composability [19]). *Let $\mathtt{f}_i$ be a set of functions such that each provides ϵ_i-differential privacy. First, the* sequential composability *property of differential privacy states that computing all functions on the same dataset results in satisfying $(\sum_i \epsilon_i)$-differential privacy. Second, the* parallel composability *property states that computing each function on disjoint subsets provides $\mathtt{max}(\epsilon_i)$-differential privacy.*

3   Analysis of Current Practices

In the following section, we review the current practice for legal data anonymization and privacy regulations. To be concrete, we illustrate the privacy risks through examples of re-identification attacks. Finally, we argue that rule-based anonymization is not sufficient to provide a strong privacy protection and discuss the (formal) issues surrounding text anonymization.

3.1 Redaction *in the Wild*

Redaction of Legal Data. The redaction process consists in removing or generalizing a set of predefined terms defined by law through a semi-manual process [48]–e.g., using "find and replace" or domain-specific taxonomies combined with named entity recognition. Furthermore, access to legal documents or even public hearings can be restricted in well-defined cases. The common practice is to replace sensitive terms, as defined below, by initials, random letters, blanks or generalized terms (*e.g.*, "Montréal" becomes "Québec"). The specific set of rules regarding protected terms and the associated replacement practice can differ between countries and courthouses [48].

According to [50], information such as names, date and place of birth, contact details of unique identifiers (*e.g.*, social security number) is to be systematically removed for any person (subject to a restriction on publication), as well as for each of his or her relatives (*e.g.*, parents, children, neighbors, employers). In some contexts, additional information such as community or geographic location, intervenors (*e.g.*, court experts, social workers), or unusual information is also removed if it can be used to identify an individual. [13] presents numerous examples of legislation putting restriction to the *open-court principle*, such as hiding the identity of victims of sexual offenses.

Paper Versus Digital. The main difference between paper and digital access is the "practical obscurity" of paper records on the one hand, and the easy accessibility of digital records, on the other. The awkwardness of accessing paper records stored in a public courthouse puts inherent limitations on the ability of individuals or groups to access those records. In contrast, digital records are easy to analyze, can be searched in "bulk" by combining various key factors (*e.g.*, divorce and children) and can potentially be accessed from any computer. Thus, traditional distribution provides "practical obscurity" [30], in that it is inconvenient (*i.e.*, time-consuming) to attend the courthouse or read case reports.

3.2 Limits of Current Approaches

In this section we provide examples of potential attacks in order to illustrate the technical difficulties of raw text anonymization. Figure 2 presents excerpts from French and Canadian opinions[7]. More examples are available in [4].

Figure 2a is anonymized according to the CNIL recommendations of 2006, which requires the last name of individuals to be replaced by its initial. However, widely available background knowledge on the "Real Madrid Club de Futbol" combined with the (real-life) pseudonyms of the "players" trivially leaks their identity.

The de-anonymization of Fig. 2b relies on the text semantics instead of background knowledge. It requires the adversary (1) to identify the link (X) between "M. [...] Abdel X" and "the use of the name 'X' to designate a drink", and

---

[7] We translated them using DeepL (https://www.deepl.com).

the association Real Madrid Club de Futbol and several players of this team, Zinedine Z., David B., Raul Gonzalès B. aka Raul, Ronaldo Luiz Nazario de L., aka Ronaldo, and Luis Filipe Madeira C., aka Luis Figo

(a) CA Paris, 14 févr. 2008, n° 06/11504

the American company Coca Cola Company markets drinks under the French trade mark "Coca Cola light sango", of which it is the proprietor; that M. [...] Abdel X, relying on the infringement of his artist's name and surname, has brought an action for damages against the Coca Cola Company [...] On the ground that Abdel X maintains that, as an author and screenwriter, he is entitled to oppose the use of the name "X" to designate a drink marketed by the companies of the Coca Cola group.

(b) Cass. 1re civ., 10 avr. 2013, n° 12-14.525, Bull. 2013, I, n° 72.

X, born [...] 2017; Y, born [...] 2018 the children and C; D the parents

Applications are submitted for X, aged 1 year, and Y, aged 2 months. The Director of Youth Protection (DYP) would like X to be entrusted to her aunt, Ms. E, until June 25, 2019. As for Y, that he be entrusted to a foster family for the next nine months. The father has two other children, Z and A, from his previous union with Mrs. F. The mother has another child, B, from her union with Mr. G.

(c) Protection de la jeunesse — 201518, 2020 QCCQ 10887

Fig. 2. Excerpts of legal decisions

(2) to infer that the drink is called "sango", thus leading to the conclusion that X = "sango". While this attack may not be easy to automatize due to the hardness of detecting the semantics inference, it is, however, trivial to perform for a human (*e.g.*, by crowdsourcing it).

Figure 2c could be attacked through a combination of attributes and relationship. This opinion from the Youth court involves children and, as such, follows the strictest anonymization rules of the SOQUIJ. However, an adversary can extract an extensive relationship graph which could be matched over a relationship database (*e.g.*, Facebook).

Besides the content of legal documents, stylometry [47] can also be used to identify authors (*i.e.*, magistrates) by their writing style. Mitigation for this kind of attack exist [20,60] but their output is only machine readable (i.e., they do not fulfill the readability requirement, but are of intereset when considering "massive" processing in Sect. 4). Similarly, it is possible to exploit decision patterns to re-identify judges, as done for the Supreme Court of the United States [32].

3.3   Reasons for the Failure of Rule-Based Redaction

Reviews of current practices for tackling the privacy of legal documents in Sect. 3.1 has highlighted the widespread use of rule-based redaction, in which a set of patterns is defined as being sensitive and is either removed or replaced. However, as shown in Sect. 3.2 (1) privacy can be violated even in "simple" instances and (2) identifying information remains in most cases. In other words, rule-based redaction does not provide any sound privacy guarantee.

We observe that it suffers from the following main difficulties. (1) *Missing rule*: many combinations of quasi-identifiers can lead to re-identification and the richness of the output space offered by natural language (*i.e.*, what can be expressed) can hardly be constrained to a set of rules. (2) *Missing match*: The current state of the art about relationship extraction and named-entity recognition makes it hard to ensure that all terms that should be redacted will be detected, in particular because of the many possible ways to express the same idea (*e.g., circumlocution*).

Although these observations make the rule-based redaction difficult, it is important to note that attacks, *e.g.*, re-identification, remain simpler than protection. Indeed, an adversary has to find a single attack vector (*i.e.*, a missing rule or a missing pattern) whereas the redaction process needs to consider all the possibilities.

4   Multimodal Publication Scheme

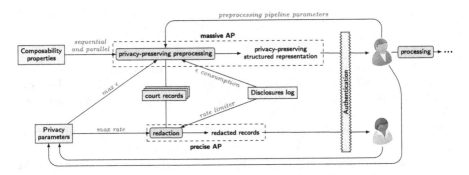

Fig. 3. Multimodal publication architecture

In Sect. 2, we have shown that the publication of legal documents serves two distinct and complementary purposes: (1) the traditional objective of transparency and case law, and (2) the modern objective of legal technologies of providing services to citizens and legal professionals. These two purposes obey to different utility and privacy requirements. More precisely, the traditional use case requires human-readable documents while legal techs need a machine-readable format for automated processing. Moreover, transparency and case law involve the access to opinions on an individual basis (*i.e.*, one-at-a-time), similarly to attending a hearing in person. In contrast, legal technologies rely on the access to massive legal databases. This difference in cardinality (*i.e.*, one versus many) entails different privacy risks. In particular, the massive processing of legal data requires the use of a formal privacy framework with composability properties (see Sect. 2). All this suggests the inadequacy of any *one-size-fits-all* approach. As a consequence, we propose that the organization in charge of the publication of legal decisions consider two modes of publication[8]: the *precise access mode* and the *massive access mode*.

Precise Access Mode. To fulfill the "traditional" use case, the precise access mode provides full access to legal decisions that are only redacted using the current practices. This access mode is designed for the transparency and case law usages, and is to be used typically by individuals (*e.g.*, law professionals, journalists and citizens). Similar to the "traditional" paper-based publication scheme, in the precise access mode [23], a user has access to full and partial documents. While the current practice of redacting identifiers could be combined with more automated approaches such as [24,54]. The aim of this mode is to provide strong utility first. It allows browsing, searching and reading documents similar to the websites currently publishing legal documents (*e.g.*, Legifrance or CanLII).

To prevent malicious users from diverting the precise access mode for performing massive accesses, users must be authenticated and their access must be restricted (*e.g.*, rate limitation or proof of work [18]). The access restrictions of a given user can be tuned depending on his trustworthiness (e.g., strength of the authentication, legally binding instruments implemented). The main objective of the restricted access mode is to make it difficult to rebuild the full (massive) database.

Massive Access Mode. The massive access mode gives access only to pre-processed data resulting from privacy-preserving versions of the standard NLP pipelines available on the server, *i.e.*, aggregated and structured data extracted from or computed over large numbers of decisions, as required for the "modern" use case. It should be compatible with most legal tech applications that traditionally use a database of legal documents (see Sect. 2). Note that the perturbations due to privacy-preserving data publishing schemes have usually less impact (in terms of information loss) when applied after aggregation (i.e., late in the pipeline, see Fig. 3 or [53, Figure 1]), at the cost of a loss of generality of the output.

Users need to be able to tune the pre-processing applied. For the sake of simplicity, we assume that the user (*i.e.*, legal tech developer) provides the parameters for a given NLP pipeline (see Fig. 3). These parameters can be for instance the maximum number of features or n-grams range to consider. In order to avoid limiting the massive access mode to the current implementation state of its NLP libraries, more complex implementations can be considered (1) by generating synthetic *testing* data in a privacy-preserving manner (*e.g.*, PATE-GAN [28]) or (2) by relying on a full pre-processing pipeline that embeds privacy-preserving calls to the server (*e.g.*, through a privacy-preserving computation framework such as Ektelo [62]).

The massive access mode must also authenticate users in order to monitor the overall privacy guarantees satisfied for each user based on his disclosures log

---

[8] The technical protection measures can be strengthened by usual legal instruments (e.g., non-disclosure agreements).

and on the composability properties of the privacy-preserving data publishing schemes used. As a result, the data is protected using authentication and strong privacy definitions.

Finally, another potential need is the annotation of documents, which is the addition of metadata to terms, sentences, paragraphs or documents such as syntax (*e.g.*, verb), semantic or pragmatic (*e.g.*, implicature). This step is crucial in NLP, and is usually done manually, for example through crowdsourcing. Crowdsourcing-specific approaches for privacy-preserving task processing [31] require splitting the task (*i.e.*, annotation) between non-colluding workers before aggregating the result in a secure way (*e.g.*, on the platform).

System Overview. Figure 3 outlines an abstract architecture for our privacy-preserving data publishing system for legal decisions. Our objective is not to provide exhaustive implementation guidelines, but rather to identify the key components that such an architecture should possess. The precise and massive access modes are both protected by the `Authentication` module. The `Authentication` module can be implemented by usual strong authentication techniques (*e.g.*, for preventing impersonation attacks). Authentication is necessary for enforcing the access control policy through the `Access Control` module and for maintaining for each user his `Disclosure Log`. The log contains all the successful access requests performed by a user. It is required for verifying that the overall privacy guarantees are not breached, *e.g.*, the rate limitation is not exceeded, or the composition does not exceed the tolerated disclosure. Finally, the `Privacy Parameters` contain the overall privacy guarantees that must always hold, defined by the administrator (*e.g.*, rate limit or higher bound on the tolerated disclosure). The user may additionally be allowed to tune the privacy parameters input by a privacy-preserving data publishing scheme (*e.g.*, the fraction spent in the higher bound on the ϵ differential privacy parameter) provided it does not jeopardize the overall privacy guarantees.

5   Conclusion

In this paper, we analyzed the needs for publishing legal data and the limitations of rule-based redaction (*i.e.*, the current approach) for fulfilling them successfully. We proposed to discard any one-size-fits-all approach and outlined a straw man architecture balancing the utility and privacy requirements by distinguishing the traditional, one-to-one, use of legal data from the modern, massive, use of legal data by legal technologies. Our proposition can easily be implemented on current platforms.

Acknowledgments. We thank the reviewers for their careful reading of the manuscript and their constructive remarks. This work was partially funded by the PROFILE-INT project funded by the LabEx CominLabs (ANR-10-LABX-07-01). Sébastien Gambs is supported by the Canada Research Chair program, a Discovery Grant (NSERC) and the Legalia project (FQRNT).

References

1. Abbasi, A., Chen, H.: Writeprints: a stylometric approach to identity-level identification and similarity detection in cyberspace. ACM Trans. Inf. Syst. (TOIS) **26**(2), 7 (2008)
2. Alarie, B., Niblett, A., Yoon, A.H.: How artificial intelligence will affect the practice of law. Univ. Toronto Law J. **68**(supplement 1), 106–124 (2018)
3. Aletras, N., Tsarapatsanis, D., Preoţiuc-Pietro, D., Lampos, V.: Predicting judicial decisions of the European court of human rights: a natural language processing perspective. PeerJ Comput. Sci. **2**, e93 (2016)
4. Allard, T., Béziaud, L., Gambs, S.: Online publication of court records: circumventing the privacy-transparency trade-off (2020)
5. Arrington, M.: AOL proudly releases massive amounts of private data. TechCrunch (2006). https://social.techcrunch.com/2006/08/06/aol-proudly-releases-massive-amounts-of-user-search-data/
6. Ashley, K.D., Brüninghaus, S.: Automatically classifying case texts and predicting outcomes. Artif. Intell. Law **17**(2), 125–165 (2009)
7. Ashley, K.D., Walker, V.R.: Toward constructing evidence-based legal arguments using legal decision documents and machine learning. In: Proceedings of the Fourteenth International Conference on Artificial Intelligence and Law, pp. 176–180 (2013)
8. Bailey, J., Burkell, J.: Revisiting the open court principle in an era of online publication: questioning presumptive public access to parties' and witnesses' personal information. Ottawa L. Rev. **48**, 143 (2016)
9. Bentham, J., Bowring, J.: The Works of Jeremy Bentham, vol. 4. William Tait, Edinburgh (1843)
10. Brüninghaus, S., Ashley, K.D.: Using machine learning for assigning indices to textual cases. In: Leake, D.B., Plaza, E. (eds.) ICCBR 1997. LNCS, vol. 1266, pp. 303–314. Springer, Heidelberg (1997). https://doi.org/10.1007/3-540-63233-6_501
11. Calamur, K.: In a first for Britain, a secret trial for terrorism suspects. NPR (2014). https://text.npr.org/s.php?sId=319076959
12. Chen, B.C., Kifer, D., LeFevre, K., Machanavajjhala, A.: Privacy-preserving data publishing. Found. Trends Databases **2**(1–2), 1–167 (2009)
13. Conley, A., Datta, A., Nissenbaum, H., Sharma, D.: Sustaining privacy and open justice in the transition to online court records: a multidisciplinary inquiry. Md. L. Rev. **71**, 772 (2011)
14. Custis, T., Schilder, F., Vacek, T., McElvain, G., Alonso, H.M.: Westlaw edge AI features demo: KeyCite overruling risk, litigation analytics, and WestSearch plus. In: Proceedings of the Seventeenth International Conference on Artificial Intelligence and Law - ICAIL 2019, pp. 256–257. ACM Press, Montreal (2019)
15. Dale, R.: Law and word order: NLP in legal tech. Nat. Lang. Eng. **25**(1), 211–217 (2019)
16. Declaration on free access to law (2002). http://www.worldlii.org/worldlii/declaration/
17. Dwork, C.: Differential privacy. In: Bugliesi, M., Preneel, B., Sassone, V., Wegener, I. (eds.) ICALP 2006. LNCS, vol. 4052, pp. 1–12. Springer, Heidelberg (2006). https://doi.org/10.1007/11787006_1
18. Dwork, C., Naor, M.: Pricing via processing or combatting junk mail. In: Brickell, E.F. (ed.) CRYPTO 1992. LNCS, vol. 740, pp. 139–147. Springer, Heidelberg (1993). https://doi.org/10.1007/3-540-48071-4_10

19. Dwork, C., Roth, A., et al.: The algorithmic foundations of differential privacy. Found. Trends® Theoret. Comput. Sci. **9**(3–4), 211–407 (2014)
20. Fernandes, N., Dras, M., McIver, A.: Generalised differential privacy for text document processing. In: Nielson, F., Sands, D. (eds.) POST 2019. LNCS, vol. 11426, pp. 123–148. Springer, Cham (2019). https://doi.org/10.1007/978-3-030-17138-4_6
21. Fleuriot, C.: Avec l'accès gratuit à toute la jurisprudence, des magistrats réclament l'anonymat. Dalloz Actualité, February 2017. https://www.dalloz-actualite.fr/flash/avec-l-acces-gratuit-toute-jurisprudence-des-magistrats-reclament-l-anonymat
22. Fouret, A., Perez, M., Barrière, V., Rottier, E., Buat-Ménard, É.: Open Justice. Technical report, Direction interministérielle du numérique (2019). https://entrepreneur-interet-general.etalab.gouv.fr/defis/2019/openjustice.html
23. Hartzog, W., Stutzman, F.: The case for online obscurity. Calif. L. Rev. **101**, 1 (2013)
24. Hassan, F., Sánchez, D., Soria-Comas, J., Domingo-Ferrer, J.: Automatic anonymization of textual documents: detecting sensitive information via word embeddings. In: 2019 18th IEEE International Conference on Trust, Security and Privacy in Computing and Communications/13th IEEE International Conference On Big Data Science And Engineering (TrustCom/BigDataSE), pp. 358–365 (2019)
25. Jaconelli, J.: Open Justice: A Critique of the Public Trial. Oxford University Press on Demand, Oxford (2002)
26. Jacquin, J.B.: Terrorisme: la peur des magistrats. Le Monde, January 2017. https://www.lemonde.fr/police-justice/article/2017/01/19/terrorisme-la-peur-des-magistrats_5065242_1653578.html
27. Joachims, T.: Text categorization with Support Vector Machines: learning with many relevant features. In: Nédellec, C., Rouveirol, C. (eds.) ECML 1998. LNCS, vol. 1398, pp. 137–142. Springer, Heidelberg (1998). https://doi.org/10.1007/BFb0026683
28. Jordon, J., Yoon, J., van der Schaar, M.: PATE-GAN: generating synthetic data with differential privacy guarantees. In: 7th International Conference on Learning Representations, ICLR 2019, New Orleans, LA, USA, 6–9 May 2019. OpenReview.net (2019)
29. Joulin, A., Grave, E., Bojanowski, P., Mikolov, T.: Bag of tricks for efficient text classification. arXiv preprint arXiv:1607.01759 (2016)
30. Judges Technology Advisory Committee: Open courts, electronic access to court records, and privacy: discussion paper. Technical report, Canadian Judicial Council (2003). http://publications.gc.ca/collections/collection_2008/lcc-cdc/JL2-75-2003E.pdf
31. Kajino, H., Baba, Y., Kashima, H.: Instance-privacy preserving crowdsourcing. In: Second AAAI Conference on Human Computation and Crowdsourcing (2014)
32. Katz, D.M., Bommarito II, M.J., Blackman, J.: A general approach for predicting the behavior of the supreme court of the united states. PLoS One **12**(4), e0174698 (2017)
33. Kim, M.Y., Rabelo, J., Goebel, R.: Statute law information retrieval and entailment. In: Proceedings of the Seventeenth International Conference on Artificial Intelligence and Law - ICAIL 2019, pp. 283–289. ACM Press, Montreal (2019)
34. Kim, Y.: Convolutional neural networks for sentence classification. arXiv preprint arXiv:1408.5882 (2014)
35. Kort, F.: Quantitative analysis of fact-patterns in cases and their impact on judicial decisions. Harv. L. Rev. **79**, 1595 (1965)

36. Lai, S., Xu, L., Liu, K., Zhao, J.: Recurrent convolutional neural networks for text classification. In: Twenty-Ninth AAAI Conference on Artificial Intelligence (2015)
37. Liu, P., Qiu, X., Huang, X.: Recurrent neural network for text classification with multi-task learning. arXiv preprint arXiv:1605.05101 (2016)
38. Mandal, A., Chaki, R., Saha, S., Ghosh, K., Pal, A., Ghosh, S.: Measuring similarity among legal court case documents. In: Proceedings of the 10th Annual ACM India Compute Conference, Compute 2017, pp. 1–9. Association for Computing Machinery, New York (2017)
39. Marques, M.R., Bianco, T., Roodnejad, M., Baduel, T., Berrou, C.: Machine learning for explaining and ranking the most influential matters of law. In: Proceedings of the Seventeenth International Conference on Artificial Intelligence and Law, pp. 239–243. ACM (2019)
40. Marrero, M., Urbano, J., Sánchez-Cuadrado, S., Morato, J., Gómez-Berbís, J.M.: Named entity recognition: fallacies, challenges and opportunities. Comput. Stand. Interf. **35**(5), 482–489 (2013)
41. Martin, P.W.: Online access to court records-from documents to data, particulars to patterns. Vill. L. Rev. **53**, 855 (2008)
42. Mikolov, T., Sutskever, I., Chen, K., Corrado, G.S., Dean, J.: Distributed representations of words and phrases and their compositionality. In: Advances in Neural Information Processing Systems, pp. 3111–3119 (2013)
43. Minocha, A., Singh, N.: Legal document similarity using triples extracted from unstructured text. In: Rehm, G., Rodríguez-Doncel, V., Moreno-Schneider, J. (eds.) Proceedings of the Eleventh International Conference on Language Resources and Evaluation (LREC 2018). European Language Resources Association (ELRA), Paris, France, May 2018
44. Mokanov, I., Shane, D., Cerat, B.: Facts2Law: using deep learning to provide a legal qualification to a set of facts. In: Proceedings of the Seventeenth International Conference on Artificial Intelligence and Law, pp. 268–269. ACM (2019)
45. Možina, M., Žabkar, J., Bench-Capon, T., Bratko, I.: Argument based machine learning applied to law. Artif. Intell. Law **13**(1), 53–73 (2005)
46. Nallapati, R., Manning, C.D.: Legal docket classification: where machine learning stumbles. In: Proceedings of the 2008 Conference on Empirical Methods in Natural Language Processing, pp. 438–446 (2008)
47. Neal, T., Sundararajan, K., Fatima, A., Yan, Y., Xiang, Y., Woodard, D.: Surveying stylometry techniques and applications. ACM Comput. Surv. (CSUR) **50**(6), 1–36 (2017)
48. Opijnen, M., Peruginelli, G., Kefali, E., Palmirani, M.: On-line publication of court decisions in the EU: report of the policy group of the project "building on the European case law identifier" (2017). SSRN 3088495
49. Organisation for Economic Co-operation and Development (ed.): The call for innovative and open government: an overview of country initiatives. OECD, Paris (2011)
50. Plamondon, L., Lapalme, G., Pelletier, F.: Anonymisation de décisions de justice. In: XIe Conférence sur le Traitement Automatique des Langues Naturelles (TALN 2004), pp. 367–376. Bernard Bel et Isabelle Martin. (éditeurs), Bernard Bel et Isabelle Martin. (éditeurs), Fès, Maroc, May 2004
51. Praduroux, S., de Paiva, V., di Caro, L.: Legal tech start-ups: state of the art and trends. In: Proceedings of the Workshop on MIning and REasoning with Legal texts collocated at the 29th International Conference on Legal Knowledge and Information Systems (2016)

52. Quaresma, P., Gonçalves, T.: Using linguistic information and machine learning techniques to identify entities from juridical documents. In: Francesconi, E., Montemagni, S., Peters, W., Tiscornia, D. (eds.) Semantic Processing of Legal Texts. LNCS (LNAI), vol. 6036, pp. 44–59. Springer, Heidelberg (2010). https://doi.org/10.1007/978-3-642-12837-0_3

53. Rastogi, V., Hong, S., Suciu, D.: The boundary between privacy and utility in data publishing. In: VLDB (2007)

54. Sanchez, D., Batet, M., Viejo, A.: Automatic general-purpose sanitization of textual documents. IEEE Trans. Inf. Forensics Secur. 8(6), 853–862 (2013)

55. Siegel, D.J.: CARA: an assistance to help find the cases you missed. Law Prac. 43, 22 (2017)

56. Sweeney, L.: K-anonymity: a model for protecting privacy. Int. J. Uncertain. Fuzziness Knowl.-Based Syst. 10(5), 557–570 (2002)

57. Tan, S., Adebayo, J., Inkpen, K., Kamar, E.: Investigating human+ machine complementarity for recidivism predictions. arXiv preprint arXiv:1808.09123 (2018)

58. Thenmozhi, D., Kannan, K., Aravindan, C.: A text similarity approach for precedence retrieval from legal documents. In: FIRE (Working Notes), pp. 90–91 (2017)

59. University of Houston Law Center: How to brief a case. Technical report, University of Houston Law Center (2009). https://www.law.uh.edu/lss/casebrief.pdf

60. Weggenmann, B., Kerschbaum, F.: SynTF: synthetic and differentially private term frequency vectors for privacy-preserving text mining. arXiv preprint arXiv:1805.00904 (2018)

61. Yousfi-Monod, M., Farzindar, A., Lapalme, G.: Supervised machine learning for summarizing legal documents. In: Farzindar, A., Kešelj, V. (eds.) AI 2010. LNCS (LNAI), vol. 6085, pp. 51–62. Springer, Heidelberg (2010). https://doi.org/10.1007/978-3-642-13059-5_8

62. Zhang, D., McKenna, R., Kotsogiannis, I., Hay, M., Machanavajjhala, A., Miklau, G.: EKTELO: a framework for defining differentially-private computations. In: Proceedings of the 2018 International Conference on Management of Data, SIGMOD 2018, pp. 115–130. Association for Computing Machinery, New York (2018)

63. Zhang, X., Zhao, J., LeCun, Y.: Character-level convolutional networks for text classification. In: Advances in Neural Information Processing Systems, pp. 649–657 (2015)

64. Zheng, J., Guo, Y., Feng, C., Chen, H.: A hierarchical neural-network-based document representation approach for text classification. Math. Probl. Eng. 2018 (2018)

Challenges in the Digital Representation of Privacy Terms

Beatriz Esteves[✉][iD]

Ontology Engineering Group, Universidad Politécnica de Madrid, Madrid, Spain
beatriz.gesteves@upm.es
https://oeg.fi.upm.es/

Abstract. This paper aims to describe a research project focused on the digital representation of information related to the privacy and data protection domain. Currently, privacy policies are used by data controllers as a tool to achieve compliance with data protection regulations such as the EU GDPR, instead of being a privacy instrument at the disposal of both controllers and data subjects. On the other hand, data subjects lack the tools to effectively establish preferences when it comes to the processing and disclosure of their personal data, as well as to easily exercise their rights. In this regard, this paper discusses the challenges of the implementation of a service based on decentralised Web technologies and Semantic Web standards and specifications to facilitate the communication between data subjects and data controllers in the light of the GDPR. The main challenges that this service intends to address are linked to the exercising of GDPR-related rights and obligations, the negotiation of privacy terms and the governance of access to personal data stores. A case study in the healthcare and genomics domain will be explored to experiment with the developed tools. Early-stage results related to the implementation of semantic policies for the representation of GDPR rights and obligations are presented.

Keywords: Personal data · Privacy · Data protection · GDPR · Semantic web · Policy languages · Intelligent agents

1 Introduction

With the wide spread of technologies in every aspect of our day to day life, the amount of data available has reached a critical level and the legal and ethical implications of its exploration has been under debate for quite a few years. In particular, the increasing amount of personal data, that is being produced by users of Web services and applications, has raised greater concerns for privacy, which has led to increased legislative actions. Adding to this situation, the vast

This work has been supported by the European Union's Horizon 2020 research and innovation programme under the Marie Skłodowska-Curie grant agreement No 813497 (PROTECT).

V. Rodríguez-Doncel et al. (Eds.): AICOL-XI 2018/AICOL-XII 2020/XAILA 2020, LNAI 13048, pp. 313–327, 2021.
https://doi.org/10.1007/978-3-030-89811-3_22

majority of personal data is centralised in large data silos and only within the reach of a few, such as government institutions, banks or large telecommunications or technological companies. Thus far, users have been willing to exchange part of their privacy for personalised services based on the collection of their personal data, however due to the recent cases of personal data breaches, they are looking for alternative methods to regain control over their own data and ensure its protection.

In this context, the European Union launched a new regulation focused on the 'protection of natural persons with regard to the processing of personal data and on the free movement of such data', the so-called General Data Protection Regulation (GDPR) [36]. Thus, when the GDPR entered into force on the May 25th, 2018, companies had to adapt their personal data driven businesses, as well as enforce new data protection processes, to be compliant with the new enforced GDPR obligations. On the other hand, users were left in overload with the amount of complex technical information about the processing of their personal data and accompanying GDPR rights and lack the tools to deal with the high number of consent forms and privacy terms that the technology era has brought to them as internet users.

From this standpoint, it is evident the need for an interoperable framework that can support individuals, the data subjects, and companies, the data controllers, navigating the GDPR. If they are to interoperate, then the need for them to speak the same language becomes evident and the use of Semantic Web technologies is only logical, as they promote interoperability and extendability, based on open standards and specifications. In this context, the development of common vocabularies and data models to specify GDPR's privacy-related terms would favour all entities involved, in particular the data subjects in (i) the management of their privacy preferences or in (ii) the exercising of their rights; and the data controllers to achieve compliance with their obligations regarding (i) the information that needs to be provided to the data subjects about the processing of their data, (ii) replies to the data subjects' right-related requests or (iii) the maintenance of records of processing activities and the management of personal data breaches. Therefore, in the scope of this research, the notion of 'privacy term' is used to refer to the items of information that need to be specified to represent concepts related to the preferences and rights of data subjects and to the policies and obligations of data controllers regarding privacy and personal data protection. For instance, the 'right to be forgotten', the 'purposes' for processing personal data or the 'legal basis' for processing sensitive data are examples of privacy terms.

In addition to this, decentralised systems could be the answer for data subjects to retake control over the privacy of their own data as they can be used to control whom has access to which data, without intermediaries, and can address the issue of data portability, making it easier to exchange data between different applications from distinct companies, without having to recreate every piece of information when using a new service. Moreover, these decentralised, user-owned data systems can be the next step towards the implementation of a negotiation system for privacy terms between data subjects and data controllers. Currently, when a data subject starts using a new service or application, the privacy policies and terms of the services are presented to them and, in most cases, they have to accept it to utilise said service, even if they do not agree with all of the terms. With a decentralised system in place, data subjects and data controllers can negotiate which data can be disclosed and specify fine-grained conditions in which the exchange can occur.

This paper is organised as follows: Sect. 2 reviews the state of the art on Semantic Web technologies, machine-readable policy languages and data protection vocabularies. Section 3 identifies current challenges in the digital representation of privacy terms, in the context of the GDPR, and discusses possible solutions to address them and Sect. 3.1 describes a case study in the healthcare domain. Section 4 provides a description of the target architecture to deal with this challenges and a few initial results of the implementation of the project and Sect. 5 concludes the paper.

2   State of the Art

Berners-Lee, Hendler and Lassila [9] first defined the term *'semantic web'* as an extension of the World Wide Web with the main vision of moving from a *'Web of Documents'* to a *'Web of Data'*. Today, most Web content is produced for human consumption, not to be processed automatically by computer programs. The Semantic Web wants to change the paradigm for a Web that can be shared and reused across applications, companies and the Web community in general. Therefore, Semantic Web's main objectives are to achieve interoperability and extendability through the development and use of open Web standards. The World Wide Web Consortium (W3C), founded and led by Berners-Lee, oversees the development of these Web standards in cooperation with academia and industry partners, with the main objectives of developing common formats for data interoperability and promoting a language to document how data relates to real-world objects. The Resource Description Framework (RDF) is the W3C standard for expressing information about resources and for data interchange on the Web [34]. With RDF, resources can be described with sets of triples: the subject and the object define two resources and the predicate specifies the relationship between these resources, which are identifiable through an Universal Resource Identifier (URI). RDF triples form a Web of information which can be used to add machine-readable information to Web services, to build distributed social networks based on RDF descriptions of people, to enrich datasets by connecting them to third party datasets and more. The Web Ontology Language (OWL), a W3C Recommendation for ontology development, builds on RDF to define Web ontologies while ensuring greater interoperability, helping reveal inconsistencies in ontologies and finding new relationships between concepts [1].

A recent project that makes use of Semantic Web's open standards to build an interoperable platform for decentralised applications is the Solid project [3]. Solid applies standards, such as the Linked Data Platform [35] and the Web Access Control (WAC) [2], to support interoperability between applications and enable data portability between different services. Using these features, the users can select which services can have access to their personal data, can reuse that data across numerous applications and can have a personal data store, a Solid Pod, which can be deployed on a public server or self-hosted on a personal server.

Naturally, the Semantic Web also has been the playground for the establishment of many machine-readable policy languages since the 1990s, as the main purpose of said languages is to govern the access of systems to resources while conveying user preferences and the privacy practices of Web services. As such, they seem perfectly suited for establishing policies related to the processing and disclosure of personal data. The Platform for Privacy Preferences (P3P) specification became a W3C Recommendation in 2002 for websites to disclose privacy protocols that can be automatically read and interpreted by user agents, however, due to a limited adoption, it became obsolete in 2018 [16]. Khandelwal et al. [22] implemented AIR (Accountability in RDF), a language focused on generating explanations for its inferences and actions that supports rule nesting and reuse. Passant and Sacco established a lightweight ontology to design fine-grained privacy preferences to govern the access to data within RDF documents - the Privacy Preference Ontology (PPO) [32]. LegalRuleML is an OASIS Committee Specification that extends RuleML with formal features to represent and reason over legal norms, guidelines and policies [26]. The Open Digital Rights Language (ODRL) is a W3C Recommendation for the establishment of policy expressions that provides an interoperable information model, vocabulary, and encoding mechanism to disclose information about permissions, prohibitions and duties related to a resource [19]. More details about relevant machine-readable policy languages are presented at https://protect.oeg.fi.upm.es/sota/languages.

Nonetheless, these machine-readable privacy languages are not sufficient to represent all the information items that data controllers must disclose in their privacy policies and therefore the development of vocabularies for the data protection domain has increased, as they can be used to close this gap on the representation of privacy and data protection concepts. PrOnto (Privacy Ontology for legal reasoning) is a legal ontology, developed by Palmirani et al. in the light of the GDPR, with the main goal of modelling the relationships between privacy agents, data types and processing operations [27]. The GDPR Provenance Ontology (GDPRov) is a Linked Data ontology focused on representing the provenance of consent and of collection, usage and storage of data, with the main goal of documenting GDPR compliance [30]. GDPRtEXT is an open data resource, that extends the European Legislation Identifier (ELI) ontology, with the aim of connecting GDPR concepts, through the SKOS ontology, to their respective GDPR chapter, article, point or recital [29]. DPV is a Data Privacy Vocabulary with the main purpose of providing a taxonomy of terms to annotate and classify the handling of personal data in accordance with the GDPR and other data protection regulations [31]. More details about relevant data protection ontologies and vocabularies are presented at https://protect.oeg.fi.upm.es/sota/ontologies.

Figure 1 provides a timeline of the analysed privacy-related policy languages and data protection and privacy vocabularies. The timeline indicates the date when the language or vocabulary was first published. The assigned colours also indicate whether the solution is based on GDPR concepts and whether it has open access and actively maintained resources. The following solutions were analysed: P3P [16] and APPEL [15], ODRL [19], XPref [4], AIR [22], DPRO and DPKO [12], S4P [7], POL [10], PPO [32], LegalRuleML [26], A-PPL [5], P2U [20], DPO [6], SPL and SPLog [23], GDPRov [30], DPF [25], IMO and PMO [24], DPV [31], GDPRtEXT [29], Cloud [17], PrOnto [27] and GConsent [28].

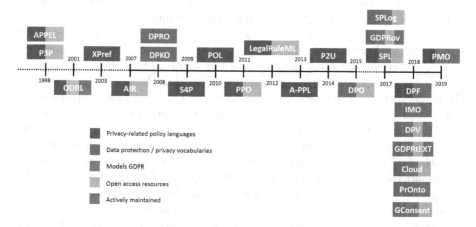

Fig. 1. Timeline of privacy-related policy languages and vocabularies from 1998 to 2019. The solutions are categorised in relation to modelling GDPR concepts, being open access and actively maintained [19]

3   Challenges in the Representation of Privacy Terms in the Light of GDPR

The domains of data privacy and data protection generate several distinct problems concerning the digital representation of concepts, especially with regard to the interoperability between data subjects, data controllers and the other entities involved. For instance, with the enforcement of the GDPR, data subjects are now in need of assistance in relation to the digital representation of consent terms, while data controllers are in need of tools to better handle the compliance with GDPR-related obligations. To this respect, three challenges, regarding the exercising of rights and obligations (C1), the negotiation of privacy terms (C2) and the governance of access to personal data stores (C3), are identified and discussed in order to determine possible solutions.

C1 *Exercising of rights and obligations in the light of the GDPR:*
Chapters III and IV of the GDPR define the *'Rights of the data subject'* and the obligations of the *'Controller and processor'*, respectively. In order to be

invoked and granted, these rights and obligations generate the need to represent information that must be communicated from one entity to another so that controllers and processors are in compliance with the GDPR. For instance, if a data subject invokes its right of access, it needs the means to represent the request and the data controller requires the means to represent certain pieces of information, such as the *purposes for processing*, the *recipients* to whom the personal data may be disclosed or the existence of the *right to lodge a complaint with a data protection authority (DPA)*. On the other hand, data controllers also need data models to handle compliance requests, i.e. upon request of a DPA, the data controller must be able to reply with a record of their processing activities, which requires the need to represent concepts such as the *identity and contact details of the controller and their data protection officer* or the *technical and organisational measures* applied to ensure the privacy of the data subjects' personal data.

C2 *Negotiation of privacy terms:* Currently, there is a lack of open source, machine readable and standardised models to support data subjects in the formulation of their privacy preferences, as well as data controllers in the establishment of their privacy policies and terms of services. Furthermore, both data subjects and data controllers lack a negotiation mechanism to support them in establishing the conditions under which the disclosure of the data subject's personal data is allowed. With the development of a privacy negotiation mechanism, both parties can communicate until they reach a consensus on what personal data can be disclosed, as well as on the fine-grained conditions under which this exchange can occur.

C3 *Governance of access to personal data stores:* Nowadays, one of the main challenges in the governance of personal data resides in the centralisation of its storage. So far, large technological companies drive their business in the exploration of this type of data, since data subjects are willing to share their data and risk their privacy in exchange for personalised services. However, recently, due to high profile personal data breaches originating from these companies, more Web users are looking for alternative solutions to regain control over their own data. In this context, the emergence of decentralised technologies can be an adequate solution to face this challenge, as they allow data subjects to directly control who has access to what data and under what conditions, and directly move their data between different services, without the need of having an intermediary.

Consequently, this paper proposes the use of decentralised Web technologies and open standards and specifications to address these challenges. More specifically, the following hypotheses were established to support the development of solutions to face these challenges:

(i) Data protection vocabularies and machine-readable policy languages can be used and extended for the representation of fine-grained privacy preferences and privacy policies for the processing of personal data.

(ii) These vocabularies can also be used to implement a service to facilitate the representation of machine-readable information related to GDPR rights and obligations.

(iii) W3C standards and specifications can be applied to improve the communication and data interoperability between the distinct stakeholders involved on the data protection domain.

(iv) These standards can also be used to establish more fine-grained access control conditions and to manage the access to decentralised file systems.

(v) Semantic Web vocabularies can be used as the basis for establishing a privacy negotiation process in which data subjects and data controllers need to communicate until a consensus is reached on what are the acceptable privacy terms for both parties for the disclosure of personal data to occur.

In Sect. 3.1, the particular challenges of representing privacy terms in the health domain are presented and discussed, as this case study will be used to test the tools developed in this work.

3.1 Case Study: Machine-Readable Policies for Healthcare and Genomics

The GDPR defines *'personal data'* as any information that can be used to directly, or indirectly, identify an identifiable natural person or *'data subject'*. Special categories of personal data, such as health or genetic data, are enumerated and require additional measures of protection. Specifically, health data is specified in the GDPR as any personal data that can be related to *'the physical or mental health of a natural person, including the provision of health care services, which reveal information about his or her health status'*; and genetic data is defined as *'personal data relating to the inherited or acquired genetic characteristics of a natural person which give unique information about the physiology or the health of that natural person and which result, in particular, from an analysis of a biological sample from the natural person in question'*. In particular, for the processing of these special personal data, the information provided to data subjects should include, not only one of the legal basis defined on Article 6 of the GDPR for the *'lawfulness of processing'* of personal data, but also a separate condition for the *'processing of special categories of personal data'*, defined on the Article 9 of the GDPR. Moreover, the European Data Protection Board (EDPB) released in February 2021 a *'response to the request from the European Commission for clarifications on the consistent application of the GDPR, focusing on health research'* [11]. This document provides new guidelines for the application of GDPR in the processing of health-related data for scientific research purposes, more specifically on compliance with data protection and ethical obligations, on the further processing of previously collected health data, on the application of the notion of broad consent for the processing of special categories of personal data, on the transparency of processing activities and on adequate safeguards to protect data subjects' privacy.

In addition to this, with the evolution of health-related data collecting services, such as fitness trackers, and the emergence of the Internet of Things (IoT), an increasing rate of health-related data is being gathered in the cloud and shared through the Web. In particular, a variety of data governance challenges, closely related to the challenges C1, C2 and C3 defined in the previous section, have emerged in this sector:

- *Interoperability of diverse data sources:* in the health sector it is necessary to conjugate a diversity of data formats since often the data comes from disparate sources, i.e. medical devices, smart-phone applications or medical records written by humans [21]. This issue fits in with the identified C1 challenge, as it impacts, for example, the right of access and portability of data subjects.
- *An increasing number of Artificial Intelligence algorithms for healthcare:* health-related data is often highly variable and stored in an unstructured manner and, as such, it recurrently faces issues related to data quality and accuracy, which can lead to poor decision making and unsatisfactory research results [33]. This challenge is aligned with GDPR's Article 5.1(d) related to the *'accuracy'* of personal data and in line with the identified C1 challenge, as it impacts, for example, the right of restriction of processing of the data subject.
- *An increasing rate of Web consultation services:* a growing number of health-care services are turning to virtual doctor consultations and remote medical guidance to improve patient care and to make health sector more accessible to everyone [37]. This issue fits in with the identified C2 and C3 challenges, as data subjects need to be able to negotiate which data they want to share and need to have control over their data to be able to share it.
- *The emergence of genetic data:* nowadays, with the possibility of sequencing the human genome, genetic data is within reach, although the cost and concerns about the privacy of processing these high-density data sets are still an important issue and their implications for public health continue to be debated [8,18]; services such as ancestry.com, which provide information on the genetic and health data of family trees, also raise the question of who has the right to be in control over the access to such sensitive data, in line with the identified C3 challenge, since it reveals information not only about the data subject, but also about its relatives.

A possible solution to overcome these challenges may be to give data subjects control over their health data. Efforts are underway to do this using decentralised systems, such as distributed ledgers. In this specific case study, Semantic Web open standards and specifications will be adopted, investigated and extended for the governance of health and genetic data. These standards will serve as the building blocks of a service that can be used to assist data subjects to govern the access to their personal data, facilitate the negotiation of privacy terms between data subjects and data controllers and support the management of GDPR rights and obligations.

For instance, the implemented service should be able to:

- Allow data subjects to govern the access to their data generated by smart bio-sensors, for instance used to monitor chronic diseases such as diabetes, which is stored in their personal database.
- Allow data subjects to share data between multiple applications: i.e. share an exam they did on a public hospital with a different private practitioner or move the data collected by a fitness application to a different one.
- Enable the direct transfer of personal data between data controllers in conformance with GDPR's *'right to data portability'*.
- Enable the negotiation of privacy terms between data subject and data controller in the context of a particular service.
- Allow data subjects to share their genetic data for the purpose of academic research with University X.
- Provide data subjects with a tool to check if the privacy terms of a fitness app comply with GDPR's *'right to be informed'*.
- Provide data controllers with a mechanism to represent the information necessary to comply with their obligations: i.e. provide a tool that allows a pharmaceutical company to represent the necessary terms of a *'data protection impact assessment' (DPIA)* for a large-scale study to cure Alzheimer.

4   A Target Architecture to Deal with GDPR Challenges

To address the challenges identified in the previous section, this paper proposes the development of a set of services, based on Semantic Web technologies, with the main goal of facilitating the communication between data subjects and data controllers in the light of the GDPR. The target architecture of this service aims to model the relations of GDPR stakeholders and to support the communication between them, to improve the means of handling data subjects' rights-related requests and to assist data controllers in the fulfilment of their obligations. A new stakeholder, an intelligent privacy assistant, will be integrated in the architecture to assist in the automation of tasks for both data subjects and controllers. These privacy assistants should be able to:

- assist data subjects in the setting of general and fine-grained privacy preferences.
- support data controllers in the establishment of privacy policies that are in compliance with the terms specified in the GDPR, related to the information that needs to be provided to the data subject when their personal data is being processed.
- negotiate the access to personal data by reasoning over the privacy preferences and policies set by data subjects and data controllers, respectively.
- handle the representation needs that the information flows generated by the GDPR rights and obligations generate, i.e. when a data subject wants to invoke its right of access, the data subject privacy assistant needs to communicate this request to the data controller and the data controller privacy assistant needs to reply the relevant information.

To implement these agents, this paper proposes the use of ODRL, DPV and GDPRtEXT for the establishment of data subjects' preferences and to extend the WAC specification to manage access to Solid Pods. The same semantic technologies will also be used to semantically represent the information that the data controllers need to provide to the data subjects for the handling of requests generated by the subjects' rights and also to support them on the management of their obligations.

Therefore, the following actions can be established for the design of these agents and to validate the hypotheses described in Sect. 3:

(i) define the system's architecture and determine the specific services to be implemented, in particular regarding the privacy agents;

(ii) identify the terms that need to be modelled to represent data subjects' preferences and controllers' privacy policies;

(iii) extend the chosen semantic technologies for the purpose of defining these preferences and privacy policies;

(iv) establish a privacy negotiation mechanism to assist in the establishment of privacy terms that satisfy both data subjects and controllers;

(v) analyse the concepts that need to be modelled to represent GDPR rights and obligations;

(vi) extend the chosen semantic technologies where necessary to accommodate the representation needs generated by GDPR rights and obligations;

(vii) validate the developed representation models with legal and ethical experts;

(viii) deploy a modifiable version of a Solid server and client;

(ix) understand how to use ODRL and DPV to extend WAC, to allow for more fine-grained access control to Solid Pods;

(x) extend DPV for the specific case of processing health-related and genetic data;

(xi) test developed services with the identified health-related scenarios.

In Sect. 4.1, initial results of the implementation of the proposed work are presented.

4.1   Early-Stage Results

Thus far, an analysis of the existing open-access specifications on the domain of privacy and data protection was performed. In addition to this, a conceptual analysis was executed to understand the concepts that need to be modelled to support the communication between GDPR stakeholders. Also, the relevant privacy-related policy languages and data protection ontologies were studied in relation to these representation needs generated by the invoking and granting of GDPR rights and obligations.

Moreover, a lightweight ontology, the GDPR Information Flows (GDPRIF)[1], was also specified to model the relationships between GDPR stakeholders, the

---

[1] https://protect.oeg.fi.upm.es/def/gdprif.

items of information that need to be communicated between them, GDPR rights and obligations and the events responsible for triggering those rights and obligations, i.e. a request of a particular data subject right, an occurrence of a personal data breach or a compliance monitoring process started by a DPA. GDPRIF's rights and obligations are therefore modelled as ODRL policies complemented by DPV, DPV-GDPR and GDPRtEXT. A complete list of these generic policies is available at https://protect.oeg.fi.upm.es/sota/gdprif, as these structured policies can be useful for data controllers to more easily fulfil their obligations when they are triggered. Two example policies are displayed and discussed in Listings 1.1 and 1.2.

Listing 1.1 specifies a data controller's obligation to respond to a data subject's request related to his right of access (*dpv-gdpr:A15*). To fulfil its obligations, in this case, the data controller has to inform the data subject about the purposes of processing (*gdprif:I6*) and the categories of personal data being processed (*gdprif:I17*), about the source of the personal data, if not directly collected from the data subject (*gdprif:I18*), about any recipients to which the personal data can be disclosed (*gdprif:I9*), about the retention period (*gdprif:I11*), about the existence of the data subject's rights (*gdprif:I12*) and the right to lodge a complaint (*gdprif:I14*), about the existence of automated decision-making (*gdprif:I16*) and about the safeguards related to a data transfer to a third country (*gdprif:I20*). The data controller also has the obligation to give the data subject a copy of the personal data (*gdprif:I21*).

Listing 1.1. Data controller's obligation to grant access to the data subject's personal data.

```
@prefix rdf: <http://www.w3.org/1999/02/22-rdf-syntax-ns#> .
@prefix rdfs: <http://www.w3.org/2000/01/rdf-schema#> .
@prefix odrl: <http://www.w3.org/ns/odrl/2/> .
@prefix dpv: <http://www.w3.org/ns/dpv#> .
@prefix dpv-gdpr: <http://www.w3.org/ns/dpv-gdpr#> .
@prefix gdprtext: <https://w3id.org/GDPRtEXT#> .
@prefix gdprif: <https://protect.oeg.fi.upm.es/def/gdprif#> .

gdprif:RA
    rdfs:seeAlso gdprtext:RightToAccessPersonalData ;
    a odrl:Policy ;
    odrl:obligation [
        a odrl:Obligation ;
        gdprif:isRequestedBy [ a dpv:DataSubject, odrl:Party ] ;
        gdprif:isAnsweredBy [ a dpv:DataController, odrl:Party ] ;
        odrl:action odrl:inform ;
        odrl:target gdprif:I6, gdprif:I9, gdprif:I11, gdprif:I12,
            gdprif:I14, gdprif:I16, gdprif:I17, gdprif:I18,
            gdprif:I20, gdprif:I21 ;
        odrl:constraint [
            a odrl:Constraint ;
            odrl:leftOperand odrl:event ;
            odrl:operator odrl:eq ;
            odrl:rightOperand dpv-gdpr:A15 ;
        ]
    ] .
```

Listing 1.2 specifies a data controller's obligation to inform a data subject about the existence of joint controllers and their responsibilities in relation to the processing of the personal data, in the cases where is a contract between joint controllers is established (*gdprif:ContractBetweenJointControllers*). To fulfil its obligations, in this case, one of the joint controllers has to give information to the data subject about the identity and contact details of the all the joint controllers (*gdprif:I31* and *gdprif:I32*) and about the responsibilities of each joint controller (*gdprif:I33*).

Listing 1.2. Joint controllers' obligation to inform the data subject about the existence of joint controllers and their responsibilities in relation to the processing of the personal data.

```
@prefix rdf: <http://www.w3.org/1999/02/22-rdf-syntax-ns#> .
@prefix rdfs: <http://www.w3.org/2000/01/rdf-schema#> .
@prefix odrl: <http://www.w3.org/ns/odrl/2/> .
@prefix dpv: <http://www.w3.org/ns/dpv#> .
@prefix gdprtext: <https://w3id.org/GDPRtEXT#> .
@prefix gdprif: <https://protect.oeg.fi.upm.es/def/gdprif#> .

gdprif:JC
    rdfs:seeAlso gdprtext:LiabilityOfJointController ;
    a odrl:Policy ;
    odrl:obligation [
        a odrl:Obligation ;
        odrl:informingParty [ a gdprtext:JointController, odrl:Party ] ;
        odrl:informedParty [ a dpv:DataSubject, odrl:Party ] ;
        odrl:action odrl:inform ;
        odrl:target gdprif:I31, gdprif:I32, gdprif:I33 ;
        odrl:constraint [
            a odrl:Constraint ;
            odrl:leftOperand odrl:event ;
            odrl:operator odrl:eq ;
            odrl:rightOperand gdprif:ContractBetweenJointControllers ;
        ]
    ] .
```

These initial results should now be validated by legal and ethical experts in the domain of privacy and data protection, and especially in the GDPR.

5   Conclusions

Recently, the privacy and data protection domains have been given a great deal of importance by European and international authorities. Regulations, such as the GDPR, and the more recent proposals for the European regulation of data governance (Data Governance Act [14]) and of privacy and electronic communications (ePrivacy regulation [13]) are examples of this trend. In this context, there is a clear need to develop interoperable technologies that support the digital representation of information, while meeting EU's legal, ethical and social expectations.

This paper identified three challenges related to the representation of privacy terms, regarding the exercising of GDPR rights and obligations, the negotiation of privacy terms and the governance of access to personal data stores, and proposed the usage of decentralised Web technologies and Semantic Web specifications to address them. These technologies will be integrated in a privacy-preserving architecture, which will include the implementation of an intelligent privacy assistant to support in the automation of tasks of data subjects and controllers. In this regard, ODRL, DPV and GDPRtEXT emerge as adequate solutions that support the described needs to represent information related to rights and obligations, as well as preferences for access control to personal data stores, in the form of machine-readable policies. Such policies can then be used in a reasoning system for the negotiation of privacy terms.

References

1. OWL 2 Web Ontology Language Document Overview, 2nd edn. (2012). https://www.w3.org/TR/owl2-overview/
2. WebAccessControl (2019). https://www.w3.org/wiki/WebAccessControl
3. Solid (2020). https://solidproject.org/
4. Agrawal, R., Kiernan, J., Srikant, R., Xu, Y.: XPref: a preference language for P3P. Comput. Networks **48**(5), 809–827 (2005). https://doi.org/10.1016/j.comnet.2005.01.004
5. Azraoui, M., Elkhiyaoui, K., Önen, M., Bernsmed, K., De Oliveira, A.S., Sendor, J.: A-PPL: an accountability policy language. Research report (2014). http://www.eurecom.fr/en/publication/4372/download/rs-publi-4372.pdf
6. Bartolini, C., Muthuri, R.: Reconciling data protection rights and obligations: an ontology of the forthcoming EU regulation. In: Workshop on Language and Semantic Technology for Legal Domain (2015)
7. Becker, M.Y., Malkis, A., Bussard, L.: S4P: a generic language for specifying privacy preferences and policies. Technical report, Microsoft Research (2010). https://www.microsoft.com/en-us/research/wp-content/uploads/2010/04/main-1.pdf
8. Belle, A., Thiagarajan, R., Soroushmehr, S.M.R., Navidi, F., Beard, D.A., Najarian, K.: Big data analytics in healthcare. BioMed Res. Int. **2015**, 1–16 (2015)
9. Berners-Lee, T., Hendler, J., Lassila, O.: The Semantic Web. Scientific American (2001)
10. Berthold, S.: The Privacy Option Language - Specification & Implementation. Research report, Faculty of Health, Science and Technology, Karlstad University (2013). http://kau.diva-portal.org/smash/get/diva2:623452/FULLTEXT01.pdf
11. Board, E.D.P.: EDPB document on response to the request from the european commission for clarifications on the consistent application of the GDPR, focusing on health research (2021). https://edpb.europa.eu/our-work-tools/our-documents/other/edpb-document-response-request-european-commission-clarifications_en
12. Casellas, N., et al.: Ontological semantics for data privacy compliance: The NEURONA project. In: 2010 AAAI Spring Symposium, pp. 34–38. Intelligent Information Privacy Management, AAAI (2010). https://ddd.uab.cat/pub/artpub/2010/137891/aaaisprsymser_a2010n1iENG.pdf
13. Commission, E.: Proposal for a REGULATION OF THE EUROPEAN PARLIAMENT AND OF THE COUNCIL concerning the respect for private life and the protection of personal data in electronic communications and repealing directive 2002/58/EC (regulation on privacy and electronic communications) (2017). https://eur-lex.europa.eu/legal-content/EN/TXT/?uri=CELEX%3A52017PC0010
14. Commission, E.: Proposal for a REGULATION OF THE EUROPEAN PARLIAMENT AND OF THE COUNCIL on European data governance (data governance act) (2020). https://eur-lex.europa.eu/legal-content/EN/TXT/?uri=CELEX%3A52020PC0767
15. Cranor, L., Langheinrich, M., Marchiori, M.: A P3P Preference Exchange Language 1.0 (APPEL1.0) (2002). https://www.w3.org/TR/2002/WD-P3P-preferences-20020415/
16. Cranor, L., Langheinrich, M., Marchiori, M., Presler-Marshall, M., Reagle, J.: The Platform for Privacy Preferences 1.0 (P3P1.0) Specification (2002). https://www.w3.org/TR/P3P/

17. Elluri, L., Joshi, K.P.: A knowledge representation of cloud data controls for EU GDPR compliance. In: 2018 IEEE World Congress on Services (SERVICES), pp. 45–46. IEEE (2018). https://doi.org/10.1109/SERVICES.2018.00036.https://ieeexplore.ieee.org/document/8495788/

18. Grishin, D., et al.: Accelerating genomic data generation and facilitating genomic data access using decentralization, privacy-preserving technologies and equitable compensation. Blockchain in Healthcare Today 1, 1–23 (2018). https://doi.org/10.30953/bhty.v1.34

19. Iannella, R., Villata, S.: ODRL Information Model 2.2 (2018). https://www.w3.org/TR/odrl-model/

20. Iyilade, J., Vassileva, J.: P2U: a privacy policy specification language for secondary data sharing and usage. In: 2014 IEEE Security and Privacy Workshops, pp. 18–22. IEEE (2014–05). https://doi.org/10.1109/SPW.2014.12http://ieeexplore.ieee.org/document/6957279/

21. Jaleel, A., Mahmood, T., Hassan, M.A., Bano, G., Khurshid, S.K.: Towards medical data interoperability through collaboration of healthcare devices. IEEE Access 8, 132302–132319 (2020). https://doi.org/10.1109/ACCESS.2020.3009783

22. Khandelwal, A., Bao, J., Kagal, L., Jacobi, I., Ding, L., Hendler, J., et al.: Analyzing the AIR language: a semantic web (production) rule language. In: Hitzler, P., Lukasiewicz, T., et al. (eds.) RR 2010. LNCS, vol. 6333, pp. 58–72. Springer, Heidelberg (2010). https://doi.org/10.1007/978-3-642-15918-3_6

23. Kirrane, S., et al.: A scalable consent, transparency and compliance architecture. In: Gangemi, A., et al. (ed.) ESWC 2018. LNCS, vol. 11155, pp. 131–136. Springer, Cham (2018). https://doi.org/10.1007/978-3-319-98192-5_25

24. Lioudakis, G., Cascone, D.: Compliance Ontology - Deliverable D3.1. Project deliverable (2019). https://www.bpr4gdpr.eu/wp-content/uploads/2019/06/D3.1-Compliance-Ontology-1.0.pdf

25. Martiny, K., Elenius, D., Denker, G.: Protecting privacy with a declarative policy framework. In: 2018 IEEE 12th International Conference on Semantic Computing (ICSC), pp. 227–234. IEEE, January 2018. https://doi.org/10.1109/ICSC.2018.00039. http://ieeexplore.ieee.org/document/8334462/

26. Palmirani, M., Governatori, G., Athan, T., Boley, H., Paschke, A., Wyner, A.: LegalRuleML core specification version 1.0 (2020). https://docs.oasis-open.org/legalruleml/legalruleml-core-spec/v1.0/legalruleml-core-spec-v1.0.html

27. Palmirani, M., Martoni, M., Rossi, A., Bartolini, C., Robaldo, L.: PrOnto: privacy ontology for legal reasoning. In: Kő, A., Francesconi, E. (eds.) EGOVIS 2018. LNCS, vol. 11032, pp. 139–152. Springer, Cham (2018). https://doi.org/10.1007/978-3-319-98349-3_11

28. Pandit, H.J., Debruyne, C., O'Sullivan, D., Lewis, D., et al.: GConsent - a consent ontology based on the GDPR. In: Hitzler, P., et al. (ed.) ESWC 2019. LNCS, vol. 11503, pp. 270–282. Springer, Cham (2019). https://doi.org/10.1007/978-3-030-21348-0_18

29. Pandit, H.J., Fatema, K., O'Sullivan, D., Lewis, D., et al.: GDPRtEXT - GDPR as a linked data resource. In: Gangemi, A., et al. (ed.) ESWC 2018. LNCS, vol. 10843, pp. 481–495. Springer, Cham (2018). https://doi.org/10.1007/978-3-319-93417-4_31

30. Pandit, H.J., Lewis, D.: Modelling provenance for GDPR compliance using linked open data vocabularies. In: Society, Privacy and the Semantic Web - Policy and Technology (PrivOn 2017), co-located with ISWC 2017, vol. 1951 (2017). http://ceur-ws.org/Vol-1951/PrivOn2017_paper_6.pdf

31. Pandit, H.J., et al.: Creating a vocabulary for data privacy. In: Panetto, H., Debruyne, C., Hepp, M., Lewis, D., Ardagna, C.A., Meersman, R. (eds.) OTM 2019. LNCS, vol. 11877, pp. 714–730. Springer, Cham (2019). https://doi.org/10.1007/978-3-030-33246-4_44

32. Passant, A., Sacco, O.: Privacy preference ontology (PPO) (2013). http://vocab.deri.ie/ppo#

33. Raghupathi, W., Raghupathi, V.: Big data analytics in healthcare: promise and potential. Health Inf. Sci. Syst. **2**(1) (2014). https://doi.org/10.1186/2047-2501-2-3

34. Schreiber, G., Raimond, Y.: RDF 1.1 primer (2014). https://www.w3.org/TR/rdf11-primer/

35. Speicher, S., Arwe, J., Malhotra, A.: Linked data platform 1.0 (2015). https://www.w3.org/TR/ldp/

36. Union, E.: REGULATION (EU) 2016/679 OF THE EUROPEAN PARLIAMENT AND OF THE COUNCIL of 27 April 2016 on the protection of natural persons with regard to the processing of personal data and on the free movement of such data, and repealing directive 95/46/EC (general data protection regulation) (2016)

37. Wang, Y., Liu, Y., Shi, Y., Yu, Y., Yang, J.: Virtual hospital apps in China: a systematic analysis from the perspective of user evaluation. JMIR mHealth and uHealth (2020)

The Use of Decentralized and Semantic Web Technologies for Personal Data Protection and Interoperability

Mirko Zichichi[1]([✉]), Víctor Rodríguez-Doncel[1], and Stefano Ferretti[2]

[1] Ontology Engineering Group, Universidad Politécnica de Madrid, Madrid, Spain
`mirko.zichichi@upm.es`, `vrodriguez@fi.upm.es`
[2] Department of Pure and Applied Sciences, University of Urbino Carlo Bo, Urbino, Italy
`stefano.ferretti@uniurb.it`

Abstract. The enactment of the General Data Protection Regulation (GDPR) has been the response of the European Union to the growing data-driven economy backed up by the largest companies in the world. It provides the data protection and portability needed by individuals that "unconsciously" generate personal data for "free" services offered by providers that lack transparency on their use. Meanwhile, the rise of Distributed Ledger Technologies (DLTs) offers new possibilities for the management of general purpose data, hence being suitable for handling personal data in a trustless scenario. These decentralized technologies bring a new concept of contract called smart because of its ability to be self-executable. DLTs and smart contracts, together with the use of Semantic Web standards, allows the creation of a decentralized digital space controlled entirely by an individual, where his personal data can be stored and transacted.

Keywords: GDPR · Personal data · Distributed ledger technologies · Smart contracts · Semantic web

1 Introduction

With the introduction of the General Data Protection Regulation (GDPR) [5] in 2018, operations carried out regarding the management and the movement of personal data have radically changed. Data privacy of European Union's Citizen has been empowered through a series of rights that provide data protection and portability. GDPR can be seen as a necessary response to the challenges posed by technological advances brought about mainly by Big Tech companies, which generate huge amounts of data without sufficient safeguards for individuals. A huge business, indeed, lies behind the trade of personal data and several companies make consistent profits operating in this sector. GDPR and current literacy help the individual to understand how their personal data is often generate unconsciously and where, how or why the data is being collected, but still,

V. Rodríguez-Doncel et al. (Eds.): AICOL-XI 2018/AICOL-XII 2020/XAILA 2020, LNAI 13048, pp. 328–335, 2021.
https://doi.org/10.1007/978-3-030-89811-3_23

further work is needed to let them develop the necessarily practical and interpretive skills [15]. More efforts are needed to reach both transparency and a balance between privacy and data sharing.

Even if GDPR requires data controllers, i.e. entities that collect and manage individuals' personal data, to release to their users the complete dataset they collected on them, upon request, there are currently no standards for this kind of requests and there is the tendency to hinder the progress of these, causing the entire process to become almost useless. These data controllers usually store this personal information in corporate databases, but they can become data providers to other parties if the individual agrees –and even if the individual does not agree they are obliged to act as data providers in extraordinary cases, e.g. national security. As of today, when these data transactions happen there is no transparency on the individual's data usage.

Meanwhile, between the many technologies that regards general-purpose data management and storage, Distributed Ledger Technologies (DLTs) are raising as powerful tools to avoid the control centralization. The current use of DLTs is in financial (i.e. cryptocurrencies) and data sharing scenarios. In both cases there are several parties that concur in handling some data, there is no complete trust among parties and often these ones compete to the data access/ownership. Such features suit perfectly with the process of moving the data sovereignty towards users and releasing them more influence over access control, while allowing anyone else to be able to consume this data with transparency. This can be made possible through smart contracts, the new concept of contract that brought a second blockchain revolution.

The purpose of this paper is to present a vision of how to integrate the use of decentralised technologies and Semantic Web standards, with the aim of supporting the design of methods and systems that help individuals to assert their rights to protect their personal data and at the same time promote their portability and economic exploitation.

2   State of the Art

One of the most remarkable novelties in GDPR is the concept of *data portability*, which defines the right to have data directly transferred from one data provider to another making a step towards user-centric platforms of interrelated services [6]. This relates to the concept of data interoperability that embodies the complex network of users interaction based on personal data flow. To this scope, it is fundamental the use Semantic Web [1] standards, that bring structure to the meaningful contents of the Web by promoting common data formats and exchange protocols. The form of its most successful incarnation is Linked Data: data published in a structured manner, in such a way that information can be found, gathered, classified, and enriched using annotation and query languages.

One of the most recent approach that involves the use of distributed technologies and Semantic Web integration in social networks is the Solid project [17]. Led by the creator of the Web Tim Berners-Lee, the project was born with the purpose of giving users their data sovereignty, letting them choose where their data resides and who is allowed to access and reuse it. Solid provides us a strong reference for our work because it uses Semantic Web technologies to decouple user data from the applications that use this data. Data is, indeed, stored in an online storage space called Pod, a Web-accessible storage service, which can either be deployed on personal servers or on public servers.

A great variety of solutions, instead, involve the use of DLTs for the management of general purpose data (for example applied to media contracts [11]) and personal data and few of them are in compliance with GDPR. A DLT is a software infrastructure maintained by a peer-to-peer network, where the network participants must reach a consensus on the states of transactions submitted to the distributed ledger, to make the transactions valid. The role of DLTs is to provide a trusted and decentralized ledger of data preserving immutability, traceability, transparency and pseudonymity. The concept of DLT is the natural extension of the "blockchain" concept, because it includes those technological solutions that do not organize the data ledger as a linked list of blocks. Ethereum [2] enhanced the technology allowing the creation of a powerful tool: smart contracts. These contracts are self-managed structures that enable a decentralized computation, thus eliminating the presence of single point of failures.

Related works for the management of Personal Data using DLTs include various proposals and many of them are also focused on GDPR [4,7,8]. However, these studies do not address DLTs and smart contract challenges, presenting only a conceptual approach. A more technical approach can be found in [18]. Meanwhile, Smart contract based data access control has been thoroughly studied in literature [12,21] and still many scenarios are conceivable.

3   Moving Data Sovereignty Towards Users

It is still not possible to feasibly ensure to individuals the sovereignty of their personal data, nor the possibility of a appropriate data interoperability for data consumers. One of the key problems is that individuals do not have control or even knowledge of the transfers that happen with their personal data. Data providers, indeed, store and maintain this data differently through several data silos, hampering their free exchange and economical exploitation. In this situation, individuals that may be good willing to offer their data for social good or that may simply make direct profit from it, do not have the power to do so [20]. The main concern for individuals, then, is to invert this trend. The solution we propose, that supports the right of individuals to the protection of their personal data, data interoperability, economic exploitation and social good, is based on the following principles: i) avoid the concentration of personal information and its opaque transfers it is needed a system that allows store and transact personal data in a controlled, transparent and non-centralized manner; ii) favour personal data interoperability, hence to facilitate cross-domain application and services, a set of common languages and protocol must be used; iii) the individual needs to be directly involved in the access control to his or her data, defining both high-level policies and fine-grained preferences.

4   A Proposal Based on DLTs and Semantic Web

We present a solution that involves the use of Decentralized Technologies together with Semantic Web technologies to satisfy the principles posed in the previous section. In order to go further, it is needed to specify the types of personal data that concern this solution (but also the general case [15]). Personal data is defined as any piece of information that can identify or be identifiable to a natural person. Digital personal data, in particular, is generated by the interaction of a user with a software or a hardware in form of numbers, characters, symbols, images, sounds, electromagnetic waves, bits, etc. [10].

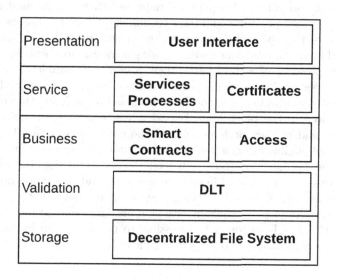

Fig. 1. Layered architecture

4.1   (Decentralized) Personal Information Management Systems

The type of data that is usually given in input by an individual to the online platform and services include self-tracking information, social networks sites data, and generally, information gathered from smartphone sensors. The use of distributed ledgers and related technologies can be used in this case serve as the basis to build novel smart services and to promote social good for what concerns individuals' personal data [7,18,22]. Online services' users produce various kind of data coming from their vehicle or smartphone, that actually concerns the same features exploited by social networks, e.g. user's location and activities. An approach that is considered in line with the GDPR [18], may be to use a composition of different services, such as:

- a DLT providing traceability and verifiability, without storing the data [22];
- Decentralzed File Storages to store data "off-chain" in an encrypted way;
- Smart Contracts to control data access [11,21];
- Cryptography techniques, such as Zero Knowledge Proof [9], to guarantee data protection;

In particular, the use of DLTs to represent and transact with personal data would grant data validation and access control, as well as no central point of failure, immutability and most importantly traceability. Moreover, it is possible to use decentralized file systems, that allow continuous data availability. These properties are necessary in order to associate each individual to the digital space that will contain his personal data and that will be used to attend the requests of data providers and data consumers. Crucial is the use of smart contracts, since they provide a new paradigm where unmodifiable instructions are executed in an unambiguous manner during a transaction between two parts. Without the presence of a third party, then, a user may completely control the access to his personal data, being sure that his decisions on how and when to access his data are always observed. Every process is completely traced and permanently stored in the blockchain. For what concerns the expression of legal requirements and privacy preferences, and the compliance with GDPR, smart contracts unintelligibleness (i.e. how their instructions, expressed in a programming language, become a contract) still needs deep investigation, but some works already address this issue [3]. Finally, the use of "suitable" cryptographic techniques, such as Zero Knowledge Proof, may allow to prove that an individual possesses a certain property without revealing his data. For instance, using Zero Knowledge Proof of Location [19] is possible to prove that an individual finds himself in a certain zone without revealing his exact location.

4.2   Data Flow Through Smart Contracts

The more the data is centralized in "silos" (not communicating between each other), the more individuals lose control over their personal data information. We envision the use of a unique digital space for each data subject, where data flow is ruled and data providers and consumers can meet to transact. A possible solution to this can be achieved through decentralization and shared standards. This approach aim to shift the control from centralized platform to users for the access to user generated personal data. This infrastructure would also give the opportunity to build a Personal Data Marketplace where individuals can decide to sell or to set access rules in order to provide data for the social good [20,22].

The use of smart contracts can be may be crucial to regulate user's data flow. The interesting aspect is that smart contracts, i.e. instructions executed by a decentralized virtual machine, allow two parties, e.g. data provider and data onsumer, to reach an agreement in the process of the data flow. Nevertheless, smart contracts can be programmed to satisfy at least the following reasoning tasks: (i) determine if a policy satisfies the legal requirements [3]; (ii) determine if a data request can be satisfied according to the individual's preferences [21].

On the other hand interoperability can be best achieved if a network of ontologies is used to model the personal data life-cycle and their actors.

4.3   Semantic Web Based Policies

The smart contracts must be thus represented in a language that favours reasoning and a language that eases interoperability. Fortunately, the W3C has published over the last twenty years a set of specifications to describe resources which simultaneously addresses these two design goals: those of the semantic web.

Whereas these specifications were born to represent data in the web, their use has gone beyond and today many applications run totally offline but using the semantic web specifications. In the most spread paradigm, information is represented using RDF (Resource Description Framework). In this framework, resources are identified with URIs and described with collections of triples. The precise meaning of each resource can be formally established with OWL ontologies. An ontology is a formal representation of knowledge through a set of concepts and a set of relations between these concepts, within a specific domain. Through the use of these ontologies it is possible to convey the meaning of data, hence to facilitate cross-domain applications and services. Ontologies in these scenarios effectively act as data models, eventually complemented with RDF Shapes[1], which further impose restrictions on the data that are easy to be evaluated.

Whereas new ontologies can be created whenever necessary, there is a set of *de facto* standard ontologies which should be reused whenever possible. For example, there are ontologies to describe the basic personal contact information, such as vCard[2], to describe basic geographical information[3] or to represent computer policies[4] or contracts [16]. Other vocabularies and ontologies have recently appeared in the domain of privacy and data protection [13,14].

The two advantages of 'interoperability' and 'reasoning' can be now well illustrated: first because the aforementioned ontologies are recommended by the W3C and thus universally understood. Second, reasoning with the information represented using these data models is easy because they are mapped in a formal language. An individual may want to say: whenever I am in the province of Lombardy, I want my data not to be transacted. If properly connected to other datasets, the system knowing that the individual is in Milano will infer that the individual is also in Lombardy and should not tranfer the data.

5   Vision and Conclusions

After having explained how a unique digital space can be built, it is fundamental to explain its use. The main idea is that this infrastructure can lead personal

---

[1] https://www.w3.org/TR/shacl/.
[2] https://www.w3.org/TR/vcard-rdf/.
[3] https://www.w3.org/2003/01/geo/.
[4] https://www.w3.org/TR/odrl-vocab/.

data flow towards a "safe" place where the individual can enforce his rights. There are different actors behind the successful implementation of this vision. First of all, the individual is obviously favoured because he assumes the full control over such digital structure. Then, all the actors behind the decentralized structure are incentivized by the use of the technology specification itself, e.g. monetary retribution. Finally, the main actors who use the space both to provide and gather data, i.e. data providers and consumers, are the one to which focus on. In particular, GDPR requires data providers to release personal data to data subjects, but this does not implies the use of the digital space. The use of common standards provided by Semantic web is a necessary incentive, but not sufficient. Hence both providers and consumers must be incentivized by the data market that generates behind the digital space.

GDPR has brought an important evolution for what concerns individuals' personal data protection, providing them rights to contrast data abuse. However, the data flow that occurs behind the scenes between data providers and consumers and the creation of data "silos" prevent the execution of transparent processes at the eye of data subjects. A possible approach may be the one where each individual maintain his personal digital space in which his personal data is stored and transacted. This can be achieved through decentralized technologies in the form of DLTs, decentralized file systems and smart contracts that provide transparency and data access control, and through semantic web technologies in the form of linked data that provide data portability. Further methods may protect the privacy of individuals while new methodologies for the analysis of such systems may bring to light new GDPR interaction models, e.g. to understand possible actors and manners to infer data.

Acknowledgment. This work has been partially funded by the EU H2020 MSCA projects LAST-JD-RIOE with grant agreement No. 814177 and PROTECT with grant agreement No. 813497.

References

1. Berners-Lee, T., Hendler, J., Lassila, O., et al.: The semantic web. Sci. Am. **284**(5), 28–37 (2001)
2. Buterin, V., et al.: Ethereum white paper (2013). https://github.com/ethereum/wiki/wiki/White-Paper
3. Cervone, L., Palmirani, M., Vitali, F.: The intelligible contract. In: HICSS, pp. 1–10 (2020)
4. Chen, Y., Xie, H., Lv, K., Wei, S., Hu, C.: Deplest: a blockchain-based privacy-preserving distributed database toward user behaviors in social networks. Inf. Sci. **501**, 100–117 (2019)
5. Council of European Union: Regulation (Eu) 2016/679 - directive 95/46
6. De Hert, P., Papakonstantinou, V., Malgieri, G., Beslay, L., Sanchez, I.: The right to data portability in the GDPR: towards user-centric interoperability of digital services. Comput. Law Sec. Rev. **34**(2), 193–203 (2018)
7. Faber, B., Michelet, G.C., Weidmann, N., Mukkamala, R.R., Vatrapu, R.: Bpdims: a blockchain-based personal data and identity management system. In: Proceedings of the 52nd Hawaii International Conference on System Sciences (2019)

8. Farshid, S., Reitz, A., Roßbach, P.: Design of a forgetting blockchain: A possible way to accomplish GDPR compatibility. In: Proceedings of the 52nd Hawaii International Conference on System Sciences (2019)
9. Feige, U., Fiat, A., Shamir, A.: Zero-knowledge proofs of identity. J. Crypt. 1(2), 77–94 (1988). https://doi.org/10.1007/BF02351717
10. Kitchin, R.: The Data Revolution: Big Data, Open Data, Data Infrastructures and Their Consequences. Sage, Los Angeles (2014)
11. Kudumakis, P., et al.: The challenge: from mpeg intellectual property rights ontologies to smart contracts and blockchains. Signal Process. Mag. 37(2), 89–95 (2019)
12. Di Francesco Maesa, D., Mori, P., Ricci, L.: Blockchain based access control. In: Chen, L.Y., Reiser, H.P. (eds.) DAIS 2017. LNCS, vol. 10320, pp. 206–220. Springer, Cham (2017). https://doi.org/10.1007/978-3-319-59665-5_15
13. Palmirani, M., Martoni, M., Rossi, A., Bartolini, C., Robaldo, L.: PrOnto: privacy ontology for legal reasoning. In: Kő, A., Francesconi, E. (eds.) EGOVIS 2018. LNCS, vol. 11032, pp. 139–152. Springer, Cham (2018). https://doi.org/10.1007/978-3-319-98349-3_11
14. Pandit, H.J., O'Sullivan, D., Lewis, D.: An ontology design pattern for describing personal data in privacy policies. In: WOP at ISWC, pp. 29–39 (2018)
15. Pangrazio, L., Selwyn, N.: Personal data literacies: a critical literacies approach to enhancing understandings of personal digital data. New Media Soc. 21(2), 419–437 (2019)
16. Rodríguez-Doncel, V., Delgado, J., Llorente, S., Rodríguez, E., Boch, L.: Overview of the mpeg-21 media contract ontology. Seman. Web 7(3), 311–332 (2016)
17. Sambra, A.V., et al.: Solid : a platform for decentralized social applications based on linked data (2016)
18. Truong, N.B., Sun, K., Lee, G.M., Guo, Y.: GDPR-compliant personal data management: a blockchain-based solution. IEEE Trans. Inf. Foren. Sec. 15, 1746–1761 (2019)
19. Wolberger, L., Fedyukovych, V.: Zero knowledge proof of location (2018). https://platin.io/yellowpaper
20. Zichichi, M., Ferretti, S., D'Angelo, G.: A distributed ledger based infrastructure for smart transportation system and social good. In: IEEE Consumer Communications and Networking Conference (CCNC) (2020)
21. Zichichi, M., Ferretti, S., D'Angelo, G., Rodríguez-Doncel, V.: Personal data access control through distributed authorization. In: 2020 IEEE 19th International Symposium on Network Computing and Applications (NCA), pp. 1–4. IEEE (2020)
22. Zichichi, M., Ferretti, S., D'Angelo, G.: A framework based on distributed ledger technologies for data management and services in intelligent transportation systems, pp. 100384–100402. IEEE Access (2020)

Author Index

Printed in the United States
by Baker & Taylor Publisher Services